The explanation of organic diversity

The explanation of organic diversity

The comparative method and adaptations for mating

Mark Ridley

Oriel College, Oxford

CLARENDON PRESS · OXFORD
1983

Oxford University Press, Walton Street, Oxford OX2 6DP
London Glasgow New York Toronto
Delhi Bombay Calcutta Madras Karachi
Kuala Lumpur Singapore Hong Kong Tokyo
Nairobi Dar es Salaam Cape Town
Melbourne Auckland
and associated companies in
Beirut Berlin Ibadan Mexico City Nicosia

Oxford is a trade mark of Oxford University Press

Published in the United States
by Oxford University Press, New York

© Mark Ridley 1983

All rights reserved. No part of this publication may be reproduced,
stored in a retrieval system, or transmitted, in any form or by any means,
electronic, mechanical, photocopying, recording, or otherwise, without
the prior permission of Oxford University Press

British Library Cataloguing in Publication Data

Ridley, Mark
  The explanation of organic diversity
  1. Adaptation (Biology)
  I. Title
  574.5'22   QH546
ISBN 0-19-857597-1

Printed in Great Britain by
St Edmundsbury Press, Bury St Edmunds, Suffolk

Preface

Comparative biology possesses two well-founded principles—design and history, adaptation and phylogeny—to explain the diversity of life. The comparative biologist can put these two together, in any particular case, and produce a hypothetical explanation. But then he is faced with a difficulty. If he goes on to try to test his hypothesis, he finds that comparative biology lacks a set of soundly-based, formal techniques. So while the comparative biologist can produce a hypothesis, which is probably of the right general kind, it is difficult to discover whether it is right or wrong, and whether it should be discarded. Comparative biology needs methods, methods which are as rigorous as the experimental and statistical methods of other sciences.

This work is concerned with the development of method. It will not remove the main difficulty from comparative biology, but I do still dare to offer it to the public. The main problem of the comparative method, I believe, is the recognition of independent trials of a hypothesis. Once independent trials have been recognized, they can be counted; and then the application of formal statistical techniques is straightforward. In the first part of the book I am going to argue that some of the techniques developed by cladistic taxonomists can (whatever their original value) be re-applied in the study of adaptation. First we need to know the distribution of the states of the characters of interest among as many species as possible: then we can cladistically count the minimum frequency of independent evolution, and test our hypothesis.

My purpose is not only the development of a technique in the abstract. I shall also try to apply it to two questions about sexual behaviour. The first is the incidence of precopulatory mate guarding, that habit (found in so many crustaceans and frogs, as well as spiders, mites, and tardigrades) of pairs of males and females going round in an intimate embrace for some time before mating. The second is the incidence of homogamy (or 'assortative mating') for size. The two questions were chosen so that the hypotheses and their literatures were of such different kinds that they would present different difficulties to the method. In both cases, I believe, we finish up with a respectable, if imperfect, test; but even if the methods, tests, and hypotheses all turn out to be utterly useless, perhaps my book will retain some value as a reference work, as I have reviewed the mating habits of whole groups, particularly the Crustacea, which have never been reviewed as a whole and only sporadically in parts.

Such, in outline, are the contents of this work. Such has been the subject of my research for the last few years. While doing it, I have received help from many people, and I would like to thank at least the more important of them. My first thanks are to Richard Dawkins, who introduced me to both the study of adaptation and the comparative method; he has also supervised all this research, in innumerable conversations, and commented, to their great benefit, on earlier drafts of the chapters. I am also grateful to Professor T. R. E. Southwood, who stepped in as supervisor when Richard Dawkins emigrated, for two sabbatick terms, to the plantations. Alan Grafen read Chapters 1 and 4, which enabled me to improve them greatly as (in Chapter 4) the reader will see. The results of his modelling have also enabled me to improve Chapter 3, by clearing out many an unnecessary abstraction; indeed I was not even sure that the theory worked until I had seen Alan's equations. To Paul Harvey I am grateful for his comments on Chapter 1, and to Fritz Vollrath (who has also helped me with many otherwise unintelligible passages of German) for his comments on the section on spiders in Chapter 3. The two questions which I have tackled in Chapters 3 and 4 were inspired, more or less, by some research on the isopod *Asellus aquaticus* which Dave Thompson conducted with me before disappearing up North, and I am grateful to him too.

I would not have been able to undertake this work, as a graduate student, without the support of two munificent trusts. I started it during the three years of an E.P.A. Cephalosporin Research Scholarship at Linacre College, and completed it after transferring, with a Hayward Junior Research Fellowship, to Oriel College. To these two colleges and their benefactors I am most grateful.

Oxford M.R.
April 1983

Contents

Part I The comparative method	1
1 The explanation of organic diversity	3
Two comparative biologies	3
The study of adaptation by the comparative method	7
Independence	9
The necessity of independence	9
The recognition of independence	18
Two problems	28
Association and cause	34
Laws of evolution	40
Part II Adaptations for mating	45
2 Introduction	47
3 The evolution of precopulatory mate guarding	52
Introduction	52
Methods	58
Arthropods	64
Tardigrades	68
Crustacea	69
Branchiopoda	70
Maxillopoda	71
Malacostraca	79
Hoplocarida and Syncarida	79
Decapoda	80
Peracarida	110
Crustacea: summary	131
Arachnids	132
Spiders	135
Mites	147
Anura	153
Some generalizations	163
4 On being the right sized mates	170
Introduction	170
Hypothesis: the sexual selection of homogamy	173
Methods	174
Systematic summary	179
Plants	179
Protozoans	179
Molluscs	185

Insects	186
Crustacea	196
Arachnida	203
Fish	204
Anura	214
Birds	214
Mammals	215
Legislation	216
References	223
Index	263

PART I
The comparative method

1 The explanation of organic diversity

TWO COMPARATIVE BIOLOGIES

The greatest tradition of research on organic diversity is to be found in taxonomy. From the earliest *scalae naturae* through the great systems of the eighteenth century with their exquisite Augustan divisions of whole into parts, through the discovery of branching systems in the nineteenth century with their explanation by Darwinian evolution: here is the great tradition of comparative biology. It is a system of ordering and classification. If the kinds of characters used to order and classify are chosen by some theoretical principle, the system can be explanatory. In the usual modern taxonomies, for example, the similarities among species related in the taxonomy are shared homologies. Communal descent from a common ancestor is the explanation of the classificatory hierarchy. But the main purpose of this tradition of comparative biology has been to discover and classify, not to explain, natural categories.

There is another tradition of thought about the diversity of life. It seeks generalizations about non-taxonomic categories. Shyness and size, for example, are not criteria of major taxonomic groups, but in Pliny's *Naturalis Historia* we can read:

Nature has determined that, among the kinds of birds, the shy are more fecund than the brave: only ostriches, hens, and partridges bear very numerous broods. (*Book X, LXXIII*)

Larger animals are less fecund. Elephants, camels, and horses produce one offspring (at a time); the thistle-thrush, the smallest of birds, twelve. (*Book X, LXXXIII*).

Similar generalizations fill a chapter, written seventeen centuries later, on 'varieties in the generation of animals', in Buffon's *Histoire Naturelle*. Take the opening sentences of three of the first four paragraphs:

One can say that in general larger animals are less productive [fecund] than small.
Oviparous animals are in general smaller than the viviparous, and also produce more [at birth].
Animals that produce only a small number of offspring, acquire the main part of their growth before they reproduce; whereas those which multiply numerously reproduce before their bodies have acquired a half, or even a quarter, of their growth. (Buffon 1749, Vol. 2, pp. 306, 307, 308)

These generalizations do not belong to the taxonomic tradition. They are the raw material of a separate, explanatory tradition, which runs parallel with taxonomy throughout the history of comparative biology. Buffon goes on to discuss why the more fecund species might continue to grow after maturity whereas less fecund ones do not. His exact explanation is too involved for us to go into here, so we will shortly change examples. What are the general kinds of explanation of these non-taxonomic generalizations? There are two: design and embryological correlation (which Darwin called 'correlation of growth'). The following passage from Locke's *Second Treatise on Government* illustrates the former. He first sets out a generalization, and then explains it by an argument of design.

This Rule, which the infinite wise Maker hath set to the Works of his hands, we find the inferiour Creatures steadily obey. In those viviparous Animals which feed on Grass, the *conjunction between Male and Female* lasts no longer than the very Act of Copulation: because the Teat of the Dam being sufficient to nourish the Young, till it be able to feed on Grass, the Male only begets, but concerns not himself for the Female, or Young, to whose Sustenance he can contribute nothing. But in Beasts of Prey the *conjunction* lasts longer: because the Dam, not being able well to subsist her self, and nourish her numerous Off-spring by her own Prey alone, a more laborious, as well as more dangerous way of living than by feeding on Grass, the Assistance of the Male is necessary to the Maintenance of their common Family, which cannot subsist till they are able to prey for themselves, but by the joynt Care of Male and Female. The same is to be observed in all Birds . . . (Locke, *Second Treatise*, §79).

Aristotle's *Politics* (Book I.8) contains a similar statement (which has previously been quoted by Clutton-Brock and Harvey 1978*a*):

Again, there are many kinds of food, and therefore there are many kinds of lives both of animals and men; they must all have food, and the differences in their food have made differences in their ways of life. For of beasts, some are gregarious, others are solitary; they live in the way that is best adapted to sustain them, accordingly as they are carnivorous or herbivorous or omnivorous: and their habits are determined for them by nature in such a manner that they may obtain with greater facility the food of their choice.

The second tradition, therefore, is as old as the taxonomic tradition: they were both founded by Aristotle. But it is not as rich. It is made up of only a few passages in works on other subjects. It is easily outweighed by the hundreds of great volumes of taxonomy. The few passages themselves mainly concern characters, such as intelligence and social co-operation, chosen for their political and moral interest. But the tradition does exist, flickering in many strange places. For an example of the second kind of explanation (developmental trade-off), we can go to the splendid, rollicking *Eulogy of Baldness* by the fifth-century pagan Bishop of Cyrene, Synesius (*c.* 370–427).

Those animals who are more deprived of intelligence, are clothed with hair all over their bodies, whereas man, in as much as his lot in life is more brilliant, is the most bare of this natural burden . . . And just as man is the most intelligent, and at the same time the least hairy of earthly creatures, conversely it is admitted that of all domestic animals the sheep is the stupidest, and that is why he puts forth his hair with no discrimination, but thickly bundled together. It would seem that there is a continual strife going on between hair and brains, for in no one body do they exist at the same time. (Translation in Fitzgerald 1927, vol. II, pp. 248-9)

Synesius here hints at a developmental explanation rather than an argument of design (which would have to explain why hairiness was beneficial in the less intelligent), and thus looks forward to the great allometric era, which was, however, to find more inspiration in body size than hair.

The two comparative biologies, in their techniques, in their concepts, and in their produce, make up complementary pairs. One seeks homologies, the other analogies; one (in Gregory's terms) studies heritage, the other habitus; the one classifies divergence, the other explains convergence; and (more dangerously: Cain 1964) the one studies ancestral, the other, adaptive characters.

This book is concerned with the second of the two traditions. It is about adaptation. So we can now concentrate this historical introduction. Let us now see how this second tradition has dealt with the problem of design.

It has always been possible for non-taxonomic comparative studies to come in two forms. One may study divergence within a small group, or convergence among more widely separated groups. These two are now often distinguished as two kinds of comparative study (e.g. Wiley 1974; Altmann 1974; Hailman 1977, 1981). The meanings of the two, of course, were radically altered by the theory of evolution. The former kind came to be known as adaptive radiation. It has its classics, such as Lack's (1947) research on Darwin's finches, or the mammalian radiation on South America (Simpson 1979). The idea was taken up by ethologists, too, such as in Tinbergen's (1959) work on gulls. But in all these studies adaptation has rarely been more than a principle of interpretation. Comparative studies of convergence have had a much closer relationship with the principle of adaptation. For comparative studies of convergence can provide powerful evidence of adaptation. Take Locke's argument as an example. A character (sociality) may be an adaptation (it helps in rearing the dependent young) in one species (dogs) of a certain kind (carnivores). This proposition can be tested by comparison. If helping to rear the young is especially important in carnivores, then other carnivores too should be social. Non-carnivorous species (other things being equal) should not. Let us study lions, tigers, and hyenas. Let us study horses and antelopes. The argument predicts how they should behave. It can be tested through its prediction.

The comparative method works because natural selection can cause convergence of different species on to the same adaptation. Darwin ('by the Spring of 1838', Ospovat 1981, p. 111) understood this. But in his wake an ice age set in on the study of adaptation. Arguments from design (and research on convergence), crushed and frozen beneath phylogenetic glaciers, all but fell victim to that most terrible technique of academic annihilation: burial alive in textbooks. How many textbooks between 1930 and 1980 have their horrible little section on convergence, unthinkingly repeating the wisdom of earlier centuries!

A whole work on convergence published at the beginning of the century (Willey 1911) was little more than an anarchist tract. 'It appears [remarked Willey on p. 56] that there is more joy over one attempt at genealogy than over ninety and nine demonstrations of convergence.' Willey himself had written on *Amphioxus* and the ancestry of the vertebrates (1894); now he listed fully 99 examples of convergence, and drew the unfortunate conclusion for the study of phylogeny. He did not even mention adaptation. The comparative study of adaptation, however, was not frozen for ever. The first signs of the recent flourishing tradition of research to which this book is a contribution are in the late nineteen-fifties. Then Chance (1959) tried classifying primate societies ecologically rather than taxonomically and thinking about the categories he found. Further analysis followed of mammalian and avian social systems, by Crook (1964, 1965), Crook and Gartlan (1966), McNab (1963), Lack (1968), and Schoener (1968), to mention just the better known. Since then the comparative biology of adaptation has progressed as part of the great revival of teleological thought.

The comparative method of 1950 was indistinguishable from the comparative method of 350 BC. The discovery of natural selection had injected a new interpretative principle, but no advance in method. Over the years many an unrigorous practitioner has tried his hand at the comparative method, and they have been followed, of course, by exaggerating critics. But neither will be providing our theme. We shall not be reviewing comparative biology. We shall instead be attempting to develop comparison as a method of studying adaptation. Some thought has recently been given to the method, and, when we shortly come to examine it, we shall review a little of the past. But, in its main aims, the book looks only forwards. It has two main aims. The first, which will be achieved in the first chapter, is to incorporate the comparative biology of phylogeny into the comparative method of studying adaptation. The reason for this incorporation is that we need to know phylogeny to measure convergence. The second, which fills the second part of the book, is to apply the method to two problems in sexual behaviour.

The next two sections of this chapter will build up the comparative method by stages. We start with a crude method. It consists of a list of species, and their attributes, and some idea of why they have those attributes. The two sections will progressively refine the method until we have both a valid comparative hypothesis, and a list of independent trials of it. The first is mainly concerned with the peculiar properties of a comparative hypothesis. The second develops the method of testing the hypothesis. Once we have a complete method we shall examine some of the commoner criticisms of the comparative method. We shall ask whether they are valid at all, and we shall see whether the particular method developed here can answer some of them. The examination fills the final three sections of the chapter.

THE STUDY OF ADAPTATION BY THE COMPARATIVE METHOD

The comparative method can be logically divided into two parts: hypothesis and test. There are some peculiarities of these two in comparative biology. Take the hypothesis first. It is not the case that every hypothesis of adaptation can be tested by the comparative method. It must satisfy certain conditions. It must explain different attributes of different species as adaptations to different environments.[1] The comparative method cannot reveal anything about adaptations which are uniformly favoured in all species. Nor should the methods discussed in this book be used with hypotheses that explain differences between species by some evolutionary trend towards greater adaptedness. The correct explanation of the difference between the 'sprawling' gait of reptiles and the 'upright' gait of mammals may be that the former have retained an earlier and inferior design. But if it is, it cannot be discovered by the methods of this book. They cannot cope with 'progress'. The hypothesis must give a static explanation of diversity, the different adaptations of the different species being advantageous in their various environmental contexts.

The next important distinction is between hypotheses about how species are adapted, and hypotheses about whether species are adapted. The methods can be applied only to the former kind of hypothesis. They assume that species are adapted to their environments, and actually try to work out how they are adapted (cf. Curio 1973, pp. 1051-3). The assumption is guaranteed by the theory of natural selection. Although the techniques of the book cannot usually test whether a characteristic is an adaptation, they can be used to test one kind of hypothesis of non-adaptation. I refer to hypothetical non-adaptive differences between species. If a difference among species is non-adaptive then it will not correlate with environmental variables. Such correlations can be sought. They can be sought especially with variables that would be expected (were natural selection operating) to be correlated. If even a thorough search reveals no such correlations, the hypothesis of non-adaptive difference is more confirmed than before the test. We may find even in the last few years many such implicit tests of whether particular comparative patterns are non-adaptive, and I shall be adding some more later in this book. Taken together, they suggest a philosophy of adaptation which differs profoundly from that well-known diatribe which flew out of Harvard in December 1978 (Gould and Lewontin 1979). Although Gould and Lewontin do not follow the many philosophers who have proved that adaptation is an unscientific concept, they do argue that 'plausible stories can always be told' (p. 588), that the methods of investigating adaptation are so feeble that in practice adaptationists cannot show whether a character is an adaptation, yet alone what it is an adaptation to. They made their task

---

[1] Throughout this chapter the word 'environment' is used with very broad meaning. It may refer to internal, as well as external properties of organisms. It may refer to the species' niche. It may refer to the environment which one gene provides for another gene.

rather easier for themselves by altogether ignoring all the more advanced work on adaptation. In comparison they may find one method of testing adaptationist hypotheses, one method of testing, although only in some cases, whether differences between species are non-adaptive.

So much for the hypothesis. What about the test? The most rudimentary kinds of comparative study are those which, after looking through the literature more or less thoroughly, present a list of only the species (or some of them) which conform to hypothesis. If the hypothesis is that adaptation A is an adaptation to environment E, then the list specifies only the species living in environment E which show adaptation A. Such a list may usefully demonstrate that a trait is taxonomically widespread. Or it may suggest a hypothesis.[2] But it cannot test one. If the method does not also list the species that do not conform to the hypothesis it is analogous to selecting only the supporting part of the results from a mass of experimental data. This is the second meaning, the one for the practitioners rather than the consumers of comparison, in Maynard Smith and Holliday's (1979, p. vii) remark that 'We must learn to treat comparative data with the same respect as we would treat experimental results'. Non-systematic studies have no doubt helped to inspire critics who believe that the whole concept of adaptation is untestable. They will notice that the adaptationists only count confirmatory instances. But they would be confusing a badly applied concept with an unscientific one. Cases that refute, as well as cases that confirm, can easily be counted, and in so doing we are moving towards a scientific test. The very first requirement, then, for a test of a comparative hypothesis is that the criterion for including a species in the test is independent of whether it supports the hypothesis. This criterion may seem obvious: it may seem that it does not require statement. But it is enough to rule out the entire literature (with one exception) from before about 1960. That exception is the French biologist Etienne Rabaud. Rabaud was concerned with Cuvierian laws of the relation between anatomy and niche, such as that herbivores have flat, grinding teeth and carnivores sharp, piercing teeth. Rabaud first looked systematically at these laws in his book of 1911, and later in a longer work on *Les phénomènes de convergence en biologie* (1925). In 1925 he summarized, systematically, such standard examples of convergence as adaptations for flying and swimming. Each systematic summary led to the same conclusion: 'one can recognize no necessary relation between the form and the way of life of the organisms considered' (1925, p. 150). The laws were not universal, he concluded, but were based only on facts that had been selected because they conformed:

If comparison is limited to organisms living in analogous conditions, morphological convergence would appear, strongly, to be related to these conditions;

---

[2] So common have comparative studies of this kind been that it has been said that the comparative method can only suggest ideas, not test them (Hinde and Tinbergen 1958).

but it is evident enough that the conclusion comes before the comparison, instead of after. The result completely changes when the comparison is practised in a complete fashion, when similar forms are placed according to their manner of life, and when these are contrasted with all the forms which coincide with them. A nonpartisan observer can then well understand that a predetermined condition of the *milieu* does not control [*commandent*] a predetermined form . . . It also demonstrates the impossibility of finding, in the disposition of morphology, the least indication of the way an organ functions, or the manner of life of the organism. Everything coincides with everything else.

Raubaud is a lonely, isolated figure in the history of research on convergence. His 1925 paper is little known, the *Suppléments* to the *Bulletin biologiques de France et Belgique* its obscure graveyard: few are the libraries that possess a copy. He inspired no tradition of research on convergent adaptations. But for the fact that he presented no quantitative summary of his conclusions he had the entire comparative method of the early nineteen-sixties. It is this comparative method that, after re-stating in abstract form, we are going to develop in the next section.

What is the abstract form of a Rabaudian test? It depends on the nature of the variables. If the results come in the form of continuous variables, such as body size or home range size, they can be plotted on a graph. If they come in the form of discrete variables, such as internal or external fertilization, they can be entered in an $n \times n$ contingency table. A test of a hypothesis about two discrete variables, A and B, each of which could exist in two states (A and A', B and B'), proceeds first to count the frequency in nature of the four states (AB, AB', A'B, A'B'), and then determines statistically whether there is any association. The hypothesis then stands or falls according to the nature of the association.

Although the general form of the test is easy, it is not without problems. Most of the rest of this opening chapter will be an examination of these problems. The first problem is that of statistical independence. A hornets' nest lies concealed beneath those innocent words 'frequency in nature of the four states'. It is time to stir it up.

INDEPENDENCE

We now have a clear hypothesis and have unearthed all the appropriate facts. We next have to determine how exactly the facts are to test the hypothesis. The following two sections will establish first that the correct method is to count only independent trials, and then will explain a cladistic technique for recognizing independent trials.

The necessity of independence

Suppose that we are testing a hypothesis with a 2×2 contingency table. That carnivores have sharp, piercing teeth, while herbivores have flat, grinding teeth, is an example. We are testing for an association between two character states (shapes of teeth) and two 'environments' (herbivory and carnivory). But what units are we going to count and dispose around the four contingencies of the table? The units which have been used most often are species, and this brings us straight to the problem: taxonomic artefacts. The taxonomic artefacts which concern us have two main sources. The first is that some higher taxa, all of whose species share the same character state, have been more studied than others: the second that some contain more species than others. We shall discuss the problem in the abstract. (Real examples exist. One is discussed by Harvey and Mace (1982).)

Imagine, then, that the adaptation A is possessed by, and the environment E occupied by, all 100 species of a single family. The alternative adaptation A' is possessed by only 50 species, all of which occupy the same environment E, but these 50 species come from 10 different families. How are we to weigh the 100 closely related species against the 50 more distantly related species? The answer according to the naïve method of counting only species is easy: there are 100 of one and 50 of the other. Easy, indeed, but not worthy of trust. The method gives more weight to taxa in which more species have been studied, and which contain more species. But why should these taxa be given more weight? We do happen to name our study organisms by their species name, and this is why species have been so often picked on as the unit of counting in comparative tests. But it is not a justification at all. Why species, rather than organisms, populations, genera, families, orders? Species do not have any special status beyond a convention of reference.

A number of methods have been used to reduce the bias which speciose and intensively studied taxa can introduce. The simplest is simply to ascend the Linnean hierarchy, using genera, families, or whatever as the units in the test. Crook (1965), for example, used families as the units in his early comparative study of avian mating systems; Lack (1968) followed him, but used subfamilies. I used families as the units in a study of the incidence of paternal care, explicitly to avoid the problem of taxonomic non-independence (Ridley 1978). (A family that contains species in more than one state is entered once for each state that it has evolved.) The many species of a family may all share a character; but it is less likely that the many families of an order all do.

This is a rudimentary beginning. The difficulty may be reduced as we ascend the hierarchy, but it is not removed. Higher categories, like lower ones, can share characters from a common ancestor. They are just less likely to. And this method raises other problems. At successively higher taxonomic levels, the numbers which can appear in the test decrease, until (at the level of kingdom) the number is (at most) one. As the test becomes more conservative it becomes less revealing. The higher levels, furthermore, may be less comparable than the lower ones. A species of insects may be, in some sense, comparable to a species of mammals: but is a family of insects comparable with a family of mammals?

Hennig (1966, pp. 154-93) and Van Valen (1973) have discussed this problem, but I do not think it is soluble. With present taxonomic methods it is probably meaningless. We may just bear it in mind as another problem of using higher taxonomic levels in the comparative method.

Now let us turn to some more sophisticated methods. Two large recent comparative studies have attempted, by different techniques, to eliminate taxonomic biases. These are the papers of Clutton-Brock and Harvey (1977, see also Harvey and Mace 1982; Harvey and Martin 1983), and of Baker and Parker (1979), on primate societies and bird coloration respectively. Let us look at these two studies in turn.

Baker and Parker on bird coloration

Why are different bird species of different colours? Sexual selection? In polygynous species with maternal care, the males do tend to be brightly coloured; while in polyandrous species with paternal care, the females tend to be. In monogamous species with biparental care, the two sexes are often similarly coloured. The hypothesis satisfies our requirements: it is truly comparative. It explains different colorations as adaptations to different reproductive habits. But it had never been systematically tested before Baker and Parker (1979). Darwin's (1871) chapter on birds, although brilliant, was not systematic. Baker and Parker tested at the same time some other comparative hypotheses, hypotheses which explain bright coloration as an adaptation against predation. 'Unprofitable prey', for example, may evolve bright coloration if then predators learn more quickly not to go for them. The familiar theory of 'warning coloration' is but a special case of an 'unprofitable prey'; the bird may be 'unprofitable' for reasons other than a sickening or poisonous taste: it may be good at running away, or fighting back.

Baker and Parker, using a colour handbook, scored the brightness of coloration between 0 and 5 for nine body regions in 516 species. For each the two sexes were scored separately, breeding and non-breeding, and the juveniles made up a fifth category. $9 \times 516 \times 5$ numbers between 0 and 5 thus make up the diversity to be explained. The variables to explain it included such things as body size, geographical latitude, gregariousness, incidence of paternal care, polygamy, and more. The method employed? Multiple regression. The units? ... Species.

Species, as we have seen, are the usual unit. But in Baker and Parker we find an improvement. They justified species (p. 86). The difficulty with using species is that there may be higher taxa made up of many species all with the same colour patterns. Baker and Parker examined their evidence to see whether it included any such taxa. To be exact: they analysed the variance of coloration both within and between families. If there was significant variation within as well as between families, then their study could not be riddled with groups of non-independent units. In fact they found significant variation within families, and so felt justified in going on to use species in the main test. But we have not heard the last of that significance test.

We are concerned in this chapter with the construction of methods, not the reviewing of results. But Baker and Parker's results are of sufficient interest in a book half concerned with sexual selection for us to mention some of them here. I have listed some of the principal trends in Table 1.1. Whether in the trend with body size, with which sex sits on the eggs, with the amount of time spent guarding the eggs, with the concealment of the nest site, or the site of mating, they found a general tendency for 'unprofitable prey' to be more brightly coloured. Birds which are more mobile and can see predators from further off tend to be brighter. The overall trends, we are told, are not so well predicted by the theory of sexual selection. Baker and Parker therefore concluded that the patterns of bird coloration are adaptations against predation, not for mating.

Now, as a matter of fact I do not think that Baker and Parker's analysis decides conclusively between the two theories. But let us not get side-tracked. Let us return to our main question of what is the correct unit for a comparative test. We have seen that Baker and Parker used species. We have seen how they justified their decision. Is it a good justification?

Their method certainly improves on the naïve use of species. It probably improves on the other method we have discussed, which is to use some higher taxa. But it is still not above criticism. The variance within each family is still likely to be lower than that within higher ranking taxa. Even if there is significant variance within a family there still may be some closely related species all showing the same characteristics. The method ensures only that, in the final test, there cannot be large numbers of these species. It insures against terrible biases but does not entirely remove bias. Small biases may still creep in when the numbers of species are finally counted and compared. We shall have more to say about the kind of method used by Baker and Parker, but it will be best to wait until after we have looked at the work of Clutton-Brock, Harvey, and their various collaborators.

Clutton-Brock and Harvey on primate societies

Clutton-Brock and Harvey, and their various collaborators at Cambridge and Sussex, have produced a series of papers on the relation of mammalian social and ecological variables. Clutton-Brock and Harvey (1977) was the first, and most wide-ranging; we shall look at it in most detail.

For 100 primate species they extracted (mainly from the literature) measurements of eight social variables, such as sexual dimorphism, groups size, group sex ratio, home-range size. These 800 or so numbers made up the diversity to be explained. They divided the species among nine ecological categories (nocturnal/

Table 1.1   Main trends in bird coloration

Character	Trend
Feeding time	diurnal feeders brighter than nocturnal
Feeding place	food off ground, belly less conspicuous
Body size	larger less 'flash' colours, brighter wings, legs
Gregariousness	more gregarious brighter
Incubation site	if concealed, bird more conspicuous
Sex incubating	if only one sex, incubater duller
Paternal guarding (not incubating)	more time guarding brighter
Polygyny	increases male brightness
Lek mating	increases male brightness

Simplified and summarized from Baker and Parker (1979). These are all independent effects in the multiple regression: the effect of lekking, for instance, is independent of polygyny.

diurnal, terrestrial/arboreal, insectivorous/frugivorous/folivorous), and looked for social differences among the nine. The method: analysis of variance. The units? Genera.

The analysis of variance was 'one-way': they tested independently for differences in each of the eight social variables. This is a reasonable method if the variables are independent, which they perhaps are. However, another technique is needed for a variable, like brain size, which is closely correlated with another variable (body size) which may vary among the ecological categories. Clutton-Brock and Harvey (1980) and Mace, Harvey, and Clutton-Brock (1981), therefore, who looked for differences in the brain sizes of mammals among ecological categories, used a different method. They plotted brain size against body size and looked for consistent deviations from the overall regression in the mammals of the different ecological categories. But there is another complication. The overall regression varies according to the taxonomic level of the points it runs through. The regression for families differs from the regression for genera. Mace, Harvey, and Clutton-Brock therefore had to make two taxonomic decisions. They had to decide which level to use for the overall regression, and at which to look for deviations. They could, for example, plot a graph of species points for a single family. The regression is then for a family, and deviations are of species. They in fact plotted all the points for families, and looked for deviations of genera. These two decisions are not fully justified in the original paper, so we cannot look any further into the method here. Once again I have summarized some of the trends (from these two and other papers) in a table, Table 1.2.

Why did Clutton-Brock and Harvey use genera? Their justification is (conceptually, not historically) a development of Baker and Parker's. Clutton-

Table 1.2 Main trends in primate societies

Character	Trend
Sexual dimorphism	increases with absolute size
	increases with group sex ratio
Body size	increases with proportion of foliage in diet
	nocturnal smaller than diurnal
Population density	arboreal higher than terrestrial
	decreases with body size
Relative brain size	smaller in folivores than frugivores (same trend in small mammals)
	greater with greater home range
Relative male tooth size	terrestrial greater than arboreal
	multi-male group greater than single-male (?)
	polygynous greater than monogamous (?)

Simplified and summarized from: Clutton-Brock and Harvey (1977, 1977a, 1980); Clutton-Brock, Harvey, and Rudder (1977); Harvey, Kavanagh, and Clutton-Brock (1978); Mace et al. (1981).
(?) indicates trend less certain.
'Relative' means relative to body size, and (in 'relative male tooth size') relative to female.

Brock and Harvey analysed the variance of social variables among genera within families. They demonstrated that (for seven of eight variables) there was significant variation within families. This is some justification for using genera. But they went a step further. They performed a nested analysis of variance on genera and families. They determined that little additional variance could be found among families. Only two of the variables showed significant extra variance among families. Thus, it may seem (and we shall return to this) that genera show as wide a range of variation as do families. Clutton-Brock and Harvey therefore settled on genera. To go below, to species, would bring in many non-independent groups of congeners. To go above, to families, would add little variation.

Clutton-Brock and Harvey have used the nested analysis of variance as a technique of finding the best single taxonomic level. The aim is expressed in the following quotation from Harvey and Mace (1982, p. 347, see also Krebs and Davies 1981, p. 40):

If our upper limit of analysis was the order, then we might use species, generic or family estimates for comparing relationships between variables within the order . . . We should use the lowest taxonomic level that can be justified on statistical grounds, i.e. the lowest taxonomic level at which maximum variance is exhibited in our measured variable.

Their suggestion is to measure the variance (of the character of interest) at all taxonomic levels, and to use in the test that taxonomic level which has the most variance. But let us now look a little closer at 'most variance'. 'Most variance' is not the same as 'significant variance'. I do not pretend to statistical expertise, so let us read the words of Harvey and Martin (1983, typescript p. 4):

We caution against the uncritical use of statistical tests for analyses of the sort described here. There are usually far more species than genera, and genera than families within an order. This means that degrees of freedom for variance-ratio ('F') tests are concentrated at the lower taxonomic levels. Therefore even when the variance ratios are the same, significant differences are more likely to be revealed at lower than higher taxonomic levels. Indeed, Clutton-Brock and Harvey (1977a) based their comparison of home-range size among primates on significance tests alone, and incorrectly concluded that maximum variance is found among genera. However, partitioning the variance components reveals additional variation among families within the order which is of even greater magnitude than that among genera within subfamilies. Yet, since degrees of freedom are appreciably fewer at higher taxonomic levels, the variance ratio is not significant.

Thus the preferred level should be that at which there is most variance, not that at which the variance component is most significant. But this is a technical point. The main point for our purposes is that Clutton-Brock and Harvey, and their collaborators, employ a method which uses a single, 'best' taxonomic level.

If a single taxonomic level is to be used, the one found by the nested analysis of variance is the best. If it favours the genus, for example, then as much variation exists on average among genera as among families. Then let us look for correlations of differences among genera. The test will be no more powerful if we used higher taxonomic levels.

But our test does not have to use only a single taxonomic level. Might not a mixture of taxonomic levels give a stronger test than any single level? If we look at that taxonomic level which on average has the most variance within its groups, there may be some groups which do not contain the full range of characters, and other groups in which the full range of characters could be found at a lower rank. If the favoured level were the family, then some families might still not show as much variance as the order, while the range found in other families might be fully represented in each of its genera. The level favoured by the method is a compromise. I do not think that there is a single optimum taxonomic level for a test. I favour instead a different solution. It is a simple extension of the techniques of Baker and Parker and of Clutton-Brock and Harvey. It is to measure the variance within higher taxa, and use as test units lower categories if there is a high variance, but higher categories if there is low variance. It is to take each taxon in turn and treat it individually. We have now come far enough to discuss the method I shall be using in this book.

The methods that we have just been reviewing were all intended to remove the bias that can be introduced by groups of related species, which share the same characteristic. These groups of species bias the test because they are statistically non-independent. The successive methods may be judged as successive attempts to find an independent unit, a unit which is a single independent trial of a comparative hypothesis. If a group of species all share the same characteristic, then their common ancestor probably also shared the same characteristic. The character evolved only once in the group, but if all the species are counted, a single evolutionary trial of the hypothesis will be counted as many times as there are (studied) species descended from that common ancestor. The methods all seem to be aiming at an ideal unit, the unit of independent evolution. As we have seen, they do not achieve this aim. I shall be trying to build a method based on the recognition that the units of the test should be independent. It will be a comparative method with independent trials.

But we have a few more questions to answer before we come to the real work of construction. What is the reason why the criterion of independent evolution should be preferred? What is the difference between a method which uses independent trials and the earlier methods which (aiming at independence) use a single taxonomic level? Are only independently evolved characters adaptations?

We will answer these questions in order. The answer to the first is to be sought in first principles. Natural selection provides a causal explanation of the relation between adaptation and environment. The comparative method tests for associations which are suggested by the theory of natural selection. However, a coincidence of two characters, A and B, may not reflect a causal relation; it could be that B is associated by chance with a third variable, C, which really causes A. C may not have been thought of, or detected, so the hypothesis that B causes A would be erroneously supported. The best solution to the problem of coincidentally associated causes is to test the hypothesis many times independently. A chance coincidence of the true cause, C, in many independent incidences of the false cause B is unlikely. Thus the more independent confirmations of an association, the more confident we may be that the association is not a coincidence. The same is not true if the association is repeatedly observed in many non-independent trials. If undetected variables are confounded in one trial, they will be confounded in all in trials which are not independent of it: the multiplication of non-independent instances does not make it any less likely that they are coincidental. The multiplication of independent instances, however, provides evidence of a natural law.

Let us consider this abstract argument in an evolutionary example. Suppose that we are interested in the hypothesis that species will evolve a particular adaptation, such as paternal care, in a particular 'environment', such as when males are territorial. In fact paternal care might be an adaptation to some other 'environment', such as external fertilization. Now we look for the facts. We find a family of 10 species all with paternal care and male territoriality. By an unnoticed coincidence, they all also have external fertilization. If we count all ten, all of which have descended from a common ancestor which had paternal care, we will deceive ourselves. They are not ten trials. They are only one.

If they are counted only once, the unfortunate coincidence of male territoriality and external fertilization will only produce a single entry in the table. Single entries do not mislead: nothing can be concluded from a single association.

If male territoriality and external fertilized are correlated, this simple argument is not enough. Correlations between characters are of course very common. The multiple regressions of Baker and Parker (1979) were one method of dealing with correlated variables. The technique of Mace *et al.* (1981) was another. The whole problem has been thoroughly discussed elsewhere (Clutton-Brock and Harvey 1979, pp. 554-8; Harvey and Mace 1982; Harvey and Martin 1983). But correlations are a separate problem. Correlations may confound tests even when independent trials are used. The use of independent trials protects us only against chance coincidences.

We come to the second question. What is the difference between the criterion of independent evolution and the criterion of any particular, single, taxonomic level? Consider how characters may appear in the multiplying branches of phylogeny. Once a character has evolved, it may be retained through any number of speciations. It may be retained in only one species before the line goes extinct or a new character state is evolved. Or it may be retained in several species, which make up only a part of a genus, or in enough sufficiently distinct species to comprise a genus. It may be retained in whole families, orders, or kingdoms. Thus, when counting cases of independent evolution, the units in the test will be a mixture of taxonomic levels, and parts of levels. There is no single taxonomic level in the test. Thus does the present comparative method differ from almost all earlier comparative methods.

The third question asks about the meaning of adaptation. The method of independent trials does not count instances of adaptation. The fact that several species share a character does mean that it has probably only evolved once in them. But that does not mean that it not adaptive in all of them (Cain 1964). When a character is shared among several related species it is sometimes attributed to 'phylogenetic inertia'. But it is not always clear what this means.

Is phylogenetic inertia an adaptive or a non-adaptive process? The term was invented by Wilson (1975, pp. 33-8), and he covered under it a variety of processes, most of them non-adaptive (e.g. absence of genetic variation);[3] Gould and Lewontin (1979, p. 585) treat phylogenetic inertia as a non-adaptive force, and Harvey and Mace (1982, pp. 237, 355) refer to it as 'phylogenetic constraints', which they contrast with adaptation by natural selection. All these discussions, if I have interpreted them correctly, have a different explanation of similarities among species that are shared by descent from a common ancestor from that which I would favour. I prefer the explanation by Clutton-Brock and Harvey (1977, p. 6): 'Evidence that a trait is similar throughout a particular genus . . . does not indicate that it is non-adaptive—only that phylogenetic

---

[3] Cain (1964, p. 56) listed a similar set of processes under the title 'genetic inertia'. That term, I believe, may have been coined by Darlington and Mather (1949).

inertia (*sensu* [they say, of] Wilson, 1975) may have been involved.' Let us hope that authors who continue to use this term will make it quite clear just why the characters in question have not changed during evolution.

If a character is the same in a group of species, one possible explanation is that it is adaptive in all of them. The reason for including it only once in the test is not that it is non-adaptive, but that the group of species only provides one independent trial of a comparative hypothesis. Similarly Cain (1964) argued that homologies, characters shared by several species, should not be thought of as non-adaptive: 'the major plans of constructions shown by the older groups are soundly functional and retained merely because of that'. However, a character that is homologous in a whole group can only be counted once in a comparative method which aims to test what environments particular characters are adaptations to.

A similar argument applies to 'exceptions' to a comparative trend. Even if a comparative trend is statistically significant, there will probably be a few taxa that do not conform to it. They show the 'wrong' adaptaton to a particular environment. But we may pause on that 'wrong'. The method, as we have already noticed once, tests what adaptations are for, not whether species are adapted. The hypothesis, not the species, was mistaken. That a species is an exception to a trend is not evidence that it is maladapted; it is possible that the exceptional species evolved its character as an adaptation to some other environment than the hypothesized one. The nature of that other environment could be investigated in further comparative studies.

The recognition of independence

Independent evolution may be the ideal criterion for the comparative method. How is it to be recognized? Evolutionary transitions took place in the past; but we can only observe the present. When we find a character in two species we cannot determine, with complete certainty, whether it is shared from a common ancestor or independently acquired. So we shall accept at the outset that we are not going to have certainty. The question then becomes whether we can even make a reasonable estimate. To recognize independent evolution is to distinguish primitive from derived character states. Derived characters are independently evolved characters. We have walked into cladism.

Cladism is a philosophy of classification. It can be distinguished by the kind of characters it uses. Any kind of classification will use some kind of shared similarity to classify organisms into groups. Cladists concentrate on the distinction between primitive and derived characters. They try to classify species according to patterns of shared derived characters.[4] They try to exclude shared primitive characters from classification. A cladistic taxonomy, therefore, is not a

---

[4] Cladists follow Hennig (1966) in usually referring to primitive characters as 'plesiomorphies', derived characters as 'apomorphies', shared primitive characters as 'symplesiomorphies', and shared derived characters as 'synapomorphies'. I do not.

hierarchy of 'overall similarity', of similarity with respect to all characters. It is a hierarchy of only one kind of character: shared derived characters. In a cladogram, shared derived characters can be represented by a horizontal line between the taxa that share them. A simple cladogram might be for three species and four characters. If species 1 and 2 share two derived characters, C' and B'; if species 1, 2 and 3 share another derived character, D'; and if species 1 has a uniquely derived character, A', the cladogram of the three species is shown in Fig. 1.1.

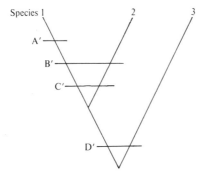

Fig. 1.1

Species 3 has the primitive character states A, B, and C, and species 2 has the primitive character state A, all of which are only implicit in the cladogram. The similarity of species 2 and 3 for character A is not represented in the taxonomy: shared primitive characters have no place in cladistic taxonomy.

If only derived characters are to be used, and primitive characters are to be excluded, techniques of distinguishing the one from the other are essential. Cladists need such techniques. We also need them. There, however, the common ground ends. We will soon be looking at the techniques of cladism; we will see how we can apply them, but our application stands the whole of cladism on its head. The cladist wants to erect and test taxonomies; we want to eliminate them. In our method cladism will only come in after we have an acceptable taxonomy. We will first look for a taxonomy in the literature. Then, when we have found one we will assume that it is true. Then we will use the techniques of the cladist to count the frequency of independent evolution of characters in the phylogeny. The function of cladism is to protect us from taxonomic artefacts. For our particular purpose we do not need a general introduction to cladism; we need only concentrate on those parts of it which may prove useful. Cladism, however, is surrounded by the kind of squabbles that no one likes to get dragged into unnecessarily. So we would do well to clarify our relationship with so controversial a system. We will pause to make quite clear which bits of it we do need, and which we do not.

We have said that we need the techniques of the cladist. We have also said that cladism is a whole philosophy of classification. The techniques are the subject of only minor controversy; it is the philosophy that is notorious. Cladists assert that only shared derived characters, and nothing else, should be represented in a classification. Other methods of taxonomy admit other kinds of shared similarity into the classification. Why do cladists prefer not to? Hennig (1966, pp. 9-27) gave one reason, although it is hardly an answer to the question because, as phylogenetic taxonomy has lapsed into cladism, it has more or less disappeared from the literature. It is however interesting enough to merit a digression. The true phylogeny of species is a real hierarchy which exists in nature. If only we can find methods of discovering the phylogenetic relations of species, we shall have a method of discovering a true, stable hierarchy. Now, a hierarchy of species based on shared derived characters is a hierarchy of phylogenetic relationships. Thus cladism does at least aim at a hierarchy which exists. (How successful it is is another question.) The main alternative to the phylogenetic hierarchy is one of morphological resemblance. Such is the aim of numerical taxonomy. Hennig reasons that there is no true hierarchy of morphological resemblance: there is no single hierarchy which the numerical taxonomist can hope to discover. The numerical taxonomist can certainly find hierarchies, but which hierarchy he finds will depend on his criterion of morphological resemblance. His hierarchy depends on his clustering statistic (whether it forms groups according to the nearest neighbour, or the centres of groups of points, or some other method): Hennig's point (translated into the modern idiom) would be that there is no true clustering statistic because there is no true hierarchy of morphological resemblance.[5] In cladism there can be true and false techniques because there really is a hierarchy to be found. (That a cladogram is not a phylogenetic tree is an important point, which was realized by cladists subsequent to Hennig (Eldredge and Cracraft 1980, pp. 137-46), but it remains true that more closely related species on a cladogram are phylogenetically more closely related. The modern distinction between 'cladogram' and 'tree' does not alter Hennig's argument.)

We do not have to make up our minds about Hennig's argument. We do not have to decide whether cladism is a sensible philosophy. It does not matter to us whether classifications are cladistic. We do need to know phylogenies (as we shall soon see); but we do not need to find them in taxonomy. We need only to be able to find them somewhere. We do not have to suppress categories such as 'fish' and 'reptiles'. Just as we can extract facts about adaptations from the appropriate literature, we can take phylogenies from the phylogenetic literature. The latter need no more be represented in the taxonomy than need the former. It would be convenient if taxonomies were cladistic, but it is only a convenience. It is not a necessity.

Techniques, valid techniques, of distinguishing primitive from derived character states are a necessity. Without these our comparative method would be

[5] Sneath and Sokal (1973, pp. 201-53) review the various clustering statistics. Johnson (1970) provides an important critique of the concept of overall similarity.

impossible. Let us now turn to the techniques themselves.[6] I should just point out that all these methods had been used before cladism ever existed. But we will call them cladistic techniques: it is in the cladistic literature that the most and the clearest discussion of them can be found.

We shall consider only two of several methods: outgroup comparison and ingroup comparison. Outgroup comparison is the most interesting. It is the main method used both by cladists and this book. Suppose that we have two related species, 1 and 2, and we are studying a character called *a* which is in state A in species 1 and A' in species 2. The question is whether A or A' is primitive. The answer by outgroup comparison is obtained by studying *a* in a related species (the outgroup). Whatever the state of *a* in the outgroup species is the primitive state. In the cladogram shown in Fig. 1.2, the primitive state for the

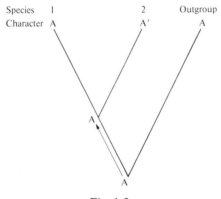

Fig. 1.2

species pair is A. We want to recognize convergence, and count its frequency. Outgroup comparison will give a minimum estimate of its frequency. It assumes that convergence is less likely than common ancestry, and only admits convergence when it has to. If A is primitive in the species pair 1 and 2, the only evolutionary event is the evolution of A' in 2. If on finding A in the outgroup we had taken A' to be primitive in the pair we would have implied that both A' and A had evolved once each in the group. Outgroup comparison thus minimizes the number of (estimated) evolutionary events.

Although the method minimizes convergence, it does not eliminate it entirely. If we study many characters, not all of them will necessarily show exactly the same pattern of sharing among the species. For example, suppose we have

---

[6] Again, this is not a general review or introduction. I have selected for detailed discussion only those techniques which may prove useful to the comparative adaptationist. Some of the more important, and more recent, discussions are: Maslin (1952), Hennig (1966), Crisci and Stuessy (1980), de Jong (1980), Eldredge and Cracraft (1980), Stevens (1980), Watrous and Wheeler (1981), Wiley (1981), Bishop (1982).

studied six characters (*a* to *f*) in three species (1, 2, and 3). Writing primes for the derived character states, the three species might be: 1(*a*, *b'*, *c'*, *d'*, *e'*, *f'*), 2(*a'*, *b*, *c*, *d'*, *e'*, *f'*), and 3(*a'*, *b*, *c'*, *d*, *e*, *f*). The cladogram is shown in Fig. 1.3. The changes in characters *a* and *c* are probably convergent.

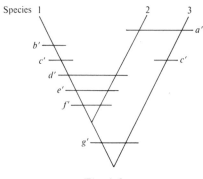

Fig. 1.3

Thus will we estimate the frequency of convergence. Thus we will count the evolutionary trials of comparative hypotheses. Our estimate is the minimum; but we are forced to use it because it is the only estimate than can reasonably be made. If we were better informed we might be able to recognize convergence by other criteria, and so make a better estimate of its frequency. But for most characters we do not have any more information than the distribution of character states among some species. If that is all we know, outgroup comparison provides the best (although minimum) estimate. How good an estimate it is will depend on the truth of its parsimonious assumption. If convergence is very much less likely than common ancestry then the estimate made by outgroup comparison will not be far wrong.

But is convergence so much less likely that common ancestery? Cain (1982) has recently criticized cladism for assuming that convergence is rare. Convergence, Cain suggests, may be very common. Suppose he is right. Suppose convergence is very common. How can we reply? We can no longer pretend that outgroup comparison estimates the frequency of convergence. It will produce an extreme underestimate: it will not estimate the true frequency of evolutionary trials. But we might still be able to use it to test hypotheses. We would no longer call it the frequency of convergence; we would call it instead an abstract statistic, which we can use to test hypotheses. We can still put the numbers obtained by outgroup comparison into a statistical test. If the test is significant we still have evidence that the trend is true. If outgroup comparison underestimates the true frequency of evolutionary events we ought to be all the more impressed by a trend which is proved by outgroup comparison.

Because it underestimates the number of trials, it makes the test more persuasive. Thus if we think, with Cain, that convergence is rife, we must stop thinking of the numbers obtained by outgroup comparison as frequencies of independent evolution. But we do not have to stop using them. They take on a new character as persuasive statistics.

We need only fall back on the psychology of persuasion against an extreme criticism. Probably most biologists (probably including Professor Cain) would agree that the estimate obtained by outgroup comparison is a useful minimum estimate of the frequency of convergence. But the estimate is statistical, not certain. Sometimes it will be wrong. Its justification is that it will more often be right than any other method.

We may notice a related criticism here. It applies specifically to our application of outgroup comparison. The characters which we shall be studying, interested as we are in adaptation, will be particularly prone to convergence. The outgroup method is most reliable for characters which are rather stable in evolution. We are going to use outgroup comparison on the characters to which it is least applicable. This is a difficulty; but the same answer can be repeated. Outgroup comparison minimizes the amount of convergence. Outgroup comparison will give a more biased estimate for characters that (in fact) show much convergence than it does for more stable characters. But the direction of the bias is always downwards. The method may not be perfect, but the imperfection only makes things more difficult for the student of adaptation. If a trend can be demonstrated despite the extreme conservativism of the technique, it must be a real one.

The use of outgroup comparison to recognize the independent evolution of characters is straightforward if the taxonomy of the species into dichotomous sister groups is known in advance. On finding a particular character state in a species we just look at the nearest outgroup. If the outgroup has the same state we do not count a case of independent evolution; if it has a different state, one entry can be made in the test.

Very few groups have been classified into simple dichotomous cladograms. We shall more often find taxa that contain several species, as shown in Fig. 1.4. What can we do with classifications like this one? Three points may be made. First, must branching be dichotomous, as most cladists demand? Their demand

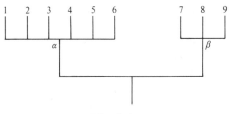

Fig. 1.4

has been criticized. The decisive issue is whether speciation is always dichotomous. One confusion should be cleared up first. If a single species splits off from an ancestral species, the ancestor itself should strictly be called a new species by the cladist regardless of whether or not it has changed (Hennig 1966, pp. 63-5; see also Hull 1978). It would be renamed by a classical evolutionary taxonomist only if it had changed considerably. Thus in insisting on dichotomous classification the cladist is only insisting that two species never split off from an ancestor at the same time; if they split off at different times the ancestor would have changed its species in the mean time. But even once this has been understood I still think that cladists are making an unnecessary difficulty for themselves. I would not be surprised if several species were diverging from one population at the same time. Anyway there is no reason why in principle only one species can ever be splitting off from another at any one time. Cladists (e.g. Platnick 1979) write that if a character analysis suggests a trichotomy we cannot be sure that subsequent analysis will not reveal a pair of dichotomies; they thus call trichotomies 'unresolved dichotomies'. But the only possible reason for believing this is that speciation has to be dichotomous. If it is not, we could just as well say that paired dichotomies are unresolved trichotomies because we do not know that subsequent character analysis will not destroy the dichotomies, suggesting in their place a trichotomy.

The second point about the classification in Fig. 1.4 is that the main reason why they exist is inadequate taxonomy. They are most often unresolved groups of smaller groups. Taxonomy is not only usually not adequate; it is also usually not phylogenetic. What should we do when we know a taxonomy is not a phylogeny but we have nothing else to go on? The answer is that we take the taxonomy as the best estimate of phylogeny that we have, and use it as if it were true. Even if it is a purely numerical taxonomy it will be a better estimate of the phylogeny than to assume that all the species in the whole group are equally related. And to assume that all species are equally related is the only real alternative. So in the absence of a phylogeny I always take the best (most nearly phylogenetic) taxonomy as a better than random estimate of phylogeny and count the frequency of independent evolution as if the taxonomy were a true phylogeny.

The third is that, with a non-dichotomous classification, we may be tempted to use the second cladistic method that I mentioned at the beginning: ingroup comparison. Ingroup comparison is another method of distinguishing primitive from derived character states. If we have two states, A and A', in the species of a higher taxon and we want to know which is primitive, we might count which is the most common in the group, and then conclude that the most common state is primitive. We would have performed an ingroup comparison. (Ingroup comparison is not so tempting with dichotomous taxa because if there are only two species, no character state can be in the majority if both are represented; ingroup comparison would be possible on a higher taxon made up of dichotomous parts.)

Most cladists (Crisci and Stuessy 1980, references on p. 121) object to ingroup comparison, because it is possible for a character to be the most common in a group but to be the derived state (Fig. 1.5). This criticism is in itself valid,

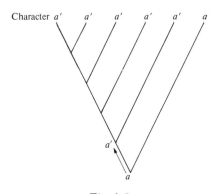

Fig. 1.5

but we are still left with the question of how to deal with non-dichotomous phylogenies. We have noticed that minimum convergence is the general principle which justifies any particular cladistic technique; we may therefore fall back on it to sort out whether outgroup or ingroup comparison is the better technique for analysing non-dichotomous taxa. We may use Fig. 1.4 again to compare ingroup and outgroup comparison. Suppose that species 7, 8 and 9 of taxon β all have character a in state A, and both states A and A' are found in taxon α: let species 1, 3, 5, and 6 be A' and 2 and 4 be A. The question is whether A or A' is primitive in α. The answer by outgroup comparison is that A is primitive. By ingroup comparison A' is primitive because it is more common. We have thus established that the two techniques can give different answers. But which technique is more conservative? In this case outgroup comparison is more conservative because in the whole group it counts only one case of the evolution of A'. Ingroup comparison makes A' primitive, so scores one case of the evolution of A', and another of the re-evolution of A in α. In fact outgroup comparison is either more conservative than ingroup comparison (as in this case, when the two methods give different answers to the question of which character state is primitive in α), or equally conservative (when they give the same answer, as they would have if A had been more common in α); there are no other possibilities. I therefore will be using outgroup comparison with both dichotomous and non-dichotomous taxonomies. The desire for consistency and the principle of minimum convergence combine to make us prefer outgroup comparison.

Finally, we will discuss two important cases where outgroup comparison cannot be easily applied. In the first it cannot be used at all. This is the case in which both the group of interest and the outgroup contain all character states

(Fig. 1.6). Outgroup comparison cannot decide which character state is primitive in the pair of species, 1 and 2. Is A primitive and has A′ evolved once or is A′ primitive, and A evolved once? Once again we may turn to the more fundamental principle of minimum convergence to try to resolve the problem. But the principle fails us. Both answers are equally conservative. In cases like this, if one of the groups contained more than two species we could turn to ingroup comparison, and argue that the more common character is primitive. If both are dichotomous, we can use either score without biasing the final test. Our methods do not lead us to prefer one course over the other.

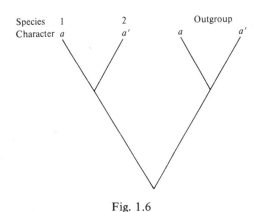

Fig. 1.6

The other difficult case is where a character reverses rapidly through a group of related species (Fig. 1.7). Here outgroup does not give the most conservative estimate. If we take species 4 as the outgroup of 3 we make A′ derived in 3; if we then take 3 as the outgroup of 1 and 2 we make A derived in 2. It would be just as reasonable (and parsimonious) to make A′ derived separately in 3 and 1, and take 4 as the outgroup of 1 and 2. We shall meet two examples of such a cladogram, one in the amphipods, and another in the toad genus *Bufo*. We will on both occasions use the nearest outgroup possible. The justification is really no more than that it makes for greater consistency of method.

Cladists use several other techniques for distinguishing primitive from derived character states. Among these techniques are embryology and palaeontology (Crisci and Stuessy 1980; Eldredge and Cracraft 1980; Stevens 1980), and there are other special techniques, such as chromosomal inversions (Williamson 1981, p. 183) which may be used in some cases. Apart from the special techniques, which can only be used in special cases, the other techniques are not as good as outgroup comparison (refer to discussions cited above). I have left them out of our discussion for another reason: I have not been able to think of how to apply them to the characters which I will be treating in the second half of this book.

Now we have completed our method of comparison. We shall discover, from the literature, the distribution of character states over phylogenies. Then, by outgroup comparison, we shall count the minimum number of independent evolutionary events. These independent units can then be counted up as independent trials of a hypothesis. They are the raw material for statistics.

Although complete, the method is far from perfect. It cannot solve all the problems of comparative biology. Its defects are numerous: the results of outgroup comparison are grossly conservative; outgroup comparison cannot even in principle be applied to all (even perfectly dichotomous) phylogenies; its

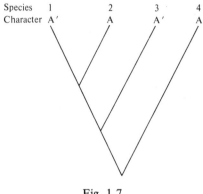

Fig. 1.7

estimates are even more wayward when based on our usual, imperfect phylogenetic knowledge.

But we do have a method, and one based on first principles. From this vantage point we can settle a number of issues which often arise in discussions of the comparative method. Some of them will be the subjects of subsequent sections, but we can now just quickly revisit a proposed distinction which we met at the beginning of the chapter. It is the distinction between two forms of the comparative method. Altmann (1974), Wiley (1974), Hailman (1978, 1981), and many others, have suggested that two main forms of the comparative method: the study of convergence of unrelated species in similar environments, and of the divergence of related species in different environments. This distinction may apply to rudimentary comparative research. But it disappears in the method that we have just built. It incorporates at once both convergence and divergence. It contains them both and is superior to them both. An observation that several species in a taxon diverge in different environments is not in itself evidence of how the species are adapted. By itself it is a single point and, as we have seen, single associations of adaptation and environment prove nothing. Only as the association is found in several taxa do we move towards a evidence that compels. And then we are moving towards a study that includes both divergence and convergence. Such are the most complete, and the most conclusive, test.

TWO PROBLEMS

The constructive work is now over. Before taking the method out to confront the diversity of nature we shall pause, for three more sections, to examine some criticisms. We shall concentrate on specific criticisms of the comparative method; readers interested in the general problems of adaptation may be referred elsewhere (e.g. Dawkins 1982, pp. 30-54). The comparative method runs into difficulty if the relation between adaptation and environment is not predictable. Indeed, as it has been formulated above, it might appear to be a method of testing only uniquely predictable relationships between adaptations and environments. The hypotheses which we have discussed have all been of the form 'in environment E, adaptation A is favoured; in environment F, adaptation B'. Bock (1976, pp. 58-9) believes that the method is only valid if there is only one adaptation for each environment: 'the comparative method depends upon the additional principle that given a particular environmental factor, or a selection force . . . there is a single optimum adaptation to it . . . If the cause-effect relationship between environmental factor adaptation is not law-like, then the comparative method of ascertaining adaptations cannot be a law-like principle'. See how he moves from 'singularity' to 'law-like'. But is a singular relation the only kind of law-like one? We shall return to that.

In nature the relationship between adaptation and environment may not be singular. More than one adaptation may be fitted to an environment; one adaptation may fit several environments. These are our 'two problems'. We will call them 'non-adaptive divergence' (or multiple adaptive solutions) and 'non-adaptive convergence.' This section is all about them.

We will first ask whether they exist in nature. We will then ask what effect they have on the comparative method. We can notice at the outset that they will make it less simple. They replace the simple relationship of adaptation to environment by one which is more complex, and less easily detected. The adaptation which a species has now depends not only on its environment but also on its history: different species entering the same environment with different initial adaptations may respond by evolving different new adaptations. Which adaptation is selected into the population will depend on the starting conditions. But some critics have gone further. Non-adaptive divergence and convergence do not just make the method difficult, we are told: they invalidate it. Bock (1976), for instance, believed that he could invalidate the comparative method by a few examples of non-adaptive convergence and divergence. 'My thesis [he announced on p. 58] is that this method is not permissible', and (he continued, on p. 59), 'the nonlaw-like nature of the principle of a single adaptive optimum can be demonstrated by showing a number of exceptions to it'. Other discussions have drawn less extreme conclusions. The following discussion is not a reply to anyone in particular. It will look at these two processes, and ask, of each, just how damaging it is: does it (as Bock believes) blow the whole system sky high? Or is it only a minor nuisance which can be ignored at so early a stage of research? Or not even a nuisance, but a part of the problem which the method was designed to solve.

Non-adaptive divergence: multiple adaptive solutions

How often do we read of multiple adaptive solutions, of multiple peaks! Most neo-Darwinian accounts of adaptation include a section on 'single solutions' and 'multiple solutions.' Simpson (1953, Chapter vi), Dobzhansky (1956), and Rensch (1959, pp. 59-68) are just three references. Bock's 1976 paper is just the extreme statement of a series of papers on (among other things) multiple adaptive solutions (e.g. Bock 1959, 1967). Many biologists, therefore, believe in non-adaptive divergence. But why? Evidence, perhaps? If there is any evidence, it should be possible to find it in the discussions of the subject. The idea of non-adaptive divergence has a broad-based pedigree, with roots in many fields of biology. Here we will look at five of the more important.

The first is classical zoology, wherein it is a commonplace. Does such a proposition need proof? 'Every naturalist' as Gould and Lewontin (1979, p. 593) say 'has his favourite illustration'. We need only start a list of old favourites, of lobsters and starfishes, which despite the great difference of their adaptations both inhabit the bottom of the sea, consuming a similar diet of deposited food; of the right-aortal circulation of birds, the left-aortal circulation of mammals; of perissodactyls which walk on odd-numbered toes, while artiodactyls walk on even-numbered ones. It is a commonplace too of the genetical theory of speciation of the moden synthesis. Again, a full proof can hardly be necessary. Shall we mention the 'adaptive landscapes' of Sewall Wright? Yes? . . . well, it is a graph of the mean population fitness in a many-dimensioned space of gene frequencies. The landscape is supposed to have many different peaks, and in a simple model at least the population will evolve up its nearest peak, and stay there unless the gene frequencies are disturbed by some nonselective mechanism. Which peak the population finishes up on will depend on which one it starts near to. It is possible, therefore, that different populations in the same environment could (from different starting points) finish up on different peaks.

Mayr's idea of a 'genetic revolution' at speciation can be formulated to give the same conclusion. Speciating populations, according to this idea, are small and may contain a non-random sample of the ancestral population's genes. This chance difference may shift the population into the neighbourhood of a new adaptive peak, which it will be pushed up by natural selection. Then natural selection will favour decreased interbreeding between the ancestral and derived populations because the intermediates are non-adaptive: speciation ensues.[7]

We turn next to Fisher's (1930) model of sexual selection. Another exposition can hardly be necessary;[8] we can go straight to its consequences. It is another process which (in theory) can produce differences between populations which are not predictable from any differences in their environments. The process starts when a character in males becomes linked with fitness; females are then selected to choose to mate with males possessing that character. The character then evolves, becoming bigger and bigger. If in different populations different male characters happened to become associated with fitness, Fisherian female choice would produce different exaggerated characters in the two populations. The true cause of the difference would be a (possibly random) historical association of characters and fitness; environmental differences need not necessarily have anything to do with which male character becomes linked with fitness.

Allometry is a fourth tradition. If some character, such as brain size, is correlated with body size, then if body size changes in evolution, a correlated change in brain size could take place. The differences between populations in brain size could be non-adaptive; they may just passively reflect past changes in body size. This argument dates from the inception of allometric studies, with Julian Huxley's book (1932, p. 214):

The evolutionary importance of the facts lies in this: that wherever we find the rule . . . holding true for a series of separate forms, we are justified in concluding that the *relative size* of the horn, mandible, or other heterogonic organ is automatically determined as a secondary result of a single common growth-mechanism, and [he adds, bursting into gleeful italics] *therefore is not of adaptive significance*. This provides us with a large new list of non-adaptive specific and generic characters.

One more theoretical tradition which has investigated multiple adaptive solutions is the theory of evolutionarily stable strategies (Maynard Smith 1982). An evolutionarily stable strategy (ESS) is one which, when it is common in the population, cannot be invaded by any rare mutant strategy. In many models there is more than one ESS. As in Wright's model (which we met above) the ESS that a population evolves to depends on which one it starts near to.

So much for the traditions of theory behind multiple adaptive peaks. The next question is whether they are concerned with something real or imaginary. Non-adaptive divergence, after all, can only mislead if it exists; and it can only be misleading in proportion to its frequency: do they exist, and how

---

[7] See Wright (1978), Mayr (1954, 1963). Wright (1980, p. 839) discusses the relation between his 'shifting balance' theory of evolution and Mayr's 'founder principle'. They are not the same.

[8] Dawkins (1976, pp. 170-1) is an exposition which is easy to understand. Harvey and Arnold (1982) explain some more recent important models by Lande and Kirkpatrick.

common are they? An answer might be sought in two sources: logical deduction from the five traditions we have examined, and empirical counting. The first possibility is hopeless. The second, as a few words will demonstrate, is fruitless: all the traditions we have examined are completely speculative. The examples from classical zoology are not rigorous: none of them have been proved to be alternative adaptations. In fact I know of no firm example of a 'multiple peak' in nature; the strongest example was the study of a chromosome polymorphism in the grasshopper *Moraba scurra* by Lewontin and White (1960, see Lewontin 1974, pp. 285-81), but this has been challenged (Wright 1978, pp. 127-45). Although we do not have any compelling evidence to guide us, the overall repeating pattern of one example after another is convincing. Multiple adaptive peaks probably do exist.

But how many are there? If they are rare we can reasonably ignore them, because they will not often spoil our studies. If they are common we shall have to take them more seriously. In fact we do not know whether they are rare or common. Having no knowledge, we can only fall back on philosophy. Research, perhaps, tends to progress faster if we do not needlessly assume complex theories. There may thus be some small justification for ignoring non-adaptive divergence at this early stage of research. Parsimony is, however, a last resort when all other arguments have failed. But fortunately we will not be compelled to rest our method on the razor edge of parsimony. Let us continue the argument.

Let us now assume that non-adaptive divergence does take place, and ask whether it is a nuisance. If it is a nuisance, it is like an invisible demon. It can mischievously and incomprehensibly move species around from one part of our table (or graph) of results to another. We might think that species are arranged in the table according to the environments they live in. But we would be wrong. They have been demonically deranged.

If the demon is completely invisible, the results become meaningless. But it may be possible that its machinations can be understood. Non-adaptive divergence is, after all, an intelligible process. We can think, for any particular character, whether a theoretical process is likely to apply. Not all characters, for example, can evolve under Fisherian sexual selection. Could the characters that we are studying have? Do they look as if they did? Does the species have the kind of properties (such as polygyny) one would expect if female choice is operating? Because we understand Fisherian female choice, we might be careful with the courtship displays of male birds of paradise. These are just the kind of characteristics for which the method could be inappropriate.

If we suspect that the difference between two characters is non-adaptive, we can incorporate our suspicion in the comparative study. We can treat the alternatives as the same character. Take, as an example, the incidence of uniparental care (as opposed to no parental care). Here, paternal care and maternal care could be treated as alternative adaptive solutions; we would simply lump them in the test. And how simple it is! In any actual comparative study there will be many obvious alternatives. Parental care might take the form of guarding a nest, carrying the eggs externally, or internally (in the mouth, or in a brood pouch), and there will be many forms of each alternative. They can be lumped in the test. It is only the unrecognized alternatives which are a potential problem. But in any particular study we can try to recognize alternatives: they are not beyond analysis. The comparative method, Bock (1976, p. 79) concluded, 'is dependent strictly upon the law-like quality of the principle that a single adaptive optimum exists for each selection force'. I disagree with him.

What if we suspect that some of our alternatives are not truly multiple, alternative solutions? We might suggest, for example, that maternal and paternal care are not alternative solutions. It is to comparison that we must turn to test our suggestion. Maternal and paternal care could be alternative evolutionarily stable strategies (Maynard Smith 1977), or one or the other might be favoured by natural selection under different particular conditions. The presence of (probably significant) correlations with these conditions (Ridley 1978; Gross and Shine 1981) lends favour to the second possibility. Likewise the suggestion that the diversity of one character (such as brain size) is purely the passive by-product of changes in another (yes, body size) can be tested. We can see whether brain size can be predicted from body size, or whether other (perhaps ecological) factors are needed to predict it. In fact body size alone does not explain differences in brain size between mammal species (e.g. Mace *et al.* 1981; see also Clutton-Brock and Harvey 1979; Harvey and Mace 1982). Thus, as we pointed out on page 7, the comparative method can test whether differences are non-adaptive.

Non-adaptive convergence

The same structure, or activity, may serve different functions in different species. For instance, in a study of the incidence of paternal care I could find no single cause that accounted for all cases; paternal care has it seems evolved in some species because of male territoriality and external fertilization (in combination with selection for parental care), but not in others (Ridley 1978). Territoriality is another example. It functions in the reproduction of some species, and in the feeding of others. Non-adaptive convergence is a natural process.

The arguments of the previous section apply again here. We need not go through them all again at length. As was non-adaptive divergence, so will non-adaptive convergence be a problem for the comparative method. It is a problem when it is unrecognized. But we can use the comparative method to test for it, and our understanding of evolution to assess its plausiblility.

A published exchange between Bock (1976) and Simpson (1978) illustrates many of these points. Bock (1976, p. 75) gave, as an example of non-adaptive convergence, the species of birds with shrike-like beaks. These beaks are long and curved, and have evolved independently in two genera: *Lanius* and *Falcunculus*. *Lanius* uses its beak to eat large insects and small vertebrates; but *Falcunculus* uses its to tear off the bark of trees. It eats the small insects beneath. The same structure has been independently evolved in different environments. Simpson (1978) pointed out that the conclusion depends on the level of interpretation. The convergence may appear adaptive if we shift the level of interpretation. Both kinds of bird use their beak to strip off hard bits surrounding soft edible bits, and in this sense, both birds use their beaks for the same kind of purpose. A similar argument might be made for the different kinds of territoriality: the requirements of feeding and of mating may have in some cases fundamental similarities which explain the convergence.

We are familiar with this kind of argument. With any particular characters we can think about whether non-adaptive convergence (or divergence) is likely. We may then use our knowledge to improve the categories in the comparative test. In the example that we have just discussed, if we trusted Simpson's interpretation more than Bock's we would not use, in a study of the evolution of beak shape, 'eats large insects and small vertebrates' as one category, and 'eats small insects under bark' as another. They would have to be replaced by a category which more accurately described how natural selection controls beak shape.

A note on random phylogenies

'Random' phylogenies are computer simulations of a particular kind of evolutionary model (Raup 1977). Each 'species' is represented in the model by a number of characters, each of which can be in one of a number of states. The species can change their character states each generation according to a specified distribution of probabilities; there are also chances of extinction and speciation. The evolution in the model is solely produced by random changes in character states: there is no natural selection. Convergence can of course occur, with a probability that can be calculated. There is convergence even though there is no natural selection. These models are a kind of null hypothesis about evolutionary change. Some authors have concluded from these models that if you find convergence you do not have to explain it by natural selection. A slightly different conclusion is appropriate here. 'Random' convergence only serves to emphasize further the importance of demonstrating comparative trends statistically. The reason for using statistics is to reject the possibility that the trend could have resulted from chance. Once again, the comparative method will be the best method of finding out how well the model, random phylogenies this time, apply to nature.

The investigation of adaptation by the comparative method has easily survived all criticisms. The relation between adaptation and environment may not be so simple as is assumed in the most naïve formulations of the comparative method. But the question is not whether the relationship is unique, but whether it is predictable. The method cannot be destroyed by demonstrating that the relationship is not unique. If the reasons for the non-uniqueness are intelligible, they can be taken into account. The comparative method can be applied whether nature is simple or complex; truth may be discovered more slowly if nature is complex, but discovered it will be. In any particular case we can think whether either of the two main problems for the comparative method—non-adaptive convergence and non-adaptive divergence—apply. If they do not we can ignore them, if they do then we may then include them in the test itself. The comparative method is the best method of analysing comparative trends, whether or not those trends are adaptive. Non-adaptive divergence and convergence are not inscrutable demons, invisibly wrecking comparative biology: they can be recognized. Once they have been found they can be treated, in our test, for what they are.

We need not stop there. Which of the alternative adaptations a species evolves is determined, as we have seen, by the initial condition of the species. Were this any more than a vague theoretical proposition, if we knew that in a certain environment a species with character A will evolve character C, while a species with B will evolve D, then we could make the comparative method more powerful still. For we could incorporate this knowledge in it. But now we are asking questions where biology can provide little guidance. I have attempted, in the cladistic techniques of the previous section, and in the next two sections, on 'cause and correlation' and on 'laws of evolution', to bring history into the comparative method. As yet we can but dimly see how history may help us to understand organic diversity. Until now history has been a sledgehammer wielded by nihilists: they have not had to think in detail about the effect of history, and they would be much less effective were they to do so. But it is now time to disarm the hammer of the adaptationists. The sledgehammer of the iconoclasts will then become a sculptor's chisel.

ASSOCIATION AND CAUSE

The comparative method is used to look for associations (between the characters of organisms) which natural selection, or some other causal principle, might be expected to give rise to. It can test hypotheses about the cause of adaptive patterns. The result of this test is an association or correlation. The mere association of two characters, A and B, does not tell us whether in nature A was the cause of B, or B of A. The question of this section is whether the cause and the effect can ever be distinguished in a pair of associated characters.

Take a real example, the tendency for monogamous species to show biparental care. Did monogamy come first, males then being selected to care for the young because that was the best way to spend the time previously spent

THE PROBLEM

looking for mates? Or did biparental care come first, whence males were selected to be monogamous, because they spent all the time that could have been spent looking for mates in looking after their young? The association itself does not tell us the direction of evolution. It is tempting to agree with Clutton-Brock and Harvey (1978a, p. 193) that 'there is no obvious way out of this difficulty'. This is not so much a criticism of the method as a limit on how much it can discover: it can discover only associations, never causes. Who has said this? And why? Is there any truth in it? Can it be answered?

We will take these questions in order. The answer to the first includes most of the modern biologists who have used the comparative method to study adaptation. Thus Clutton-Brock and Harvey wrote of their own work on primate societies: 'perhaps too clearly, it demonstrates the problems of this approach: the necessity for reliance on heterogeneous data, the problem of dependence on closely related species, and the difficulty in distinguishing between causal relationships and non-causal associations' (1978b, p. 314). Likewise, from Krebs and Davies (1978, p. 5): 'more serious is the difficulty in distinguishing cause and effect when it comes to interpreting results', which changes from a difficulty to an impossibility two pages later: 'the comparative approach, as emphasized earlier, cannot sort out cause and effect'.

Why do Clutton-Brock and Harvey, Krebs and Davies, tell us that it is difficult to tell cause from correlation? We may assume that they do not have in mind the general philosophical problem of causality. They must mean that causality is rather more difficult to infer in the comparative method than in other areas of science. And they would be right. Experiments are the usual, and the best, scientific method of discovering causes. This may be demonstrated by a familiar ethological example. The subject is communication. A putative signal, A, given by one animal, may always be followed in nature by a putative response, B, by another. How are we to tell whether A and B are indeed signal and response, rather than just being temporally correlated with some third, unidentified, variable which is the true cause of both (Cullen 1972, p. 104)? Experiments are the answer. The signal must be presented experimentally many times, and if the response still continues to follow it then we may conclude that it is being elicited by the signal. The experimental presentations are unlikely all to be correlated with that third, unidentified, variable. *Ex hypothesi* the natural presentations were, so an experimental association is much more convincing evidence of a causal association than is a natural association.

We cannot tinker experimentally with the subject matter of the comparative method. We cannot experimentally make species monogamous and then measure their evolutionary response. If we are to find out anything about the causes of comparative trends we will have to resort to less direct, less satisfactory techniques. This section is to show that however indirect, however unsatisfactory they may be, such techniques do exist. They are used intuitively; they are not often explicitly discussed: but they can be found, veiled and little-known, in the literature. Biologists do notice that it is difficult to distinguish cause from correlation; but they do not consider how it might be done, even with difficulty. In the absence of any other treatment, then, we will look at two methods. It is not a complete list: one applies natural selection; the other cladism.

The first principle is contained in the statement that, if natural selection controls evolution, not all directions of evolution are equally likely. Suppose that, from a comparative survey, we know that characters A and B are associated. We want to know whether A evolved first, and set up selection favouring the evolution of B, or the other way round. The first method is to think whether natural selection would favour B were A already present, and whether it would favour A were B already present. In some cases, one direction of change would have been favoured by natural selection, but the other would not. Take for example the correlation which we have met before, between the mode of fertilization and the sex of the caring parent. Of species with uniparental care, those with paternal care tend to have external fertilization, and those with maternal care internal fertilization. There is a theory which explains why external fertilization predisposes species to the evolution of paternal care, and internal fertilization to maternal care (Dawkins and Carlisle 1976). But could natural selection have driven the system in the other direction? Had paternal care evolved for some other reason, would that set up selection in favour of external fertilization? I think not. However, it does seem likely that had (female) viviparity evolved for some other reason, internal fertilization might be selected for: the eggs would perhaps be safer if they stayed all the time inside the female rather than being released for fertilization and then taken back into the female. Another, and more convincing example may be found in Chapter 4. The hypothesis of that chapter explains homogamy for size by the presence of three other reproductive characteristics: that larger females lay more eggs, that larger males are stronger competitors, and that mating takes a long time. For reasons which may be obvious to many readers, and which are explained in detail later (pp. 173-4), if a species has these three characters, it will be selected to become homogamous. However, there is no reason why the evolution of homogamy should set up selection in favour of any of the three characteristics. If a species mates homogamously there is no particular tendency for it to evolve so that, for example, larger females lay more eggs. Larger females laying more eggs might evolve subsequently to homogamy but this is just as likely in a species without homogamy as one with it, so no correlation with homogamy would be expected. In short, if we can demonstrate the comparative association, we can be fairly sure what the cause of the association was.

In other cases this principle does not work. It does not determine the direction of evolution. That association of biparental care and monogamy can serve again as an example. If a species were monogamous, selection would favour biparental care; if it had biparental care, selection would favour monogamy: evolution could go in either direction. What now is the solution to the problem of cause and effect? It is that each variable could be both cause and effect (or they could co-evolve) in different lineages, and it would be senseless to try to disentangle which is 'the' cause. In any particular evolutionary lineage, one of the variables may have come first; but which one came first may vary between different lineages. We cannot use our understanding of natural selection to determine which came first in which lineage; we only use it to determine that it is unlikely that one of the variables came first in every lineage: the causal arrow may have flown in both directions. And that is not a useless piece of knowledge.

This principle might be criticized. It might be said that it relies too much on the imagination of the biologist. Evolution, we are saying, can go in whatever direction we can imagine natural selection would cause. But it may be that sometimes our imagination will fail us. We will sometimes fail to imagine what is in fact true. And we will then be wrong. I would only point out that this is not so much a criticism of the comparative, as of the whole scientific method. The state of scientific knowledge is always limited by the ideas that scientists have imagined. What was it that Thomas Henry Huxley said on being told the theory of natural selection? 'How stupid of me not to have thought of that.' In 1840 he was wrong; in 1860 he was right: all the difference was made by the imagination of a hypothesis. The comparative method does indeed rely on the imagination of the biologist: the argument of this section perhaps relies more on the imagination than other areas of science. But this is a difference of degree, not of kind.

Biologists not uncommonly use the theory of natural selection in their reconstructions of evolutionary history. In the reconstruction of phylogeny, the principle dictates that all the postulated intermediates have to be functional organisms. One cannot (or, rather, should not) argue that A evolved from B if the intermediate stages would have been selected against. Sidnie Manton (1977) applied this principle with singular insistence. Her case for the polyphyly of the arthropods rests on it: arthropodan intermediates between many of the existing kinds of arthropods could not have functioned, so they could not have existed. They can be connected only by non-arthropodan intermediates. Comparative anatomy offers many comparable examples.

Classical evolutionary taxonomists often use the theory of natural selection in the reconstruction of phylogeny. That is one reason why they are concerned to reconstruct ancestral ways of life. For some reason, cladists do not generally admit functional arguments in the reconstruction of phylogeny. They scornfully write them off as 'scenarios', and 'story-telling'. But they may only be making things difficult for themselves. The method is logically possible. They could use it to enrich their technical repertory. The reconstruction of history is difficult enough without ruling out possible lines of evidence.

Mention of cladism bring us to the second method of distinguishing cause from correlation. We have already met some of the lines of evidence which cladists use in the reconstruction of history. They can be re-applied again here. The method of outgroup comparison allows us to determine which character state evolved first in a cladogram. It may sometimes be possible to use the same method to determine which of two associated characters came first in the cladogram. We have to determine the state of each character everywhere in the cladogram. Direct inspection will then reveal which change of state, if either, came first. Consider the following cladogram in which there is an association between the states (A with B, A' with B') of two characters (*a* and *b*) (Fig. 1.8). The cladogram would have to be much bigger to make the association

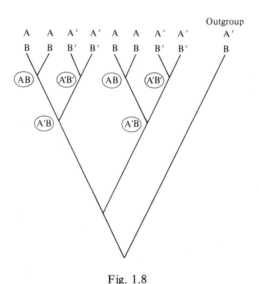

Fig. 1.8

statistically significant. We can use this smaller cladogram to sort out the arguments. The probable character state of each ancestor has been determined by outgroup comparison and then written, in a circle, on the cladogram. We need only to look at the ancestral states to see that character state A' evolved before state B'. The crude rule is that if one character in the association is taxonomically more widespread, then that character is more likely to have evolved first. Chapter 4 again provides an example. Two of the causal conditions (larger females lay more eggs, larger males are stronger competitors for mates) are taxonomically so widespread that outgroup comparison would certainly show them to have evolved before homogamy. These two, at least, we may conclude are causes rather than effects; Thus the conclusion of the cladistic method accords with the conclusion reached before by reasoning from natural selection. Because we have two techniques we can test one against the other, and when both agree the conclusion is more firmly established.

If this method is to be used, the comparative association must be imperfect. If the two character states of the association are always found together, and never found apart, it would not be possible, by this method, to determine which evolved first. In the cladogram above, it is the fact that the outgroup has the character state $A'B$ which allows us to infer that A' evolved before B'.

The literature of biology contains fewer examples of this second method than of the first. An early, if only implicitly argued, case is provided by Winterbottom (1929, p. 192). Concerning the association of extravagant displays with polygamy in birds, he wrote:

In those polygamous birds which display in common we have to decide before considering, or, rather, *pari passu* with this consideration, whether the birds were already polygamous or whether they became so later. In view of the fact that somewhat similar ceremonies may occur in monogamous birds, but in a less advanced state, I am inclined to think that the latter may be the more probable hypothesis.

The paper by Gittleman (1981) on 'the phylogeny of parental care in fish' provides a related example. He examined the transitions between four character states: no parental care, paternal care, maternal care, and biparental care. He was interested in whether some transitions are more likely than others. His evidence of a 'transition' was the presence of more than one state within a low taxonomic category (family, or genus). If a family contained some species, for example, with paternal care, and others with no parental care, Gittleman counted a 'transition' between the two. He thus had a question about the direction of evolution, and a method which, in an abstract sense, is cladistic. And what did he find? Only four of the six possible pair-wise transitions are represented in fish: paternal care to maternal care, and no parental care to biparental care are missing. Biparental care thus tends to evolve through a state of uniparental care, and paternal care can only evolve into maternal care through either biparental, or no parental, care.

Any technique of determining the evolutionary polarity of character states may also, by allowing us to fill in the ancestral character states on a cladogram, help us to determine which of a pair in an association came first. Other cladistic techniques besides outgroup comparison may be of use. I will only suggest a few further lines of argument, without trying to be exhaustive. So far, we have implicitly assumed that each species has a unique set of character states. We might extend it by observing natural, or pathological, variation within a species. Suppose that species of kind A, although they generally show character state B, sometimes also show B' in some form or other; suppose also that species with state A' are always B', and never B: B, we might suspect, is derived from B'.

Embryology might help. If one state of a character is taxonomically more widespread in the development of species, and changes into another character state later in the development of only some species while retaining its former state in the others, then it may be inferred that the developmentally earlier state is the primitive one. In *The Descent of Man* Darwin used this embryological argument in his discussion of bird coloration. He writes: 'When the young and the female closely resemble each other and both differ from the males, the most obvious conclusion is that the males alone have been modified' (1894 edn, pp. 467-8). Elsewhere he combines cladistic embryology with the theory of natural selection: 'When the adults differ in colour from the young, and the colours of the latter are not, as far as we can see, of any special service, they may generally be attributed . . . to the retention of a former character' (1894 edn, p. 463).[9]

This section is intended as no more than a preliminary discussion; it is not intended to be complete. Perhaps more arguments can be thought of. We have fetched two methods from two of the great fields of evolutionary biology: natural selection and taxonomy. Thus equipped we may in principle determine which members of a comparative association are the causes, and which the effects.

LAWS OF EVOLUTION

Explanatory generalizations about the diverse adaptations of different species are scientific laws. To be a law, the generalization must be true, and there must be some theoretical explanation of why it is true. A law is a generalization which does not just happen, contingently to be true, but results from a natural cause. How is a contingently true generalization distinguished from a law? A philosopher would look to a contrary-to-the-fact instance (a 'counterfactual'). Counterfactuals only obey laws. Suppose, for example, that we are in a room full of zoologists, and they are all English. 'All zoologists in the room are English' is a true generalization. But is it a law? Consider whether it would necessarily still be true if another zoologist came into the room. Obviously, he might not be English: the generalization just happened to be true of those zoologists in the room at the time. For a generalization to be a law, it must remain true of contrary-to-the-fact instances. (This meaning of the word 'law' is a commonplace in philosophy: e.g. Goodman 1955; Ayer 1972; Mackie 1976.) We can therefore call comparative generalizations laws if they are validly based in the theory of natural selection. The theory will supply a reason for believing that the generalization would be true of undiscovered, or unstudied species.

When discussing a law I shall not hesitate to call it what it is. I am, however, aware that many biologists do not use the word with so unhesitating an eagerness. It is a brief and vigorous word, which marks and leaves exposed any idea it touches: it thus terrifies those who are most comfortable when concealed behind a puff of scholary caveats. The word has, moreover, been driven under-

[9] Ghiselin (1969, pp. 228-9) and Kottler (1980, p. 220) discuss Darwin's analysis of the evolution of bird coloration.

ground while the crusade against reductionism, the demarcation dispute with physics, and other quixotic battles have raged overhead. In a wider context still, historical laws have suffered from their undemocratic political associations. But let us not dwell on the character and inspiration of the opposition. Let us concentrate on their reasons.

Laws of organic diversity are not customarily thought of as evolutionary laws. Evolution is usually conceived as historical change, so its laws have concerned temporal trends. The laws of Cope, of Dollo, and of Williston are of this nature. Cope's law, for example, states that the size of animals tends to increase during evolution. These are the kinds of laws which people have in mind when discussing evolutionary laws. These are the kinds of laws which criticism has been fired against. The criticisms, as it happens, apply equally to comparative laws, so they are worth looking at here. We shall proceed in two stages: looking first at the criticisms on their own terms, and then asking whether they are valid against the kinds of laws which we are more concerned with.

Simpson (1963, p. 128), with Gould (1970) following him, maintains that 'the search for historical laws is . . . mistaken in principle'. They are led to this by their conception of true scientific laws. True laws refer (according to Simpson, and Gould) to the 'immanent', not to the 'configurational' properties of the universe. Those two words need a little exegisis: 'immanent' properties are inherent, universal, and necessary; 'configurational' ones are the contingent, actual distributions of properties at any one time and place. Gravity is an immanent property; Kepler's law is configurational. Now, Simpson and Gould believe that true scientific laws refer to immanent properties of the universe. They also think that configurational, rather than immanent, properties of nature are the main concern of evolutionary biologists. It follows that evolutionary laws are 'mistaken in principle'.

They have another objection too. Evolution is historical: each evolutionary event is preceded by its own unique history. Generalization about unique events is of course not possible. 'It is further true [and I am quoting from Simpson again] that historical events are unique, usually to a high degree, and hence cannot embody laws defined as recurrent, repeatable relationships' (1963, p. 128). Historical laws are philosophically impossible. Similar reasoning about the uniqueness of historical events led Popper (1957) to his influential critique of historical laws.

However, it is not enough just to plead that historical events are unique. All events are historical, and all events are unique. If we say that a set of events can all come under the scope of a law we do not mean that all the events are identical (in the sense of Simpson or Popper), but only that they have sufficient similarity for the same cause to lead to the same effect. The history of sharks is undoubtedly different from that of dolphins; but they occupy a similar environment and that similarity is probably sufficient to explain the similiarity of their body shapes. The distribution of body shapes among species is of course a configurational property of the world, in Simpson's sense; but that will not prevent us from formulating a law about it. Simpson does concede that 'historical events may . . . be considered predictable in principle to the extent that their causes are known and are similar' (1963, p. 140). But he does not think that this provides any justification for historical laws. Proper laws are not statistical: 'The laws immanent in the material universe are not statistical in essence' (1963, p. 129). Bock (1976, p. 78) also thinks that laws cannot, without a disaster, be statistical: 'Inclusion of a chance-based mechanism has serious, and usually disastrous, consequences for a covering law in science'. The events covered by biological laws may be more easily distinguishable by their history, and more statistical in their character, than in physical laws. (They may, on the other hand, not be, but that is another matter.) But this is a difference of degree: it need not lead us to abandon our search for evolutionary laws, whether they concern temporal trends or organic diversity.

The historical uniqueness of evolutionary events can hardly be doubted. Apart from the whims of history and geography which influence the course of evolution, a number of other evolutionary processes, of which Mendelian segregation and mutation are the most important, are inherently probabilistic. We thus have some theoretical understanding of why evolutionary laws may have exceptions. But probability, exceptions, and historical constraint do not in themselves prevent the formulation of historical laws: they can be written into the laws. If a comparative generalization applies to all species except the amphibians, for example, the method has not broken down; the law is just not universally applicable. It should be rephrased 'all species, except amphibians, have adaptation A in environment E' (or whatever). Philosophers of science refer to these escape clauses as 'scope restrictions' and regard them as perfectly proper and respectable (e.g. Suppe 1978). Often scope restrictions are not written in, but only because the legislator cannot be bothered: *de minimis non curat lex*.

Probability and historical contingency can also be written into laws. Thus: 'Species with history H_1, when in environment E_1, have probability $p(x)$ of evolving x, $q(y)$ of evolving y, $r(z)$ of evolving z; species with history H_2, when in environment E_1, have probability . . . etc.' This is a perfectly good form for a law.

The term 'law' in evolutionary biology has been tainted by crude and atheoretical generalizations about the direction of evolutionary trends (Cope's law is an example). These laws have no real theoretical basis: they are non-explanatory. A worse taint still is the association with the directional theories of evolution that prevailed before the modern synthesis. An earlier generation of evolutionists had to fight against the deterministic laws of orthogenesis (Simpson 1944, 1947, 1950, 1953), for example, and they were right to do so, and to emphasize the probabilistic part of natural selection. Evolution is probabilistic, but it is not unpredictable. The evolutionary laws of organic diversity must be contrasted with the earlier laws of organic change. They are static, and refer to the distribution of character states among environments. They are firmly rooted in a theory, the theory of natural selection, which explains the fit between character and environment. No one, therefore, should be troubled by the term 'law'. Let us bring it out to show that the search for the explanation of organic diversity is on. While we remain content with the unpredictability of evolution, and admit contritely that biologists can only discover 'rules', crude generalizations to be rejoiced in when compared with the laws controlling the 'simpler' systems of physics, we shall make little progress in understanding why nature is the way it is.

PART II
Adaptations for mating

2 Introduction

The comparative method can be used to investigate any trend in the diversity of life. In the next two chapters we are going to use it on two questions about 'mating systems'. From the point of view of exercising the method developed in Chapter 1, these two were picked arbitrarily. One (Chapter 4) happens to be a review of a rapidly expanding field of research; the other (Chapter 3) is not. But the choice may not be wholly unpredictable. Two studies of mating systems are rather likely to turn up in a book on the study of adaptation by the comparative method: of all the areas of modern biology, social behaviour is the one where the comparative method has been used most to study adaptation; it is also the area which has supplied most of the critical discussions of the method.

This introduction is only a short bridge between the general methods which we have been through, and the two particular studies of mating which are to come. When we come to Chapters 3 and 4, we will need to understand the theory of mating, or (as it is usually called) the theory of sexual selection. Sexual selection was discovered by Charles Darwin; over half of his *Descent of Man* (1871) is about 'selection in relation to sex'. Sexual selection (like natural selection) consists of both general, abstract ideas, and applications to particular circumstances. We do not need to go into all the particular applications here: the only two we need to know about will be fully explained when we come to them. But to understand them we do need to know about the general, abstract theory. There is no up-to-date account of sexual selection from first principles. Most modern expositions, including the brief one below, owe a lot to Trivers (1972). Trivers' original paper is now out of date in many respects (see, for example Dawkins 1976; Emlen and Oring 1977). The paragraphs that follow are not intended as a complete modern account; they aim to provide the minimum necessary to understand the two theories which we will be testing at length.

Organisms, by natural selection over evolutionary time, come to possess those habits and characteristics which enable them to reproduce more than organisms with other characteristics. Sexual selection applies this statement to reproduction. With sexual reproduction, new organisms develop from the contributions of two parents. But the contributions of the two are not the same. One sex (the male) may contribute only a tiny sperm; the other, a large nutritious egg. From this difference, many others more or less directly follow. Imagine a simple mating system in which the two sexes meet for only a short time, mate, and then separate. Some fish, for example, may do little more. Because the male gamete is so much smaller than the female gamete, a male can make many more gametes than a female. A male, therefore, is capable of fertilizing many females. Any adaptation which enables him to meet and mate with more females will be strongly selected. The special structures and habits which enable the males of so many species to find and mate with females—large eyes to see them with, bushy antennae to smell them with; antlers or tusks to fight off other males—probably arose for this reason. *The Descent of Man* is packed with examples, and many more have been found since.

The same argument does not apply to females. There is no advantage to a female in meeting and mating with many males. A costly adaptation, such as a pair of antlers, in a female might enable her to mate more; but they will not be favoured. A female does not reproduce more by mating more; a male does. Sexual selection can, however, work on females. Males will be crowding in on every female, eager to mate with her. The female has no difficulty in finding males. She can find dozens whenever she wants. She meets so many that, if selection favoured it, she could be choosy about the kind of male that she would mate with. Why might selection favour choice? The answer is that all males may not be equally good mates. Some males may be better: they may have better genes (and so will father better sons and daughters), or defend better territories, or have some other property which would make him a desirable mate. Thus selection might favour female choice. Again, the same argument does not apply to the opposite sex. If males contribute effectively nothing (a miniscule sperm) at each mating, they will be strongly selected to mate with all-comers. A choosy male, who did not mate with certain kinds of females, would leave less offspring than a non-discriminatory male.

We have been through this argument as if males were the sex that invests little in each mating and females the sex that invests much. Thus do the sexes of many species. But the argument applies independently of which sex happens to invest more. If (as, for example, in the wading birds called jacanas) the males look after the eggs which the female lays, the direction of selection will be reversed. Now the females compete among themselves to obtain males, and the males may be choosy. The direction of sexual selection, in Trivers' terms, is controlled by the relative parental investment.

Such is the theory in the abstract. In real species in nature (rather than the simplified species we have just imagined) the possibilities of obtaining mates are affected by many other factors. These other factors are the subject of all the theory of mating systems which I have alluded to but am not going to review. We shall see in Chapters 3 and 4 how sexual selection may operate in two real cases. This chapter aims to cover just enough of the basic theory to carry us straight into the special cases. We shall need only two basic ideas: that males are selected to meet and mate with females as fast as they possibly can, and that organisms are more choosy about their mates in proportion to how much time and energy they invest in mating.

Here is an appropriate place to say a little about my procedures of research. I have two reasons to do so. One is that it may interest someone else who thought of doing a similar study (or someone who was curious about how it can be done); the second (and more important) is to enable the expert reader to

PRACTICAL METHODS

judge how complete my summaries are. The summaries do aim at a kind of completeness. They aim at a correct estimate of the minimum number of times each character state has evolved in the entire animal kingdom. I have not gone into all the character states to the same depth. Take the study of precopulatory mate guarding (Chapter 3). There are some whole groups, such as fish, in which it is probably not found at all. I have not listed all the species of fish which lack a precopula: I have not even mentioned any of them. There are other groups, such as crustaceans, in which there are many species with precopulas. I have treated the Crustacea at length, mentioning all the species which are known to have a precopula, and all the species which are known to lack one. I have discussed the crustaceans, but not the fish, which lack precopulatory mate guarding. That is the first difference: some species which lack precopulas are mentioned, others are not. There is another. All the species which possess a precopula are mentioned, but not all the species which lack one. The difference has two reasons. The important one is that we will only be discussing as much evidence as we need to estimate the frequency of evolution of the different character states. All fish are the same so they will not appear independently in the final test. There has been extensive reversal between the relevant character states in the Crustacea, so the reversals have to be studied in detail. Of course, I may have overlooked some odd species within the big groups which I believe to be completely without species which have precopulatory mate guarding: if I have, the study is defective. The second reason for looking at some character states in more depth than others is to keep the study down to size. It would be merely pedantic to list all the species of fish, cephalopods, and so on, which neither have nor are (theoretically) expected to have a precopula. In the second study (Chapter 4) the problem does not arise. There are no taxa for which huge amounts of non-independent facts are available. In Chapter 4 I have collected all the facts that bear on the question.

I have also had to decide how direct evidence must be to be included. Take the question of precopula again. Species with lengthy precopulas are usually often found, in general faunistic surveys, as sexual pairs; conversely, species without precopulas are not found in sexual pairs. Quite a common kind of paper in the literature is the 'annual life-cycle of X', which records a series of samples through the year. These papers may record the season when pairs are found, and only species with precopulas are found as pairs. If pairs are not found in such a paper, should it be used as evidence that the species lacks a precopula? I have not. As evidence, I think it is too indirect: but it is evidence of a sort. I have included only direct observations of precopula or mating without a precopula. The reason is to keep the study down to size rather than any philosophical importance of the distinction between 'direct' and 'indirect' evidence.

So far we have been discussing the sorts of decisions that have to be made to keep the study consistent, and down to a manageable size. I will finish by saying a bit about how I actually find the relevant facts. The overall purpose is to find all the relevant papers, read them, and decide how to use them in the test. We also need modern, reliable phylogenies for the groups in which the characters of interest have changed state. First the papers have to be found. Where should one look? I use four main kinds of source, in two stages. I first aim to find out the main groups which show the traits of interest. This one just has to know, from one's general zoological knowledge. If a zoological education accomplishes anything it should result in a knowledge of the main sources, which should guide the comparative biologist both to the literature on the characters that he is interested in, and also to the best phylogenies. They are usually much more reliable for the latter. I mean volumes like Hyman, Grassé, H. G. Bronn's *Klassen und Ordnungen*, and so on. For some reason these books tend to be bad on arthropods in general and crustaceans in particular, which happen to be the main groups that we are going to be concerned with. Hyman never reached the Crustacea; they are one of the few groups still not covered by Grassé, and the treatments by the great German handbooks are still sporadic. The best introduction, and one that I have used most is Kaestner (1968), in the English translation. The other main group which we will be dealing with is the Anura. A comprehensive treatise on the amphibians and reptiles of Europe (Böhme 1981-) is being published, but the volumes on anurans are not yet out. Other groups have particular volumes: Boolotian for echinoderms, Ramazzotti (1972) for tardigrades. There are also general works on mating. Ghiselin (1974) is a good, if necessarily (it was not meant to be a handbook) taxonomically uneven modern introduction. Meisenheimer (1921) is valuable: it reviews the great age of German microscopy, just as it was drawing to a close. The volumes on *Reproduction in Marine Invertebrates* (Giese and Pearse 1975-) are no help to the student of mating, and have not yet reached the arthropods.

Once I have decided on the main groups which I want to investigate in depth I move on to the second kind of source: the *Zoological Record*, reviews of the mating of particular groups, and (what I will call) windfalls. The *Zoological Record* is an excellent bibliography. One can pick up the annual volume for the Crustacea, and look up all the papers classified under reproduction. The bibliography itself is almost complete, but the subject index does not list all the references on reproduction. The trouble with subject indexes, of course, is that they do not exist for new subjects. If you want to review a 'subject' which has not been thought of as a 'subject' for a research discipline, it will not have been built into the indexes. Neither mate guarding, nor homogamy, are 'subjects.' One discovers the incompleteness of the subject index from the other two kinds of source. Some groups have been the subject of review papers, or even books, on mating. There is Hartnoll (1969) on crabs, and Wells (1977) on Anura. These reviews are valuable, but are not available for most groups. They are easy to find: one does not have to read around for long before bumping into the reference. The third kind of source is less easy to find. Some (otherwise ordinary) research papers just happen to have very useful bibliographies. Patel and Crisp

(1962), for example, under an innocent title about the reproductive cycle of barnacles, contains many references on crustaceans with precopulatory mate guarding. Jackson (1978) contains a windfall of references on precopulas in spiders, at the end of a paper on *Phidippus johnsoni*. Windfalls are a matter of luck: I bumped into Patel and Crisp in Ghiselin (1974), and Jackson (1978) while looking through recent issues of journals. A windfall is not, from its title, distinguishable from an ordinary paper, so there is no method of searching for them. I have probably overlooked some.

We now have a complete set of theoretical and practical techniques. The time has come to see whether we can answer some questions about real animals with them.

3 The evolution of precopulatory mate guarding

INTRODUCTION

Fish around with a pond net on the bottom of a freshwater stream and you may, depending on the time and place, be rewarded with pairs of the amphipod *Gammarus* or the isopod *Asellus*. The pairs consist of a larger male clasping a smaller female beneath him. The male clasps the carapace of the female in his fourth pair of legs (in *Asellus*), or his first pair of legs (in *Gammarus*). These pairs stay together for several days until the female moults; they then shift to their copulatory posture (with ventral surfaces opposed), the male releases his sperms, and they separate.

Watch a pond used for breeding by the common frog. The breeding season is short, but explosive. As they come out of winter hibernation they make their way to the breeding pond. Breeding is stimulated by rainfall, and the frogs come in great numbers; it used to be believed that they came down with the rain. Many come in pairs. The male, once again, is on top, but in frogs the male is the smaller of the two. He sits on his mate's back, clasping her around the belly by his front legs, until, some time after reaching the pond, the female spawns her eggs. The male, without moving, squirts his sperms among them. Then the sexes separate. Lengthy associations of the sexes before actual mating have been found in other groups as well, in spiders, in mites, in insects, in crabs. And we shall come to them.

The habit has collected a varied vocabulary of descriptive terms, in different languages, and for different groups: the riding-position, *Reiterstellung, chevauchée nuptiale, promenade nuptiale*, marriage clasp, as well as precopula, *Präkopula*, and *Vorkopula*, in crustaceans: the amplexus of toads: cohabitation of spiders, and (perhaps) the pupal attendance of mosquitoes. The habit is frequently mistaken for actual mating, so we can also find reference to it under such terms as mating and copulation. Because the male of the pair frequently has to defend his mate from other males, the habit may also be called mate guarding. The vocabulary also includes Parker's (1970) term, the 'passive phase'. We will use 'precopula' as a general term, drawing on other terms as appropriate. 'Precopula' is not etymologically very appropriate in groups which lack a true copulation, but we will not let that worry us.

The alternative to mating with a precopula is to mate without one: the sexes copulate (or otherwise discharge their gametes) soon after meeting. If they are not ready to mate straightaway, they separate. They do not wait together until they are ready to mate. This essay is going to be a summary of which species mate after a lengthy precopulatory association, and which do not. We will have more to say about the definition of precopula. The best method of defining terms is to make them refer to a set of entities which all may result from the same theoretical process. So we will save the discussion of definitions until after we have met the theory.

What, then, might predispose a species to evolve a precopula? The answer lies at the end of the association. Take *Asellus* as an example.[1] Its precopula culminates in the moulting of the female, which is followed almost immediately by mating. But why do they not mate until after the female moult? For the answer to this question we must look, down a microscope, at the genital openings of the female. *Asellus* has internal fertilization, and the male introduces his sperms by his second abdominal appendages (pleopods) through openings on the female's sixth thoracic segment. Just look at the difference in size of those genital openings before and after a moult (Fig. 3.1). It is probably physically

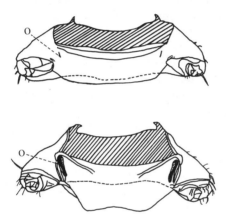

Fig. 3.1. *Asellus aquaticus* female genital openings (marked O) before (top) and immediately after (bottom) her moult. (From Maercks (1930).)

impossible for the pair to copulate before the female moults. And after the moult yet further obstructions are soon set between the genitals of the male and female. The female broods her eggs in a pouch beneath her thorax. The pouch is walled beneath by outgrowths, called oostegites, from her thoracic legs. Within 24 hours of her moult (whether or not she has been mated) she lays eggs into her pouch. Eggs and pouch combine to make an impenetrable barrier, and anyway the eggs cannot be fertilized after they have been laid. They have either been fertilized already, and will develop, or they are infertile, and will soon be resorbed. Mating, therefore, is only possible in the short period between the moulting of the female's posterior half, and the hardening of the oostegites and laying of eggs which take place within 24 hours.

[1] This section is based on Unwin (1920), who however made some errors of detail. So he is supplemented by Maercks (1930), and some observations of Dr David Thompson and myself (Ridley and Thompson 1979).

The brevity of the interval when mating is possible is the key to the problem of precopulatory mate guarding. The explanation can most easily be followed by imagining, to begin with, a population in which the males do not guard the females before mating. We may idealize the population a little more. Imagine all the males are identical, all the females are identical, and the sexes meet by bumping into each other at random. If when a female bumps into a male, she has just moulted, then they mate. If she is at any other point in her cycle they separate. Now, if the interval when mating is possible is very short, meetings at which mating takes place will occur only at a very low rate. Many mating opportunities will be missed simply for want of a meeting of the sexes at the right moment. Now let a mutant arise. It will exert its effects solely in males. The males (we will assume) can tell, when they meet a female, how long it will be until she moults. The mutant capitalizes on this ability. It causes males to wait with any female who has less than, say, one day, to go until her next moult. Only after mating with her does the male depart to look for another female. The mutant males will meet 'acceptable' females at a higher rate than do the rest of the males. The consequent decrease in his 'searching time' (average time to meet an acceptable female) may more than make up for the increase in 'guarding time' (average time spent with each female) incurred. If it does, the mutant will spread. As the mutant form increases in frequency, natural selection feeds back positively, to favour it still more. The relative advantage to guarding increases as the number of unguarded, moulting females available at any one time is depressed by the habit of longer guarding. The number of free females with only 24 hours or less to go until the moult also decreases in the population as the mutant spreads. Now there may be selection for a mutant with a longer criterion, such as 48 hours. The process would repeat itself.

Natural selection will continue to favour longer and longer criteria until an equilibrium is reached between the decrease in searching time and increase in guarding time. Eventually the guarding time may grow so long that further increase is no longer compensated by a decrease in the searching time. Whether that equilibrium is reached below the point at which males join on indiscriminately to the first female they meet depends, in this simple model, on the circumstances of sex ratio and meeting rate (Grafen and Ridley 1983). But we shall not concern ourselves with the mathematical details here. We are only going to try to predict qualitative comparative trends. To do that we need only a qualitative understanding of the theory. Precopulatory mate guarding evolves in this model because a male that will accept females earlier in their moult cycle than do the majority of males in the population opens up to himself a large pool of unexploited females. Naturally he is selected to exploit them.

This is not an original theory. Many authors have vaguely realized that a precopula may be advantageous, 'to ensure that the females are fertilized', or some such formulation. Thus have authors such as Blegvad (1922, p. 32) and Le Roux (1933, p. 29) written of *Gammarus*; Unwin (1920) of *Asellus*; Menge

(1866) and Nielsen (1932, I, p. 35) of spiders. Patel and Crisp (1961) stated the theory more precisely for crustaceans. Ghiselin (1974, pp. 187-8) mentioned that precopulas and mating at moulting are connected in crustaceans. (He also mentions another theory, that precopulas serve to protect the fragile moulting female from predators.) Seibt and Wickler (1979, and Wickler and Seibt 1980) stated and modelled a similar theory. Parker (1974, p. 172) stated the model briefly but clearly. And Alan Grafen and I have modelled it mathematically (Grafen and Ridley 1983). We did this for a number of purposes. One was to check whether the verbal reasoning which we have just been through is correct. We confirmed that it is.

Even if this theory is correct it is not the whole story. The reader may concede that if females lay their eggs only in a brief, predictable interval, then precopulas will tend to evolve. But he may not yet be satisfied. He may point out that we have only substituted one mystery for another. For why do females only lay their eggs just after their moult? This question cannot be satisfactorily answered. One theory in the literature (Yonge 1937) runs as follows. The female must brood the eggs after she has laid them. While they are being brooded, the eggs are attached to their mother's body. Thus if she moults she loses her eggs, unless they have already developed, and the larvae swum away. If the amount of time that it takes for the larvae to develop is similar to the length of the female moult cycle, the female would have to lay her eggs as soon as possible after the moult in order for them to have grown up before she must moult again. An observation supports this theory. In several crangonid prawns the moult cycle of the female is longer than normal when she is brooding eggs (Hess 1921; Yonge 1937; Scudamore 1948). If the moult cycle is adapted to some other circumstance, the females are probably selected to lay their eggs soon after their moult.

I do not know whether this theory is correct. It is not my purpose here to defend it. I would instead make a different point. I do not have to defend it. The relation between precopula and mating cycle is a hypothesis about adaptation which is, if correct, explanatory in its own right. It does raise further questions. But to raise questions which you cannot answer is not the same as to self-destruct.

The mathematical model (Grafen and Ridley 1983) can also predict precise evolutionarily stable values of the duration of mate guarding. The information in the literature is not good enough to test these precise predictions. It will prove difficult enough to infer which species practise precopulas. We will not be reviewing the literature of some tradition of research on precopulatory mate guarding. We will be summarizing scattered observations from many kinds of literatures, from faunistic and taxonomic surveys, from the technical pamphlets of applied biology, from descriptive papers on the natural history in general or reproduction in particular of odd species. We will be little helped by analytical papers on mating behaviour. With such sources of information, as a first resort, it is easiest to test only a qualitative comparative hypothesis. The theory does allow such a hypothesis. It is as follows. Precopulatory mate guarding should evolve only when the female will become receptive during a short, predictable interval. If the females of a species are receptive only during short, predictable intervals, all matings should be preceded by a precopula. If their periods of receptivity are unpredictable, and mating may take place at any time during the adult female's life, mating should not be preceded by a precopula. If the females of a species have both predictable and unpredictable periods of receptivity, matings during the former should be preceded by a precopula, but not the latter. Such is our comparative hypothesis. It is a true comparative hypothesis. It predicts trends in diversity; it is explanatory, and validly based in the theory of natural selection; it is testable. It does not stretch the model to the extremes of its predictive powers, but that is only because the available literature is limited. But before trying to test the theory, we will pause a while to examine the theory a little more. The theory, as it has been expressed so far, is deliberately simplified. It assumes that all males are identical, that all females are identical; it also omits many other factors which may be thought to affect the evolution of precopulatory mate guarding. Can the theory still stand if these assumptions are taken away?

Take first the assumption that all females are identical. It is false, of course. Different females have different fecundities, and males are naturally selected to join preferentially with more fecund females. When Manning (1975) and Thompson and Manning (1981) tested male *Asellus*, they found that the males preferred to pair with larger (more fecund) females. Le Roux (1933, p. 30n) observed that male *Gammarus* would not pair with females that had been parasitically castrated by the acanthocephalan *Polyphemus* (see also Thompson and Manning 1982, p. 285). Although the fecundity of individual females will affect whether a male pairs with her, the qualitative theory still stands. Relaxing this assumption alters the quantitative prediction, but not the theory we will be testing.

The assumption that all males are identical is also false. Different males have different sizes, and it has been shown in many species with precopulatory pairing that bigger males can take over females from smaller males (Peckham and Peckham 1889 for a spider; Potter, Wrensch, and Johnston 1976 for a mite; Davies and Halliday 1979 for an amphibian; Ridley and Thompson 1979 and references for crustaceans). Alan Grafen and I examined the effect of male takeovers in our model. Once again, there is an effect on the quantitative prediction of the actual criteria of males (Grafen and Ridley 1983, Figs. 2 and 3), but the comparative hypothesis which is the subject of this essay is hardly affected at all.

Another factor which affects the duration of precopulatory guarding is the relative mortality of individuals and pairs. If pairs have a higher mortality than individuals, pairing will tend to evolve to be of a shorter duration. (Many kinds of predator in nature preferentially eat larger items of prey, and so tend to pick on pairs.) Strong (1973) explained the difference in the duration of precopula between two populations of the amphipod *Hyalella azteca* by differences in the intensity of predation: where predation was higher, there the pairing was shorter. Wickler and Seibt (1980) considered this result theoretically. Again, our qualitative comparative hypothesis will stand except at very extreme differential predation.

Another theoretically important factor is the sex ratio (Manning 1981; Grafen and Ridley 1983). Pairing may be expected to be of shorter duration if there are more females about. Manning (1981) experimentally demonstrated this effect in *Asellus*. Earlier qualitative observations on a parasitic mite by Downing (1936) suggest a similar effect, as does the observation by Potter *et al.* (1976) that more females had males waiting with them before mating when the sex ratio was more male biased. The sex ratio is another factor which will usually only alter the quantitative duration of mate guarding. Grafen and I quoted some facts suggesting that differences in sex ratio may explain why some species have permanent monogamy while others have a limited precopula. Only when the sex ratio is extremely female biased does it upset the qualitative comparative hypothesis. If there are a sufficiently large number of females, then a precopula may not evolve even if each female is only receptive for a very short period. For our comparative analysis we will just have to hope that in nature very few species have, unknown to us, such female-biased sex ratios. At such extremes the simple prediction is no longer the real prediction of the underlying theory. If we know that the sex ratio is very biased towards females, we can modify the prediction of the theory.

The model assumed that males can tell how long females have to go until they will be receptive for mating. This assumption can only be tested directly by experimentally offering females of different stages in their moult cycles to males. If males choose the females with a shorter time until their moult, the assumption is confirmed. This has only been proved for two isopods (Shuster 1981; Thompson and Manning 1981). In other species it has been noticed that there are more pairs containing females later in their moult cycle than pairs with females earlier in their cycle; the frequency distribution of pairing durations is biased towards shorter times (e.g. Hartnoll and Smith 1977 for *Gammarus*; Cone *et al.* (1971*a, b*) for a mite; Sweatman (1957) reports the opposite for another mite). This suggests, but does not prove, that males preferentially pair with females nearer to their moult. The observation would also be made if pairs formed at random, but split after the female moult: more late females would be paired because they would have had a longer chance of being found. The ability of males to tell the time of a female in her moult cycle may be universally true in crustaceans with precopulas, and pheromones (Dunham 1978) are probably involved. This ability is important for the evolution of precopulatory mate guarding. Under one condition this assumption can also be ignored. When the evolutionarily stable male criterion is to accept any female, regardless of where she is in her moult cycle, precopulas will evolve even if males cannot tell how long a female has to go until her moult. (Such a male criterion corresponds to permanent monogamy: the male stays with a female after her moult and waits with her to fertilize her next brood.) This is the reason why, in Seibt and Wickler's (1979) model for the evolution of monogamy, precopulas (i.e. permanent precopulas) could evolve even though males were joining females at random times in their moult cycles.

So much for the model's assumptions. They are probably all false, or at best simplifications. But that need delay us no longer. Our comparative analysis can go forward without them. We are to test a sufficiently crude hypothesis that most of the fine details of the model are dispensible. We shall look at the relation between the incidence of precopulatory mate guarding and the incidence of restrictions on the timing of mating in the reproductive cycle of the female. We have now said enough about the theory. Now let us consider how to test it. It is time to return to the temporarily postponed problems of definition.

METHODS

The final test will be a simple 2 × 2 contingency table, of 'mating restricted in time' (predicting precopula), and 'mating unrestricted in time' (predicting no precopula). The four categories have been constructed from two continuous variables (time, in both cases), so we will have to decide on a criterion to divide the continuum. We must keep in mind the qualities of the literature when framing our definitions, just as we did when choosing the hypothesis. Let us just pause to consider the kind of evidence we will be using.

Biologists do not usually measure the duration of any precopulatory sexual association, or the total duration of the reproductive cycle. We may be told a rough estimate of the duration of a precopulatory association, such as 'several days', or 'a few minutes'. Some of our information will be even cruder. Single samples of a species may repeatedly contain sexual pairs: this suggests a lengthy sexual association. Whether natural pairs are detected in general faunistic samples depends on the kind of pairing, and the kind of sample. In some species the precopulatory pair are physically attached; they are easily detected. In other species the male and female only come very close to each other: if the species is aquatic and is collected by pulling a net through the water, the pairs will come up as separate individuals; if it is terrestrial, a naturalist would be much more likely to notice physically unattached pairs. In fact, it is probably commoner for the male to grip the female in aquatic species, so nature to some extent makes up for our sampling artefacts. But not entirely. The crude collecting apparatus which is dragged along the sea bottom, or grabs samples of bottom sand, must have missed any cases of pairing in which the male and female are not firmly attached to each other. Many crustaceans, for example, live in tubes, and the male may join the female in her tube before her moult. Clumsy sampling techniques would never reveal these.

So much for the evidence. How are we to interpret it? Both variables present problems: we will discuss precopulas first, and then restrictions on receptivity to mating. What questions must we ask of our evidence to convince ourselves that it shows a precopula? If sexual pairs are found, the first question is whether the pairing comes before or after copulation. If this question cannot be answered the information cannot be used in the test. Our theory applies only to precopulatory mate guarding,[2] so only precopulatory pairs can be used to test it. Next, we have to decide how long an association must be to count as a precopula. Our theory gives little guidance here. It predicts exactly how long a precopulatory association should be, so the dichotomous distinction which we are trying to impose simply does not exist in the theory. The distinction is methodological: but we still need it. It can be used to test the theory provided that the terms are clearly defined in advance. I shall take a precopula to be an association of more than about 24 hours before mating in species, such as most crustaceans, in which the reproductive cycle is weeks long. A rather shorter criterion seems appropriate for the many mites in which the moult cycle may be only two or three days long. For mites the information on the duration of precopulatory associations is almost all qualitative anyway, so I have taken any mention of an association before the female moult to be evidence of precopula. When precopulas have been measured in mites they are usually less than a day long. Within a species of *Gammarus*, the moult interval is reduced at warmer temperatures, and so, correspondingly is the duration of precopula (Fig. 3.2). The average duration of precopula decreases from 13 to 4 days over the temperature range from 6 to 17 °C. For species from warmer water (such as the amphipod family Melitidae) a shorter criterion of a precopula is appropriate. The criterion can be made objectively comparable with that for *Gammarus* by Fig. 3.2. The duration of precopula approximately halves from 10 to 17 °C, so if the criterion were one day at 10 °C (when the moult cycle is 40 days or so) then it should be about 12 hours for a gammarid which lives at about 17 °C. We will only need this adjustment for temperature once in the following review: very few tropical species have been studied.

We have implicitly defined what is not a precopula: any association before mating which lasts less than about 24 hours. It is in fact usually fairly obvious if a species lacks a precopula; the sexes just mate within a few minutes of meeting. They may court each other a little first, but its duration is much shorter than that of a precopula. There will be a grey area between short precopula and long courtship. At the outset we can hope that few species in nature fall into it. If we meet species which we cannot clearly classify, we will have to exclude from

---

[2] Males of many species guard their mates between copulations, and after copulation. It is an adaptation to prevent other males from inseminating the female, and is the subject of an excellent review by Parker (1970).

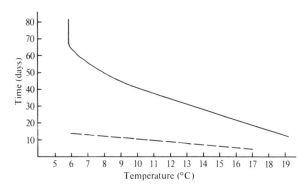

Fig. 3.2. Duration of precopula (– – –) and total moult cycle (———) at different temperatures in *Gammarus duebeni*. (Redrawn from Kinne (1959).)

the final test. At the end we can reconsider how much of a problem it has been.

The distinction between a courtship and a precopula is not simply one of time. Courtship involves more active behaviour from the male. In a precopula, the male just holds onto, or waits near, a female. In a courtship he must dance. He must try to induce the female to mate. Courtship, therefore, by definition, makes the female more ready to mate: she mates quicker if courted than if not. A precopula has relatively little effect on how long the female waits until mating. The time of mating is determined more or less externally, by the breeding season (amphibians) or the moult cycle (arthropods). Entering precopula has no effect on the seasonal cycles, but it may have a little effect on the moult cycle of a crustacean female. In some species, such as *Asellus* (Thompson and Manning 1981) and *Bufo bufo* (Huesser 1963), the time when the female lays her eggs is independent of whether a male joins her in precopula. In others, such as *Gammarus* (Kinne 1959; Hughes 1978), the female lengthens her moult cycle slightly if she is not joined by a male. The mange mite *Chorioptes bovis* (Sweatman 1957, p. 660) takes this to exceptional lengths: females will not moult until joined by a male. Isolated from males, they die without moulting. But *Chorioptes bovis* is an exception. For most species we do not know what females do if isolated from males; but in most of the few species which have been studied, the female moults and lays her eggs regardless of whether a male is around at the time or not. In precopula, a male passively waits for the female to become receptive; in courtship he actively induces her to become receptive: the distinction may have blurry edges, but it does exist.

The definition of precopula is now decided. It remains to decide what is meant by mating being restricted in time. The details of why mating may be restricted in time vary greatly between groups. The theory we explained in terms of one crustacean, *Asellus*. Like *Asellus*, the mating of many species of arthropods are restricted by the moult cycle. The exact details of the restriction vary between groups, and we will discuss them in the several sections of the systematic summary. I will review any evidence which seems relevant. Whether fertilization is internal or external is, for example, of interest. If it is external then re-insemination is essential after each moult, whereas if it is internal and the female can store sperm then it is not so essential. But the main general kind of evidence of a restriction of mating in time for arthropods is the observation that mating always follows immediately after the female moult.

Three main kinds of evidence suggest that in these species mating is only possible for a short interval after the moult. The first, and most impressive, is experimental. In a mere four species, males have experimentally been put with females at various times after the female moult (Table 3.1). These experiments determine exactly how restricted mating is. Four is a small number, but the results all agree. In all the crustaceans which mate only after the female moults in nature, and which have been experimented on, it has been found that mating is only effective during a short time after the female moult. The second kind of evidence is a crude field-version of these experiments. In many species it has been observed that females in their intermoult, or just before their moult, do not mate; but females just after their moult do. Thus, some time after their moult females must become unable to mate. The third kind of evidence is less direct. It consists of observations of how long after the moult the eggs are laid; effective mating would no longer be possible after laying. The interval, from moulting to laying, has been roughly measured in some isopods and amphipods, in which it is as little as three hours in *Gammarus*, and is less than 24 hours in *Asellus*. However, in many decapods the eggs may not be laid for many weeks after the moult, but mating always takes place in a much shorter interval. So this third kind of evidence can only suggest an upper limit to the interval of receptivity, and it may be misleadingly long. All three lines of evidence suggest that in species which only mate after the female moult, there is only a short interval after the moult when mating is possible. The quantitative details of that word 'short' vary between groups.

Just as the observation that mating only takes place at moulting is evidence of a restriction on mating in time, so the observation that mating takes place at all times in the moult cycle is evidence that mating is unrestricted. There is a clear distinction between species which can only mate in the day or two after a female moults, and those which can mate at any time through the moult cycle. (Moult cycles are typically a month or more long, although their duration varies enormously.) The former we predict to have a precopula, the latter not to. In other groups, and in some arthropods, there are other kinds of restriction on when mating can occur. In anurans we will be comparing species with short explosive breeding seasons with species with long breeding seasons. As we will see, several authors have thought about the classification of anuran breeding season. Breeding seasons come in all durations, so we will

Table 3.1 How long females are available for mating

1. Males experimentally added at intervals after the female moult

Species	Times males added	Mating success (%)	Authority
Hymenocera	3 h	100	Seibt and Wickler (1979)
	5–7 h	< 100	
Paralithodes	2–9 days	100	McMullen (1969)
	10–12 days	56	
	13–15 days	0%	
Lobster	0–3 h	77	Templeman (1934)
	3–6 h	37	
	6–20 h	26–9	
	23–85 h	0	
Cumacean	just before	0	Forsman (1938)
	at moult	?>0	
	10 h	0	

2. Female unreceptive or male not interested except after moult

Copepoda (Fuhrenbach, 1962), Natantia (L. Nouvel 1939, 1940; Needler 1945; Kamiguchi 1972; Bauer 1976), *Sphaeroma* (Forsman 1956; Daguerre de Hureaux 1966), *Typhlodromus* (El-Badry and Zaher 1961) mysids (H. Nouvel 1940; Clutter 1969; Clutter and Theilacher 1971).

3. Eggs laid certain time after moult

Species	Laying time	Authority
Pontoniinae	40 h	Hipeau-Jacquotte (1973c)
Hyalella	12 h	Wilder (1940, p. 451)
Gammarus duebeni	1 h	Le Roux (1933, p. 31)
pulex	3–4 h*	Heinze (1932, pp. 414–16)
chevreuxi	3 days	Sexton (1938, p. 39)
Asellus aquaticus	<24 h*	Unwin (1920)

*Eggs laid whether or not female inseminated.

only use in the test those that are obviously 'long' (say, more than a month) and 'short' (say, less than a fortnight). The former tend to have long precopulatory amplexuses. But the idiosyncracies of each group are best dealt with in the relevant systematic section.

Our hypothesis may appear to apply to all species of animals. Ignoring the grey area which arises in the crude test we will attempt, it should predict for every species whether or not it should have precopulatory mate guarding. I have therefore tried to summarize the evolution and loss of precopulas in the entire animal kingdom. The only large groups in which precopulas have evolved are the arthropods and the frogs and toads. So the study can immediately be cut down. We do not need to go through all the species which mate without a precopula, when none is predicted. We can cut it down further by leaving out any species to which the theory cannot be unambiguously applied. These must be distinguished from species which do not conform to the theory. They are species which neither conform nor refute. The main species in this category are those which have a permanent sexual association for some purpose other than mate guarding. In fully social species, the sexes are permanently associated, so the male must have been close to his mate for a long period before mating. Such is true of social insects, spiders, mammals, and so on. It is also true of barnacles. The male barnacle settles near a female; they are both attached to the ground. The male is close to his mate before mating, but he would have to be whether or not the female had a restricted period of receptivity to mating. The pre-copulatory positions of the sexes are constrained by variables outside our theory. I have excluded social species, and species like barnacles, from the test and summary. There are some other species which cannot be used in the test but which I have mentioned in the summary because their close relatives are of interest. Pyemotid mites are an example. Pyemotid males mate with their sisters inside their mother, or on her external surface immediately after birth. The sexes have therefore had a long precopulatory association; but, once again, it neither counts for nor against the theory.

I have, for a similar reason, excluded 'precopulatory mate guarding' in monogamous birds, and some social mammals. In some species of birds, such as blackbirds (Horn 1968), swallows (Samuel 1971; Hoogland and Sherman 1976), and magpies (Birkhead 1979), and in some mammals, such as primates (Seyfarth 1978; Packer 1978), and lions (Packer and Pusie 1982), it has been reported that the sexes come closer together for a period before and after mating. It is called 'consorting' in primates. This kind of mate guarding has sometimes (and reasonably enough) been treated as analogous to the precopulas of crustaceans and other groups. There is, however, a difference between the mate guarding of these vertebrates and the precopulas which our theory applies to. In our theory, precopulas evolve when it pays a male to wait with one female rather than searching for another closer to her time of receptivity. The male waits with her because of the difficulty of finding another female. But the male magpie, or lion, would have no difficulty in finding the female come her time for mating. In magpies the only difference is whether she is about 20 yards (when being guarded) or 40 yeards away (when not) (Birkhead 1979). Mate guarding is here an adaptation to prevent 'sneaky' males from stealing copulations; or, in primate troops, as a convention of 'ownership' of the female through her oestrous phase. It does not evolve as a trade-off between time spent guarding and time spent searching. A nearer analogy with the precopula of crustaceans is the permanent association of the sexes in these species of social mammals and monogamous birds. In crustaceans, too, we can find permanent sexual associations. In some isopods, natantian shrimps, and others, the male may stay with a female after her moult and wait with her until her next moult to fertilize her next brood. The theory can explain the permanent monogamy of these species as a simple temporal extension of the precopula (the conditions for a precopula to evolve into permanent association are not severe: Grafen and Ridley 1983). The theory can explain permanent sexual associations; but it cannot explain changes in the average distance between the sexes at different times in the permanent association. In fact numerous additional factors have probably affected the evolution of permanent sexual associations in addition to that identified in our theory (Wickler and Seibt 1980).

The mate guarding of birds and mammals presents an additional problem. The facts are few. The theory does not really predict whether the pairs of monogamous birds should stay closer together near the time of ovulation; but if some species do, then probably nearly all do. Yet it has only been confirmed in a few species. Probably it should be entered only once in the test. Similarly the predictions we could make about which primate species should (multi-male troops, perhaps?) evolve consorting are not as clear as the predictions we can make for species like *Asellus aquaticus*. The application of our comparative hypothesis is not straightforward with mammals and birds. And when application is not straightforward, we will quickly give up. As we are now about to see, there is plenty of information for species to which the theory is readily applicable.

ARTHROPODS

No species of invertebrate outside the arthropods possess an unambiguous precopula. The only other group for which it has been suggested is the echinoderms. Parker (1974) pointed out that the 'attachment phase' of some echinoids may be a precopula. I have followed up a few references through Hyman (1955, Parker's reference) and some other general works on echinoderm reproduction, and have found no conclusive evidence that the attachment phase comes before the release of gametes. Thus although they may have a precopula, the evidence is not strong enough to include them in the test. So we will confine our attention to the arthropods. During the evolution of the arthropods precopulas have been lost and gained many times, but, their outgroups all lacking it, we can assume that the common ancestor of all the arthropodan groups lacked a precopula. Unambiguous precopulas may be found within the Crustacea, the Arachnida, and (probably) the Tardigrada. We will also discuss some more ambiguous analogies from the insects, so we have to decide the sister-group relations of these four main groups. The traditional phylogeny would be as in Scheme 3.1. (The Tardigrada are often classified as a phylum separate from the Arthropoda, but that does not matter to us.) Another phylogeny would be proposed by Manton (1977). She did not believe in the 'Mandibulata' (Crustacea + Insecta in Scheme 3.1), and suspected (1977, p. 288) that the tardigrades belonged in her controversial phylum Uniramia. She therefore would have preferred Scheme 3.2. Before choosing one of these two, let us first ask whether

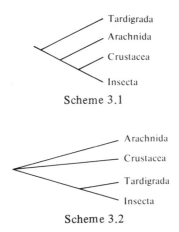

Scheme 3.1

Scheme 3.2

we have to. Maybe it will not make any difference to the final score which one we use; then we will not have to choose one. The primitive state for these groups as a whole is (by outgroup comparison) the absence of a precopula. Both Arachnida and Crustacea (as we shall see) primitively lacked one, so must have independently evolved them. This will give the same score on either phylogeny. The only possible difference stems from the tardigrades. Because the only tardigrades for which we have evidence possess a precopula we might reason that they primitively had one. In Manton's phylogeny the tardigrades are the outgroup of insects, so (because, as we shall see, the insects lack one) we might score a case of the loss of precopula. In the 'standard' phylogeny the tardigrades are not the outgroup of insects, so this loss would not be scored.

The easiest way out is to assert that the reproduction of tardigrades (which we will come to) is an absurd model for the ancestral insect. Other considerations over-ride outgroup comparison and dictate that the state of tardigrades is not the primitive state of the insects. Then we can slip back on the standard primitive state, the absence of a precopula, for the insects as well. Then the two classifications give the same score, and we do not have to choose one rather than the other.

The Crustacea, the Arachnida, and the Tardigrada each have their own section later. There are also some mating habits among the insects which are not unlike a precopula. Some of these habits have been cited as examples of precopula (Jackson 1978), or described as a 'precopulatory passive phase' (Parker 1972; Parker and Smith 1975). The analogy does appear good; but I am going to argue that it is not good enough. Furthermore, to include these habits in our summary would force us to include many more habits which are even less like precopulas. We would have to broaden our summary undesirably.

The insects cannot be dismissed without argument. We would hardly be surprised if the insects provided many examples of precopulas. They have the right general life-cycle. The females cannot be inseminated until after they emerge as adults. The transition from static, unreceptive pupa to active, receptive adult takes place suddenly. The pupae would be physically easy to guard, because they are stationary. The males of temperate species with discrete generations usually emerge before females, which should also encourage the evolution of precopulatory mate guarding. In many species the females do mate soon after they emerge. It is especially common in Hymenoptera and Lepidoptera (Engelmann 1970, p. 86). Some chironomid midge females mate immediately on emergence (Oka 1930, p. 274). In the cockroach *Diploptera dytiscoides*, Roth and Willis (1955, p. 61) tell us, 'newly emerged, teneral females are very attractive to, and are courted by, older males'. They may mate before they have fully emerged from their pupal skin. And yet these insects have no precopulas. I do not regard them as violations of the theory. The females can mate at any time during their adult lives, and they will, therefore, sometimes mate just after they emerge. The absence of a precopula is therefore predicted. But other insects do provide three kinds of possible analogies of precopulatory mate guarding.

The first is the case of 'pupal attendance' in certain mosquitoes. This habit has undoubtedly been selected for by the same process as selects for precopulas in other animal groups. We will only exclude it from the analysis because the duration is either too short, or too uncertainly known, to satisfy our definition of a precopula. In most mosquitoes, the males form swarms which attract fully adult females. Various kinds of exceptions are known (Downes 1966, 1978), of which two species interest us here. Males of the New Zealander *Opifex cancer* (Kirk 1923; Haeger and Provost 1965, p. 25) and the Floridan *Deinocerites cancer* (Downes 1966; Provost and Haeger 1967) find their mates by waiting for them on the water surface. The pupal stage of the mosquito is aquatic; males wait for pupae to surface, and when one does, the male may grab it (*Opifex*) or stay with it (*Deinocerites*). Males will fight over them, even though they are unable, initially at least, to distinguish its age, sex, or even species. If an adult female emerges from the pupa, the male mates with it immediately it emerges.

Pupal attendance before mating does not last long. It is dangerous for a young pupal *Opifex* to be caught by a male, who will be driven by lust to flay it alive. The young pupae therefore only come to the surface briefly and avoid the males on the surface. They are only attended by males just before they emerge. (No measurement exists, but my impression is that the sexes are only together before mating for perhaps a minute or two.) The duration of pupal attendance in *Deinocerites cancer* is also not clear in the literature. Downes (1966, p. 1170) mentions that males attend pupae more persistently when there is only two hours to go until emergence. Provost and Haeger (1967, p. 569) report that males stayed for more than 10 minutes when they had lights flashed at them. It seems unlikely that pupal attendance in these two mosquitoes lasts long enough to count as a precopula on our definition. However, the habit probably evolved by the process underlying our theory. The male stays with the pupa because of the predictable mating it can obtain after the emergence. Our classification of these species as lacking a precopula is not a fact of nature, but an artefact of method. We have to live with artefacts such as these: it would be difficult to define precopulas to exclude convincingly all cases of mere courtship, but to include these mosquitoes.

Once we have accepted that mosquitoes lack a precopula, we can count them in favour of our theory. The females which are available to them at the surface are all about to emerge: the younger ones swim away, or die. There is no selection for a longer period of pupal attendance. In conclusion, our first possible analogy of a precopula we will count as a predicted instance of the absence of a precopula.

The same fate awaits the second. Parker (1972), working on the dipteran *Sepsis cynipsea*, and Parker and Smith (1975), on *Locusta migratoria*, described how males may guard ovipositing females, and then mate with them after the eggs are laid. The phase they described, reasonably enough, as a precopulatory passive phase. However, it is not long enough for us to count as an instance of precopula. In *Sepsis* it lasts a few minutes (less than an hour), and in *Locusta* for either less than an hour, or less than four hours, depending on the condition of the female (Parker and Smith 1975, p. 166). Although these are not, as a matter of definition, precopulas, like the mosquitoes their behaviour was probably selected for by the same general process as favours precopulas. The female starts developing her next egg batch after laying. In *Locusta* the first male to inseminate her during the next reproductive cycle fertilizes 60 per cent of her eggs even if another male mates with her (and 100 per cent if one does not, Parker and Smith 1975). Once again there is a fairly sudden, predictable change in the state of the female: before she lays her eggs she is much less worth mating with than after. Males therefore are selected to guard females while they lay their eggs in order to be the first male to mate with her in her next cycle. We cannot predict, from the available information, whether the males would be selected to guard the females for longer. We count these species as mating, as predicted, without a precopula. (They might alternatively be conceived as being dubious, and excluded from the test: it would not alter the numbers in the test.)

A third possible insectan analogy of precopula may be found in the Hymenoptera. The relevant system of mating is as follows. The males emerge from their pupae before females. (Or at any rate adult males and pupal females co-exist in time.) The pupae dwell underground, but in recognizable areas. The males may then defend territories at the emergence sites, and mate with the females as they come to the surface. Such habits are found in many species, which have conveniently been reviewed by Alcock and his seven co-authors (1978). This habit conforms with our theory. The females become receptive at a predictable instant, and males wait for it in a precopula. We might therefore count in favour of our theory the many instances of territoriality at female emergence sites. I however prefer not to, and for two reasons. One is that the analogy is imperfect. The theory referred to waiting with a particular female, once she has been found. By waiting at the emergence site the male certainly waits for a female, and it may seem pedantic to make anything of the fact that the males cannot actually see the females they are waiting for. But there is reason in this pedantry. If we include these species in the test, where are we going to stop? Consider some of the other mating habits of hymenopterans. Males of some species, instead of awaiting females at their emergence sites, await them at their feeding sites. The males defend territories around flowers. This habit is favoured for much the same reason, and it would be an easy extension to include all cases of territoriality for mating in our study. I do believe that the theory being tested is of much wider importance than for the evolution of precopulas as narrowly defined. But it is also desirable to keep the study down to size, to define its frontiers clearly. Our ultimate goal may be to understand mating systems as a whole, in all their diversity, but we cannot achieve it in one short book. Any excuse, provided that it can be applied objectively, will do to keep the study manageable. Let us seize on the pedantic distinction between the territoriality of hymenopterans and the precopulas of other species. It will save us from a huge literature. We will have more than enough to think about without them.

So much for the insectan analogies of precopulas. They were all good analogies, but must be excluded so that our summary may be coherent. The insects provide no examples of precopulas according to the definition that we are using; with them out of our way, our quest for precopulas can proceed into the ranks of the other groups of arthropods.

Tardigrades

The tardigrades owe their obscurity to their tiny size. That their phylogenetic position is unknown we have already noticed. Not much can be said about their mating habits either. We know of two main kinds of mating. Some, particularly the terrestrial species, have internal fertilization; the male deposits his sperms in the female's cloaca. Others, particularly aquatic species, have a less conventional method of fertilization. The male places his sperms beneath the female's old cuticle just as she is moulting. The female then lays her eggs into her old skin, and there they are fertilized. The skin acts as a sack for the eggs. In this second kind, oviposition, mating, and moulting all take place at about the same time. This has been shown for four species (the nomenclature following Ramazotti 1972): *Hypsibius dujardini* (von Erlanger 1895; Hennecke 1911), *H. augusti* (van Wenko 1914, p. 192), *H. convergens* (Baumann 1961, p. 376; Greef 1866, p. 119: *Macrobiotus teradactylus* is perhaps a synonym), and *H. nodosus*, in which the female is inseminated about 24 hours before she finally casts off her skin (Marcus 1929).

We might expect them to have a precopula. Our expectation will be satisfied. In *Hypsibius augusti* (van Wenck 1914, p. 492 and Fig. 30), the males seek out and grip onto the females before she moults. The males are smaller than the female, and more than one may cling to her (Fig. 3.3). Up to nine males may cling to a single female. Van Wenck does not tell us how long the males stay with the female before her moult, but *Hypsibius augusti* looks like a predicted instance of precopula. In *H. convergens* the male also joins the female before her moult. In this species, Baumann (1961, p. 376) tells us, the male and female spend a 'long time' together before copulation. The Tardigrada, must have evolved a precopula at least once. That will be their contribution to the test. The habits of the tardigrades, and of all the other groups discussed in this chapter, are summarized in Table 3.2 (p. 160).

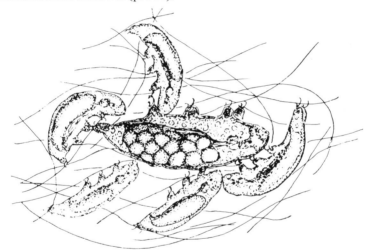

Fig. 3.3. Five male *Hypsibius augusti* in precopula with a female. (From Marcus (1929, redrawn from a blurry photograph in van Wenck 1914).)

Crustacea

Precopulatory mate guarding has been evolved and lost more among the crustaceans than among any other animal group. We are in for a long section. The habit is widespread because, in many crustacean groups, the female lays her eggs soon after her moult. Mating can then take place during the short interval between moulting and oviposition. These species, as we shall see, tend to evolve precopulatory mate guarding.

The cladistics of the large groups of crustaceans are not known with much certainty. We will use the phylogeny suggested by Siewing (1963, p. 102), which is as good and convenient as any. His phylogeny of the large groups with which we shall be concerned is shown in Scheme 3.3, and that is the order in which we shall run through them.

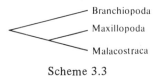

Scheme 3.3

Branchiopoda

Our libraries contain facts about all of the four branchiopodan orders (Anostraca, Notostraca, Conchostraca, and Cladocera). We start with the Anostraca. It contains the fairy shrimp *Artemia salina*. *Artemia* has often been watched, but not often carefully. The male has a large pair of cephalic horns with which he grips a female around her waist, next to her egg-sac, during the period of mating. A male and a female may go around thus paired for many hours. Precise measurements cannot be found in the literature; Sclosser (1756) was the first to describe the pairing and he just said they swim around together for 'some time', von Siebold (1877, p. 273) saw a pair together for three days, Leydig (1851, p. 298) said that pairs stay together for 'weeks' (*Wochenlang*). *Artemia* is easy to obtain, and Matthew Holley and I have made the necessary measurements. Thirty hours was the average duration of pairing at 21 °C ($n=35$). Sperm transfer and fertilization in *Artemia* take place just after a moult by the female when her eggs move from her ovaries down either side of her tail to her egg sac. Fairy shrimps conform to the theory.

The information on the other anostracans is less easy to interpret. The males of the other species also have enlarged cephalic horns, which suggests a lengthy and tenacious pairing. However, the published remarks on the duration of pairing refer only to the actual act of sperm transfer, not the full pairing. For *Chirocephalus*, Matthias (1936, p. 60) writes about a brief *'accouplement'* which is clearly distinct from the period in which the male grips the female. Valousek (1926, p. 15) writes of a copulation which lasts for but a few seconds and is preceded by the male swimming near to (but not connected with) the female; however, his remarks are too vague for us to use. In *Branchipus stagnalis* (a synonym of *Chirocephalus* according to Baird 1850) too the literature only reveals the duration of *l'accouplement*, not of the whole pairing (Matthias and Bouat 1934, pp. 331-2). And in *Eubranchipus dadayi*, when Pearse (1912) put a male with a female, 'he usually clasped her every few minutes until copulation took place', which is another statement that it is difficult to use to test our theory. In *Chirocephalus* (Prevost 1803, p. 100; 1820 p. 212), and *Eubranchipus dadayi* (Gissler, in Packard 1878, p. 422), a moult always precedes the transfer of eggs to the egg-sac, when insemination also takes place. We might predict a precopula, but can only conclude that the facts are too ambiguous for us to use.

The remaining Branchiopoda all seem to lack a precopula; but it is not always easy to decide whether or not they conform to the theory. For the Notostraca we have information on two genera. In *Triops cancriformis* (Hotovy 1937, p. 31) and in *Triops granarius* (Longhurst 1955, p. 679) copulation is rapid: it lasts less than a minute, and is not preceded by a precopula. Hotovy does not mention a moult, and Longhurst states that 'males may attempt to copulate with females in any condition', although he thought that copulation was only efficacious if it followed immediately after a moult. This last thought would suggest that *Triops granarius* is an exception to the theory: if mating is only successful just after a moult there should be a precopula. However, Longhurst presents no proof, and the evidence from *Lepidurus apus* (Desportes and Andrieux 1944, pp. 64-6) suggests that he was wrong. Copulation in this species is again quick: it lasts less than a minute (Du Réau de la Gaignonnière 1908; Le Goffe 1939, p. 42; Desportes and Andrieux 1944). The female can be mated at any time, and lays her eggs only after mating. Isolated females do not lay eggs, but they lay soon after a male is added. The loss of precopula in the Notostraca appears to conform to the theory.

The information on *Cyzicus cycladoides* (Conchostraca) is inadequate. 'The association of the sexes [*la rapprochement des sexes*] lasts only a few minutes' (Gravier and Matthias 1932, p. 1127), and a female moult always precedes mating (Gravier and Matthias 1930, p. 185). This might seem to contradict our theory, but it is not conclusive. It is neither clear how long before mating the moult comes, nor whether the sexual association is just the act of sperm transfer or the whole pairing. In most of Matthias's papers where the information is clear the measurements he gives are only for the act of insemination. *Cyzicus* must be excluded.

The final branchiopodan group is the Cladocera, which includes the familiar water flea *Daphnia*. Copulation has been seen but rarely, which itself suggests that it is quick. It was first described by Jurine (1820, pp. 108-11 and Pl. 11, Fig. 3). It is quick, and 'rarely' lasts as much as eight to ten minutes. The female mates only once even though other males try, uselessly, as the gallant Swiss said '*à pénétrer dans le sancteur des plaisirs*'. He makes no mention of a moult in a detailed description, which suggests that there is none. Likewise, in exquisitely detailed descriptions of the mating of many species belonging to 17 or more cladoceran genera, Weismann (1879) does not mention a female moult. He also implies throughout that mating is quick, and writes of one attachment of ¼ hour as amazingly long (p. 72). *Daphnia*, we may infer, conform to the theory. The Branchiopoda, in conclusion, contain species (among the Anostraca) which have a precopula when predicted, and other species (among the Notostraca and Cladocera) which lack one when predicted.

Maxillopoda

The main maxillopodan groups are the ostracods, barnacles, branchiurans, and copepods. Siewing's phylogeny of them is shown in Scheme 3.4. The first three

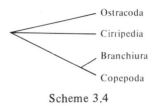

Scheme 3.4

of these will not hold us up for long: only for the copepods do we have much interesting information. I have not been able to find out anything about the incidence of precopula in the ostracods, or whether they mate immediately after a female moult. The sessile way of life of barnacles excludes them from our test. They have a rather variable relation between mating and moulting, as we have already discussed in the introduction to this chapter, and 'usually a male is found with a female' (Kaestner 1968, III, p. 199). Patel and Crisp (1961, p. 102) say that the habit of males and females of the same species settling near each other 'corresponds in function with the recognition of the female by the male and the carrying of the female until she is mature in non-sessile forms'. But, for our test, it cannot. Because they are sessile they do not have any option but to stay permanently close together, whether or not this is required by the confinement of mating to a short part of the female moult cycle. We will say no more about barnacles. The information on branchiurans is also not sufficient to include them in the test. They are fish parasites. Baird (1850, p. 253) writes that the male *Argulus* seeks out the female, and the copulatory embrace lasts 'several hours.' One might infer from Fryer's (1960) account of *Dolops ranarum* that mating is short and unconfined in the female moult cycle, as our theory predicts, but the inference is too weak to include the species in the test. The sexes only stay together for spermatophore transfer (p. 419-22), and he describes a female moult after spermatophore transfer, the sperms having migrated into the female, where they are stored, before the moult. But this information is inadequate to use in our test. Let us turn to a group which we can use in our test. Let us turn to the copepods.

Seven orders make up the Copepoda. Of the seven, I have uncovered nothing on the relevant habits of the Notodelphyoidea and Monstrilloidea, and very little on the fish parasites Caligoida and Lernaeopoida. In the Caligoida, according to Kaestner (p. 178), mating takes place just after the final moult of the female; but I have not found out if mating is preceded by a precopula. Many species of Lernaeopoida have extreme sexual dimorphism, with a dwarf male permanently attached to the female. In *Sphyrion lumpi*, for example, of 263 females examined by Squires (1966, p. 525), 12 had a dwarf male attached. Little is known about insemination, although in *Lernaea apprinacea* the female mates as a stage V copepodite, then moults, and continues her life-cycle (Bird 1968). No more seems to be known. They are unusable in our test. After a few general remarks, we shall concentrate in turn on the Calanoida, the Cyclopoida, and (especially) the Harpacticoida.

A typical copepod starts life as an egg, grows up through several naupliar stages, and then metamorphoses into a copepodite. After five subadult copepodite stages (marked by moults), a final moult takes it to the adult stage VI. Once adult the copepod does not moult again. So if mating is confined to a short period just after a moult, the moult in question is likely to be the last one; precopula would therefore be between a stage V subadult female and an adult male. Our search of the literature must attend to remarks about the stage of paired females. At the inseminatory act the male fixes a spermatophore near to the genital openings of the female; the sperms are received into a 'receptaculum seminis'. Lang (1948, p. 1513), writing about harpacticoids, states that the receptaculum seminis is ectodermal so any contents would be lost at a moult.

Williams (1907), generalizing about the duration of pairing, stated that copepods could be divided into those in which only one antenna of the male was modified for gripping the female, and those in which both antennae were so modified. Those with a single modified antenna, he thought, tend to have a short pairing, while those with both antennae modified tend to pair for longer. In many calanoids, indeed, only one of the male antennae is modified for grasping, and mating tends to be quick; and in many harpacticoids both antennae are modified (Fig. 3.4) and long pairings are known. However, in the Cyclopoids both antennae are usually modified and pairings are usually short. I have not formally tested Williams' generalization, but my strong impression is that the literature does not support it, and my weak impression is that it is wrong.

We come to the individual groups. In the Calanoida paired females are always adult, and precopula is absent. Mating has not often been seen, which itself suggests that it is quick. The only records of lengthy pairings are one between 'two specimens of *Eurytemora* [*vetox*]' which, Gauld (1957) observed, 'remained paired for several days', and another of *Pseudodiaptomus coronatus* (Jacobs 1961, p. 445) pairs of which 'may remain in copula for a few hours and sometimes for days'. But Gauld also writes of pairs of this and three other calanoids (*Centropages hamatus, Temora longicornis,* and *Acartia clausi*) which stayed together for only 'several minutes', and Jacobs' statement is too vague to support any conclusions. Jacobs also states that 'mating in *Acartia tonsa* Dana was never observed to last longer than a few seconds'. In *Eurytemora affinis* Katona (1975) tells us, 'males . . . usually only engage in copulatory behaviour only with adult females', and they pair for 'from 30 seconds to over an hour'. Blades (1977, pp. 59-60) describes a precopulatory stage in the mating of *Centropages typicus* which lasts for only a few seconds to a few hours; the mating females are adult (see also Lee 1972, p. 7). In a detailed description of copulation in *Labidocera*, Blades and Youngbluth (1979) do not mention any

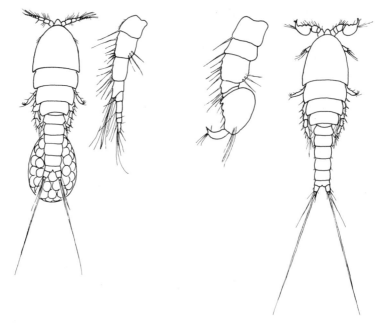

Fig. 3.4. A male (right) and female (left) *Tigriopus brevicornis*. (Redrawn by Matthew Holley from Sars (1903).) The magnified limb is the first antenna: the male uses his to grasp females in precopula.

moult by the female. *Limnocalanus gracilis* (Roff 1972, p. 156) pairs for only two or three minutes to mate when the female is adult. From this scant information we may conclude that calanoids generally mate when adult (and so are unconstrained by any moult) and have no precopula. They conform to theory.

Next, the Cyclopoida. They too will not detain us long, because we know so little about them. We know the most about *Cyclops*, particularly from the papers of Wolf (1905) and of Hill and Coker (1930). In many species of *Cyclops* there is a short mating, unpreceded by a precopula, and (if we may take silence about moulting as evidence of its absence) only between adults. Such has been reported for the following species: *Cyclops fuscus*, *C. strenuus*, *C. vernalis*, *C. viridis* (Wolf 1905, p. 125), *C. serrulatus* (Wolf 1905; Hill and Coker 1930), *C. americanus*, *C. bicuspidatus*, *C. prasinus*, and *C. varicans* (Hill and Coker 1930), and *Asterocheres lilljeborgi* (an ectoparasite of starfish, Röttger, Astheimer, Spindler, and Steinborn 1972, pp. 263-4). The posture of the pair is an important part of the evidence. The pair in these species without a precopula face each other, the ventral surface of the male facing that of the female (Fig. 3.5), as first described by Jurine (1820, Pl. I). Later (it was first described by Schmeil 1892, p. 166, and in more detail by Wolf 1905) a different kind of pairing was seen. In *Cyclops fimbriatus*, *C. affinis*, and *C. phaleratus*, Wolf saw pairs in which the male was attached behind the female, and both male and female swim in parallel, with the same dorso-ventral orientation. He described the posture as resembling that of the harpacticoids (Fig. 3.5). Now in harpacticoids (as we shall soon see) such pairs may pass several days together.

Fig. 3.5. Normal copulatory posture of copepod illustrated by *Cyclops fuscus* on the right. Precopulatory posture of a harpacticoid, *Canthocamptus crassus*, on the left. (From Wolf (1905).) (Photograph supplied by the Bodleian Library, Oxford.)

There are further hints. Rehberg (1880, p. 536) had mentioned paired immature females in *C. fimbriatus*; Schmeil (1892, p. 166 n4) questioned Rehberg's taxonomy, stating that he had seen inseminated immature females of other species of *Cyclops*, but not of *C. fimbriatus*. Wolf did not specify the stage of his paired females. Holmes (1909) tells us that *C. fimbriatus* swim around in pairs 'for a long time prior to copulation', and he compares their mating habits to amphipods (which have a long precopula followed by mating after the female moult). It remained for Hill and Coker to investigate the matter thoroughly. Hill and Coker observed this parallel pairing in *C. modestus* too, but in this species it was not the final mating posture. After the pair had spent some time ('in several instances the male held this position for more than ten hours' p. 212), the female moulted, and the pair then changed their grip to the normal cyclopid mating position with the male and female facing each other. The actual spermatophore transfer lasted only five minutes. The parallel pairing position is a precopula. The several remarks about paired immature females, and moulting during pairing, in *C. fimbriatus* (Rehberg 1880, p. 536; Hill and Coker 1930) combine to suggest that this species also has a precopula and mating after the female moult. The only observations of paired immature females, and of moulting during pairing, are in the species with the parallel pairing posture. The evidence for a lengthy pairing in these species is slim (the only two vague remarks are quoted above), but these species appear to have a longer pairing than other *Cyclops*. The other *Cyclops* species mate only when adult (at least it has never been said to be otherwise) and is not preceded by a precopula. In *C. serrulatus* both mating postures have been observed (Hill and Coker 1930). We may conclude that *Cyclops* conforms to our theory. In some species there is a precopula followed by mating after the female moult; in other species mating is between adults and is not preceded by a precopula. In some species both kinds of mating take place. The evidence for a precopula from any single study is not convincing, but examined as a whole, it is.

The literature on the harpacticoids, although larger than that for other copepods, is no more conclusive. We shall be concerned with only nine of the families (dispersed among six of the superfamilies) of Lang's (1948) classification. Precopulatory pairing between an adult male and a subadult female is widespread but not universal. Unfortunately it is not clear whether the long precopulas are only found before matings with females as they moult to adulthood. The literature neither contradicts nor proves this crucial generalization. We will work through the facts in the order of the following classification:

superfamily 1	family Ectinosomidae
superfamily 2	family Harpacticidae
	Tachidiidae
superfamily 3	family Peltidiidae
	Tisbidae
superfamily 4	family Thalestridae
superfamily 5	family Canthocamptidae
superfamily 6	family Cletodidae
	Laophontidae

For the Ectinosomidae we have only a single illustration of a pair of *Pseudobradya*, by Willey (1931, Tav XXI), who states in his text that the illustrated female was subadult. For the Harpacticidae we have information on *Harpacticus gracilis, H. uniremus, Tigriopus brevicornis*, and *T. japonicus*. All four pair up reportedly some time before the female's final moult; later she moults, a spermatophore is transferred, and the pair separate. Williams (1907) writes of an 'apparently normal' pair of *Harpacticus gracilis* which stayed together for 24-39 hours. The male *Tigriopus brevicornis*, like many other harpacticoids, has antennae modified to grip the female (Fig. 3.4). Fraser (1936, p. 526) writes of pairs which swum around together for 'a few minutes to several days'; he implies that the female of the pair is normally a stage V subadult, but 'fully grown females can often be seen mating'. What we want to know is whether that variation in pairing duration is related to the stage of the paired female; our want, at the time of writing, must remain unsatisfied. *Tigriopus japonicus* stay together for two to four days and mate after the female moult (Ito 1970, p. 476, who tells us that some stage IV females were paired). Lang (1948, p. 1513) states that female *Harpacticus* and *Tigriopus* of a variety of stages, including adults, may be seen mating. We can conclude that the Harpacticidae have a correctly predicted precopula. For the Tachidiidae we have a single paper by Haq (1972), on *Eupertina acutifrons*. 'Newly moulted females readily copulated', he found (p. 226), but contrary to Williams (1907) he found no precopula followed by mating after the female moult. Males only 'very rarely' clasped subadult females. Haq tells us nothing about the duration of pairing, so we cannot decide which of the four boxes of our test to put *Eupertina* in, and we will ignore it in the test.

Claus's (1863, p. 71) observations on the Peltidiidae were the first of paired subadult female copepods: 'I often saw males of Peltidiidae in the afore-mentioned grip-position with immature [females], united to the gripped female before her last moult'. In 1863 he considered the possibility that females were inseminated before their last moult; but in a later paper (1889, p. 6) he said that sperm transfer took place only once the female was adult. The later theory is more likely because, as Lang (1948) argued, sperm transferred before the female's last moult would be lost at the moult. As we turn to the Tisbidae, we find a single paper by Johnson and Olson (1948), on the littoral *Tisbe furcata*. Their observations are suggestive but not conclusive. They saw pairings of 'a few hours up to one or more days' (p. 327; Lang 1948, p. 1514 gives similar figures), but the paired females were generally adults. 'On only one [of 87] occasion was there evidence of moulting during the clasping period' (p. 327). Moults, of course, are easy to overlook, but we may conclude that pairing with subadult female are rare in this species. Again, we would like to know whether that variation in pairing duration is related to the stage of the female; again the literature does not tell us.

The females of *Diathrodes cystoecus* secretes herself a capsule of mucus cement, on seaweed. The males tour these capsules, and on finding a female, seizes hold of her. 'This clasping position may be maintained for hours or days, although rarely the latter' (Fahrenbach 1962, p. 345). Insemination takes place within 24 hours of the female's final moult, and eggs are laid within another 24 hours (pp. 315 and 342). Males sometimes unsuccessfully try to mate with stage V copepodite females or ovigerous females (p. 346). This species probably conforms to the theory.

Observations have been published on three genera of Canthocamptidae. Schmeil (1894, Taf. III, Fig. 1) illustrated a pair of *Canthocamptus trispinosus* in what looks like a precopulatory posture, although he gave no information on the pair. Wolf (1905, p. 126) saw long pairings in *Canthocamptus staphylinus* (Fig. 3): 'I isolated several pairs, and can thus state that 4-6 days passed before the spermatophore was stuck on'. He did not, however, state whether the females were adult or subadult. Complementary information is available for *Athyella* (Willey 1925, p. 125; 1931, p. 603) and *Mesochra* (Giesbrecht 1882; Willey 1931, Tav XXI, Fig. 10): in these paired females are subadult, but we are not told how long they pair for.

Pairing of subadult females has also been documented in the final two families of our list, the Cletotidae and Laophontiidae. For the former we have only a single illustration by Willey (1931), Tav XXI, Fig. 3) of *Cletocamptus bermudae* together with his textual assurances that the female of the pair is subadult. Willey similarly depicts and describes three species of Laophontiidae: *Laophonte lunata*, *L. sigmoides*, and *Platychelipus littoralis*. Lang (1948, p. 1513) states that the pairing of subadult females is the rule for the family, although he may only be generalizing from Willey. We now also have the paper by Lasker, Wells, and McIntyre (1970) on *Asellopsis intermedia*. 'Adult males [they tell us] were found clasping copepodites of all stages from Stage II to V, but never clasping adult females' (p. 154); stage V females were the most frequent partners. Clasping lasted 'several days at least' (p. 154). Their evidence that spermaphores were transferred only after the final female moult is suggestive, if not conclusive. We may conclude that *Asellopsis intermedia* has a precopula, which ends with mating after the female moult.

Let us now consider the harpacticoids as a whole. A conclusive study would report the stage of the paired females, the duration of pairing, and the timing of spermatophore transfer. Fahrenbuch (1962) and Williams (1907) more or less achieve all this, but Williams only asserts it in a short note. The paper of Lasker *et al.* (1970) is reasonably conclusive, and those of Fraser (1936) and Ito (1970) slightly less so. For the harpacticoids, we cannot found our conclusions on single excellent studies. Single papers are frustrating: many record paired subadult females, but few also report the duration of pairing. Whenever we are told the duration of pairing—'several days', '5-6 days'—we are left ignorant of the stage of those enduring females. Whenever the stage of females (in lengthy pairings) has been reported, whether by Claus, by Giesbrecht, by Ito, by Williams or Willey, Fraser or Fahrenbuch, Lang or Lasker, Wells and McIntyre, they have been subadults. Only Haq, and Johnson and Olson stand out against this throng, and even they found some paired subadult females. And in none of the species with paired subadult females were the pairings said to be brief. Few papers, indeed, record the duration of pairing at all, but Williams, Fraser, Wolf, Johnson and Olson, Fahrenbuch, Ito, and Lasker, Wells and McIntyre all remark on pairs which lasted for more than a day. We can at least conclude that some harpacticoids mate after the final female moult, after a precopula. We can, however, be less sure of whether some do not also mate when adult, without a precopula. It looks likely, and the theory would predict it; but there are insufficient facts to be sure.

How, then, shall we enter the copepods in our test? The calanoids have a short mating and can mate any time during the adult female's life. Most cyclopoids do likewise, but some may have a precopula and mating after the final female moult. Many harpacticoids have a precopula and mating after the final female moult. The counting for the test depends on the ancestral state. The nearest outgroups for which we have some kind of relevant information are, using the phylogeny of Siewing (1963, p. 102), the Branchiura, the Cirripedia, and the Ostracoda. As we have seen, little is known about these, but we may reasonably suppose that they lack a precopula, for the existing literature betrays no sign of one. The nearest outgroup for which we know whether a precopula is present or absent is the Branchiopoda; and both states are found in that group. The primitive state of the copepods is therefore obscure, but we will suppose, because no sign of one can be found in other maxillopods, that they primitively lacked one. The calanoids, and most cyclopoids, would then have inherited their mating habits from a common ancestor of like habits. Precopula, and mating after the final female moult, would have re-evolved twice, once in *Cyclops*, and once in the ancestor of the harpacticoids. We will send these two to the test.

Malacostraca

The next, the last, and the largest branch of the Crustacea is the Malacostraca. The cladogram of the four main groups, following Siewing (1963, p. 102), is shown in scheme 3.5.

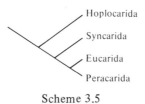

Scheme 3.5

Hoplocarida and Syncarida

The first two groups will not detain us. Nothing is known about the mating of the Syncarida. The Hoplocarida contain the mantis-shrimps, which, despite a rich tradition of ethological research, yield but one paper on mating habits. It is by Dingle and Caldwell (1972), and concerns *Gonodactylus bredini*. The sexes usually live in separate retreats. When a female becomes receptive for mating, it is accomplished within a few minutes: there is no precopula. And, as we would suppose, 'female receptivity was not connected with the molt' (p. 423).

The male does stay with the female after mating for about 10 days, until she lays her eggs. She then evicts him.

We may move straight to the final sister groups, Eucarida and Peracarida. On these two groups there is a very large literature indeed. Their members mate variously with and without a precopula, so the single paper on *Gonodactylus* is valuable in demonstrating that the absence of a precopula is the primitive state. The Eucarida and Peracarida we will take in turn. The superorder Eucarida is made up of two orders, the Euphausiacea and the Decapoda. Nothing is known about the mating of krill, so we can go straight onto the decapods.

Decapoda

We will work through the decapodan groups in the order of the classification shown in Scheme 3.6.

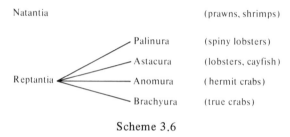

Scheme 3.6

Natantia

We will make a few general remarks about the suborder, and then work through its three sections (Penaeidea, Caridea, and Stenopodidea). The first general remark is that insemination has followed a female moult whenever it has been observed. However, as Nouvel and Nouvel (1937, p. 209) remarked, there does not seem to be a precopulatory phase in which the male carries the female (*une appariage*), like that found in many other crustacean groups. They thus initially look like an exception. But, as we shall see, the observations are inadequate to test the theory. Fertilization is external. After the female's moult the male attaches a spermatophore to the ventral surface of the female at a region, often called the thelycum, designed to receive it (Heldt 1931; Nouvel and Nouvel 1937; Palombi 1939; L. Nouvel 1939, 1940; Hudinaga 1942; Hoglund 1943; Needler 1945; Burkenroad 1947; Lloyd and Yonge 1947); it had been thought that, in the planktonic prawn *Lucifer*, the thelycum (spermatheca) was internally connected to the oviducts. Hartnoll (1968a, who also reviews earlier writings) has demonstrated that it has no such internal connexion; it has external fertilization like any other natantian.

The time and mode of fertilization we know with reasonable certainty; but we know much less about the association of the sexes for mating. Discoveries subsequent to Nouvel and Nouvel's (1937) review do not change their conclusion that male natantians do not carry the female before her moult. But that is not the end of the matter. Nouvel and Nouvel gave only one reference to (and that only a stray comment in a faunistic survey: Heller 1864) shrimps living together in pairs. We know of many more examples now. Furthermore, the observations of mating without a precopula were nearly all made by placing a male in a tank with a recently moulted female. This method was, sensibly enough, being used to observe sperm transfer. But we would hardly expect a precopula in those circumstances. A few of these papers also record that males take no interest in females at other stages of their moult cycle, or even treat them aggressively (L. Nouvel 1939, 1940; Needler 1931; Kamiguchi 1972; Bauer 1976); but again the observations were taken after a male had been added to a female in a tank. Aggression is exactly what one would expect, especially if the species naturally live in fairly permanent pairs. No observations yet rule out the possibility that males stay near females before their moult without being in physical contact. (If they physically clasp each other, we would expect someone to have seen it; but no one has.) Several of the species (alpheids, crangonids) studied by Nouvel and Nouvel, and of which they said that there is no precopula, have subsequently been found to live in permanent pairs. Who dares to say that further appropriate studies of natantians, in nature or in more or less natural laboratory conditions, will not reveal more? To discover whether a species has a precopula without physical contact, rather more careful observations are required than for precopulas in which the male grips the female. The Natantia are not the last group for which we will point this out.

Let us now turn to the groups. We start with the Penaeidea. Relevant facts are available only for the genus *Penaeus* (Penaeidae). In both *Penaeus japonicus* (Hudinaga 1942, pp. 307-9) and *P. caramote* (Heldt 1931), fertilization is external and mating follows a female moult. Burkenroad (1939, p. 21), found that in *P. setiferus* 'of 259 females from all hauls [from nature] closely examined, 145 . . . showed signs of having been impregnated . . . Soft- and paper-shelled individuals showed no traces of impregnation, so that mating is in this species apparently not an accompaniment of the molt'; these results are too indirect to persuade. Heldt and Burkenroad did not watch pairs before the female moult; but Hudinaga did. Penaeids tend to live in the bottom sand, but come out at night to creep around and mate. Hudinaga placed his prawns in artificial tanks. He then himself stood in the tanks and watched the prawns during the early hours of the morning. His vigil was rewarded five times. 'The male follows the female' before she moults, for a short time (which was difficult to measure, but lasted 3-7 minutes). This is too short a time to count by itself as a precopula. For the penaeids then we may conclude that mating is confined in time to a period just after the female moult, but that good evidence of a precopula is lacking.

Next, the Caridea. Information is available for six of the families in four of the superfamilies (out of totals of 11 and six respectively). In the Pandalidae (which are protandric hermaphrodites) mating has been seen in *Pandalus danae* (by Needler 1931) and in *P. platyceros* (by Hoffman 1973). In both species mating followed soon after a female moult, and in neither was there any record of precopulatory mate guarding. Needler, however, only ever put recently moulted females with males, so he had no chance of seeing a precopula; and Hoffman only incidentally observed mating during an unrelated study. So the absence of a precopula is not proven. Let us move on to the next superfamily, which contains both the Alpheidae and the Hippolytidae. We will take the snapping shrimps (Alpheidae) first. All the species which have been appropriately studied have been found to live in pairs, in the niches and cavities of coral reefs. Coutière, in his great monograph on the alpheids (1899), tells of a number of species which he found living in couples at Djiboiti, baix de Tadjourati (in the Red Sea). He mentions nine species of *Alpheus* (pp. 489-508), two of *Synalpheus* (pp. 492 and 497), and *Arete dorsalis* (p. 509). Other authors have found further species living in pairs, eight more species *Alpheus* (Cowles 1913, p. 122; Fishelson 1966; Magnus 1967; Patton 1974, p. 234; Castro 1974, p. 400; Schein 1975, pp. 93-4; Knowlton 1980), and *Synalpheus charon* (Castro 1971, p. 399). Mating and laying are associated with the female moult (Knowlton 1980, p. 164). For the Hippolytidae there are no natural observations, and no records of species living in pairs. Both Bauer, studying *Heterocarpus pictus* (1976, p. 418), and *H. paludicola* (1979, p. 162), and L. Nouvel (1940), studying *Lysmata seticaudata* (a protandric hermaphrodite), placed recently moulted female with males, to observe mating, which is confined to this period of the female moult cycle.

Our third superfamily contains the Crangonidae, sand-dwelling prawns from the intertidal to the deep sea. The transfer of the spermatophore has been watched in two species, the common prawns *Crangon vulgaris* (Lloyd and Yonge 1947, p. 641; Havinga 1930, p. 70) and *C. crangon* (L. Nouvel 1939). Mating follows just after the female has moulted; in *C. crangon* the female cannot be inseminated more than about 12 hours after her moult. The observations on these two species appear to have been made by putting a male with a recently moulted female, so they tell us little about the natural association of the sexes. MacGinitie (1937, p. 1035; 1935, p. 706) has observed three other species of this family in nature and found them all living in sexual pairs. The three are: *Crangon californiensis, C. dentipes,* and *Betaeus longidactylus.* Insemination has not been seen in these three species though we can presume that it follows the female moult like in all related species. Lengthy pairing, too, is found in *Hymenocera picta* (Gnathophyliidae). Seibt and Wickler (1979, p. 172) added males to females one, three, five, and seven hours after the moult and found that mating was generally not successful after five or more hours after the moult. Mating normally takes place within an hour after the moult. *Hymenocera* evidently conforms to the theory.

The final superfamily contains the Palaemonidae, which are littoral, brackish, and freshwater prawns. The family divides into two subfamilies, Palaemoninae and Pontoniinae: the studies, mirroring the taxonomy, divide into faunistic surveys in nature (of Pontoniinae) which describe sexual associations, and laboratory observations (of Palaemoninae) which describe sperm transfer. Only two species have been studied in both ways. Insemination, like in other natantians, consists of the sticking of a spermatophore onto the outside, ventral surface of the female just after a moult. It has been described for the following species: *Palaemon elegans* (Hoglund 1943, who called it *Leander squilla*), *Palaemon paucideus* (Kamiguchi 1972), *Palaemonetes vulgaris* (Burkenroad 1947), *Palaemonetes varians* (Antheunisse, van den Hoven, and Jefferies 1968, p. 260), *Macrobrachium amazonicum* (Guest 1979, p. 142), and the giant (20 cm or more long) Indian freshwater prawn *Macrobrachium rosenbergii* (Bhimachar 1965, p. 5). Rao (1967) put males of *M. rosenbergii* with females only just after their moult; the pairs mated, but more interesting is his incidental observation that the male stayed with the female after mating. Perhaps the Palaemoninae have monogamous species? The reported observations on the subfamily are not adequate to rule out a precopulatory sexual association in any species. Many other species (but all of Pontoniinae) have been observed in nature, and have been found living in pairs: *Pontonia flavomaculata* which inhabits the mantle cavity of the ascidian *Phallusia* (Heller 1864, p. 51), *Pontonia domestica* (Herrick 1895, p. 73) and *Anchistus custos* (Jacquotte 1963, pp. 59 and 62; Johnson and Liang 1966, pp. 443-3; Hipeau-Jacquotte 1974*b*), which both live in the bivalve *Pinna*, *Conchodytes biunguiculatus* and *Paranchistus ornatus* (Jacquotte 1963; Hipeau-Jacquotte 1974*b*), and four species mentioned by Castro (1971, pp. 396-7), *Periclemenes imperator*, *Harpiliopsis depressus*, *Conchodytes tridacnae*, and *Stegopontonia commensalis*.

Only Hipeau-Jacquotte (summary 1974) has brought the two kinds of study together. As we have seen, she has recorded pairing in three species. In two of them (*Anchistus custos* and *Conchodytes biunguiculatus*) she has also confirmed that mating, and egg-laying follow immediately after the female moult. The eggs are laid about 40 hours after mating, and are laid whether or not the female mates. These two species, at least, conform to the theory. The rest of the subfamily presumably do as well, although they would not all appear independently in the test. For the Palaemoninae, we have hardly any observations of the behaviour prior to insemination. Nataraj (1947, p. 91) writes of courtship 'fencing' an hour before insemination in *Palaemon idae*. The observations of Ruello, Moffitt, and Phillips (1973, p. 198) on the Australian *Macrobrachium australiense*, in aquaria, are also suggestive. The male (of a pair in the tank) built a nest and courted the female several days before her moult. She only actually entered the nest about half an hour before her moult. After the moult, the pair mated, and a day later the female left the nest again. But none of our information on the Palaemininae is good enough to use to test the theory.

Of the natural history of the third natantian section, Stenopodidea, little is known. *Spongicola venusta* lives in pairs in the venus basket sponge *Euplectella* (Kaestner 1968, III, p. 320), and *Stenopus hispidus*, a cleaner of fish, also lives in pairs (Brooks and Herrick 1891, p. 327; Johnson 1969). Yaldwyn (1968, p. 379) saw a 'nuptial dance' by a male to a female who had just moulted in the same aquarium; but the pair never actually mated (p. 388). Only the timing of the dance is suggestive. Analogy with other natantians is, of course, much more powerfully suggestive; but analogical reasoning is not allowed in this comparative study.

The facts about the mating of the Natantia, commercially interesting though they no doubt often are, are difficult to bring into confrontation with our theory. We may reasonably conclude that fertilization is external in the whole group, and takes place soon after the female moult: this has been confirmed in all the species which have been appropriately studied. The facts on the association of the sexes, however, are less conclusive. In all the species which have been studied in nature, the sexes have been found living in pairs. We may be tempted, with one eye on our theory, to generalize this observation. A weaker generalization would be that all species have some form of pairing, although without physical clasping, before insemination. But this would be a rather radical reinterpretation of the mating of the Natantia: no-one has ever suggested it before. Perhaps people have been over-influenced by the behaviour of the Natantia in aquaria. These observations in aquaria prove nothing: males and females have only ever been put together just after the female moult. The suggested generalization is reconcilable with (although not suggested by) the remarks of laboratory observers. With so much uncertainty, we will rely on only the species whose sexual association has been studied in nature and the timing of whose mating is known. There are only four such species, and they all have lengthy pairing (*Alpheus armatus, Hymenocera picta, Anchistus custos*, and *Conchodytes biunguiculatus*), which is predicted, because insemination only follows the female moult. They could all appear as at most one entry in the test. All the facts on the Natantia are summarized in Table 3.2 (p. 160).

Reptantia

The four sections of the Reptantia are the Palinura and Astacura (often combined in the Macrura), the Anomura, and the Brachyura. We will, for convenience, first consider together the mode of fertilization in the two macruran groups, and then separately their sexual associations for mating. After that we can proceed to the Anomura, and to the Brachyura.

Fertilization in the Macrura The classification of the groups with which we will be concerned in this section is shown in Scheme 3.7.

Until recently, it was thought that all the Macrura, like the Natantia, had external fertilization. A spermatophore is stuck on to the external ventral surface of the

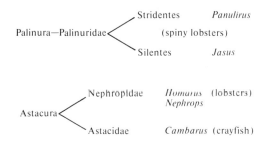

Scheme 3.7

female at mating. Then, it was thought, when she lays her eggs, sperms are released from the spermatophore and fertilize the eggs on their way from the oviducal openings on the sixth thoracic segment to be brooded on the legs of the abdomen. In the final year of this *ancien régime* Berry and Heydorn (1970) described (indirect) observations of external fertilization in five genera, *Panulirus*, *Palinurus*, *Puerulus*, *Linuparus*, and *Jasus*. Then came an important paper by Silberbauer (1971) on *Jasus lalandii*. Silberbauer discovered an internal connection between the seminal receptacle (where the spermatophore is deposited) and the oviduct, and he demonstrated histologically that sperms were present in the oviducts after mating. *Jasus lalandii* has internal fertlization. The spermatophore is attached, then releases its sperms internally, and is soon lost. The sperms are stored until the eggs are laid some 15–32 days after mating.

Silberbauer's result has since been extended to lobsters. Sperms have been found in the oviducts of female American lobsters *Homarus americanus* (refs. in Aiken and Waddy 1980), and Farmer (1974) argued for a similar mode of fertilization in the Norway lobster ('scampi') *Nephrops norvegicus*. Experts (Silberbauer 1971; Aiken and Waddy 1980 for review), however, do not extend the result to the Stridentes or to the Astacidae. It may be recalled that the original proof of external fertilization in *Jasus* was identical to that for *Palinurus* and *Panulirus*. Careful external observation suggested, in *Jasus* at least, external fertilization where it was not. Could not the observations of the Stridentes have been in error as well? Perhaps a histological investigation like Silberbauer's would reveal internal fertilization in the Stridentes too; a histological study is certainly needed to prove external fertilization. However, the fact that *Jasus* have a clear internal seminal receptacle whereas the Stridentes do not is a good reason for not generalizing the result (Silberbauer 1971, p. 41).

So much for the site of fertilization. What about its timing? When in her moult cycle does the female mate? In the Palinura, as we shall see, mating is confined to a short period after the female moult in *Jasus*, but not in the Stridentes; in the Astacura, mating is confined to a short period after the female moult in the Nephropidae, but not in the Astacidae. Let us look at the details.

In *Jasus*, according to Silberbauer (1971, p. 25),'mating is most likely to occur when the female is soft-shelled, during a brief period following ecdysis, some 15-22 days before the shell fully hardens', and according to Von Bonde (1936, p. 9) mating follows within 'about two hours after' the female moult. Mating is not so confined in the Stridentes. In *Panulirus longipes cygnus*, mating occurred all through the moult cycle in a study by Chittleborough (1976, p. 502); although Sheard (1949, p. 13) states, without proof, that mating must follow a female moult to be successful. In *Panulirus argus*, mating is 'usually' (Kaestner 1968, III, p. 321) and (in one study) on 10 occasions out of 10 (Sutcliffe 1953, p. 176), with a hard female; neither Heydorn (1969) nor Berry (1970) mention a moult in their descriptions of mating in *Panulirus homarus*.

The lobsters (Nephropidae) resemble *Jasus*. They have similar internal fertilization, and mating is again confined to a short period after the female moult. Females can store matings through a moult, and a single insemination can fertilize more than one egg batch (Aiken and Waddy 1980, p. 245); an electrophoretic study suggested the multiple paternity of a brood of one wild female (Nelson and Hedgecock 1977). The earliest, but indirect, observations on the American lobster did suggest that mating took place throughout the female moult cycle (Herrick 1895, p. 35), but direct observations subsequently undid the suggestion. Templeman (1934, p. 424; 1936) found that nearly all mating took place within two days of the female moult, and experimentally showed that females do not normally mate more than about five days after the moult. The extensive observations of Hughes and Matthiessen (1962, p. 419) revealed no matings more than two days after the moult. Atema *et al.* (1979) also found that mating followed moulting. For an anatomical explanation we can turn back to Templeman (1934, p. 424; 1936, p. 225). When the spermatophore is first placed in the female's receptacle it is a gelatinous mass, but after 9-10 hours it hardens and physically prevents the deposition of any more spermatophores. Only the paper of Dunham and Skinner-Jacobs (1978) stands out against this stream of observation and theory. They experimentally put 12 pairs of lobsters together when the female was hard-shelled, during her inter-moult. Seven pairs mated. But this result need not contradict the earlier observations. Mating probably normally takes place after the female moult, but if a female is isolated and does not then mate, she may mate later. Even if she has already been inseminated, another male may stick a spermatophore on, but it will probably not result in fertilization, being either too late, or prevented by a previous spermatophore, or both. All that matters to us is whether mating normally takes place just after the female moult, and the facts suggest that it does. Fewer, older facts suggest a similar conclusion for the European lobster *Homarus gammarus* (Anderton 1909, p. 19). In *Nephrops norvegicus* too mating follows the female moult (Farmer 1974, p. 172).

In the crayfish (Astacidae), the spermatophore is not deposited immediately after the moult. The external attachment of the spermatophore in the crayfish[3] was observed by Gerbe (reported by Coste 1848), and external fertilization was described by another worker in Coste's laboratory, Chantran (1870, 1872). Chantran (1870) also showed that the female was hard at mating: the female only moults once while adult, in August or September, but she mates at sometime between November and January. Subsequent observation has all agreed that the female crayfish mates when hard (Andrews 1895; 1904; Herrick 1911, p. 301; Templeman 1934, p. 423; Yonge 1947, p. 511; Tack 1941; Smith 1953; Mason 1970; Ingle and Thomas 1974, p. 536; Ameyaw-Akumfi and Hazlett 1975).

In summary, we may state that mating in the Palinura is confined to the period after the female moult in *Jasus*, but not in *Panulirus*; in the Astacura it is confined in the Nephropidae, but not in the Astacidae. The theory clearly predicts some kind of precopula in *Jasus* and in the Nephropidae, but not in *Panulirus* and the crayfish. How well do the animals measure up to it?

Sexual associations for mating   Berry (1970, pp. 4-5) described a precopulatory phase, lasting 'for 5-13 hours', in the mating of *Panulirus homarus*. The duration is too short to count as a precopula, and from Berry's description it sounds more like a courtship than a precopula. The precopulatory phase seems to be spent in coaxing the female out of her shelter, to make mating possible. There is no mention of anything like mate guarding in Chittleborough's (1976) account of mating in *Panulirus longipes cygnus*.

There is a hint, but not a definite description, of a precopula in *Jasus lalandii*. Silberbauer (1971) tells us that males show 'responses' to females before the moult. The group of females that attracted males he describes as moulting over the next two weeks (pp. 25-6). Males, it therefore appears, 'respond' to females up to at least two weeks before their moult. However, from Silberbauer's description, we cannot decide conclusively whether *Jasus* has a precopula. The male 'response' he describes as 'an increase in activity and aggression accompanied by a raised stance . . . ' (p. 25): this may tantalize, but it will not convince, the student of mate guarding.

We turn now to the Astacura (lobsters and crayfish). The papers before Atema *et al.* do not betray the slightest hint of a precopula in the American lobster. Nor is there any reason why they should, for they all only describe the act of insemination. The history of lobster research may serve as a parable for the prawns and shrimps which we have recently reviewed. From Herrick (1895) to Templeman (1934), from Templeman (1936) to Hughes and Matthiesson (1962), no precopula was seen; but as soon as tanks were set up, containing lobsters under approximately natural conditions, and the lobsters were allowed

---

[3] Neither Gerbe (or, rather, his boss Coste), nor Chantran state the formal name of the species that they observed. They call it *l'écrevisse*. It was probably *Astacus fluviatilis*.

to mate when they wanted to (not only when they were put together), and they were watched before mating, then a precopula was discovered. Atema *et al.* (1979, pp. 283-8) described the following habits. A few days before her moult the female leaves her shelter in search of a male; she then joins the male in his shelter and cohabits with him for some time (up to seven days); the female then moults, the pair mate, and they then stay together for a further period (of up to seven days); the female eventually departs. *Homarus americanus* has a precopula. The paper on the European lobster *Homarus gammarus* by Anderton (1909, p. 19) does not mention a precopula. The single female sought a male after she had moulted, the pair mated, and then cohabited, 'off and on, for several weeks'. But this was only one female, in a paper on artificial cultivation. Even if there had been a precopula it might not have been seen, or mentioned. We do not use observations of this low quality to test our theory. Farmer (1974) also does not mention a precopula in *Nephrops norvegicus*; but then his methods were not designed to reveal one.

We have predicted the absence of a precopula in the crayfish, and none of the papers on the mating of crayfish describe one (Andrews 1895, 1904; Tack 1941; Smith 1953; Mason 1970; Ingle and Thomas 1974). Once again, the observations are in most cases inadequate to prove the absence of a precopula. Only the paper of Ingle and Thomas (1974) is sufficiently detailed that we can conclude that, if there were a precopula, they would have noticed it. Their species, *Austropotamobius pallipes*, we can count as an instance of the predicted absence of a precopula.

Field observations on mudshrimps (Thalassinioidea, Callianassidae, another superfamily of the Astacura) suggest that many species have lengthy sexual associations. They live in burrows in the sand, and have often been seen in pairs: *Callianassa affinis* (MacGinitie 1937, pp. 1034-5) lives in pairs, *C. gigas*, in one observed instance, lived in a pair (MacGinitie 1935, p. 712), *C. filholi* lives in groups of one male with two or more females in burrow (Devine 1967, p. 109), and *Upogebia pugettensis* lives in pairs (MacGinitie 1930, p. 43; 1935, p. 707); *C. californiensis* does not live in pairs (MacGinitie 1937, p. 1034, the references she cites do not mention pairing), and nor does *Callianassa major* (Rodrigues 1976, p. 88). But we do not know when mating takes place during the female moult cycle. Because other Astacura vary in the time of mating, we must exclude the mudshrimps from the test.

The habits of the Macrura are summarized in Table 3.2 (p. 160); we will count their contribution to our test later.

Anomura The Anomura, the hermit crabs and their relatives, mainly inhabit shallow water and intertidal areas. We will be concerned with six of the families, scattered through all four of the superfamilies. It will turn out that most of the facts are too inconclusive to use in the formal test; but we can still confront the theory with the facts, and see how well they fit. The classification of the six families, and the order in which we will consider them, is:

Coenobitoidea	1	Diogenidae
Paguroidea	2	Paguridae
	3	Lithodidae
Galatheoidea	4	Galatheidae
	5	Porcellanidae
Hippoidea	6	Hippidae

We will consider a few general facts before working through the individual groups. The hermit crabs, which make up the Diogenidae and Paguridae, are the most studied anomurans. It has been generally accepted since the nineteenth century (Brandes 1897 for *Pagurus prideauxi* and *Galathea strigosa*; Mayer 1877; see Block 1935, pp. 252-3, for *Diogenes pugilator*), that in hermit crabs the spermatophore is attached externally near the female genital openings at the base (coxae) of the third thoracic legs (pereiopods). When the eggs are fertilized is less certain. They may be fertilized on their release to be brooded (Brandes 1897, p. 350; Block 1935, p. 253). Or sperms may migrate from the spermatophore into the female, and fertilize the eggs inside the female. It has been said that the eggs are physically incapable of being fertilized after they are released: chemical changes in the covering of the egg dictate that it is fertilized before release (Mayer 1877, p. 201; Alcock 1905; Kamalaveni 1949, p. 124-5). It does not matter much to us whether fertilization is internal or external. The theory applies a little more forcefully if it is external, but this is not essential. Kamaleveni's observations, on *Clibanarius olivaceous*, did not prove external fertilization (she relied on the chemical argument for this); but they are a little suspect. In nature, hermit crabs do not come completely out of their shells to mate, but Kamaleveni took hers out; furthermore, Hazlett (1966, p. 4) informs us that the copulatory posturing of Kamaleveni's hermit crabs 'may have been aberrant'.

We can sit on the fence between internal and external fertilization, but we have to know when mating takes place in the female reproductive cycle. This is a most vexing problem. Earlier biologists variously observed mating immediately after the female moult (Block 1935, p. 253; Bott 1940; Kamaleveni 1949, pp. 124-5) or not necessarily after the female moult (Matthews 1959, p. 252). Hazlett has since more thoroughly observed several species, and has conveniently tabulated his results (Hazlett 1973, p. 681). Of 12 species of Paguridae, he classified only one as mating only after the female moult, two 'usually', in seven there was 'no' association, and in two it was 'variable'; of 14 species of Diogenidae, six only mate after a female moult, three 'usually' do, in four there was 'no' association, and in one it was 'variable'. From these figures there appears to be no strong link between mating and moulting; perhaps there is a tendency but it is very variable. A likely interpretation of much of the variation in Hazlett's table is that all species have a similar, variable tendency to mate after the female moult, but the play of chance throws up some observations of some species showing no association, while in others the association is variable. However, Hazlett's observations do suggest that there are some differences between species in addition to variation within species. In some of the species, the timing of insemination was the same in many observations. The play of chance is only a plausible explanation of the apparent differences between species if the sample sizes are small; not all of them are.

The reason why many other crustaceans only mate after the female moult is that, as we have seen, the eggs are laid just after the moult. But oviposition may not regularly follow the moult in hermit crabs (Hazlett 1966, 1970). Often the females do release their larvae, then moult, then mate, then oviposit; but this sequence is not universal. Females may release their larvae and oviposit without a moult between (Mayer 1877, p. 202 provides another instance). Mayer's and Hazlett's observation raise two possibilities. Either the matings with hard females take place when females are about to lay eggs (without moulting); or mating may be unrelated to oviposition. Which is true is crucial for the theory: the theory demands the former if there is a precopula, and the latter if there is not. But the crucial observations have not been made. Stranger things are to come. In other crustaceans with external fertilization, mating is necessary after each moult because the externally attached spermatophore is cast with the exoskeleton. So we may be surprised by the mode of fertlization suggested by Hazlett (1966, p. 134) for *Pagurus bonairensis*: the spermatophore is attached while the female is still carrying a brood, she then releases her larvae, and then moults, before finally laying and fertilizing her next brood. Why are the sperms not lost at the moult? The answer is not known, but the female must retain the spermatophore somehow. The hermit crab's habit of living in shells may make it easier to keep the old exoskeleton, or spermatophore, than it would be for many species. There would then be a kind of *de facto* sperm storage over the moult. Whatever the solution to these problems, we may for the present concude that mating and moulting are not strongly associated in many hermit crabs, and that the early reports of mating only after moulting were due to insufficient observations. Maybe there are differences between species, but more observations are needed to prove it. The apparent variability in this timing of mating is matched in the observations on the Porcellanidae (Molenock 1975) and the Hippidae (MacGinitie 1938; Efford 1967; Subramoniam 1977). But it seems well confirmed that mating only follows moulting in the Lithodidae (refs. below). The observations on *Galathea* (Pike 1947) are less extensive.

Now the review of the anomuran families can begin. The facts about precopulatory behaviour in the two families of hermit crabs are, like the facts about mating, not easy to interpret. We will start with the Diogenidae. *Diogenes pugilator*, which has a short breeding season of less than 15 days where it was watched by Block (1935), has a lengthy precopulatory association. The male of the pair drags the female around, gripping her by her first ambulatory legs, or by her shell. Block picked up some of these pairs and took them back to his laboratory, where they stayed together for about another day or two (p. 253). The pairing always ended with the moult of the female, followed by mating, in both Block's and Hazlett's (1968a) studies. This well-studied species appears to conform to the theory exactly. Males have been reported dragging females about before mating in other species (Matthews 1956, p. 306 for *Dardanus punctulatus*; Hazlett 1966; pp. 32, 39, 54-5, 58 for four more species; Hazlett 1972, pp. 809-15 for another 11) but the duration of this phase is unclear. A period of 5-6 hours, although 'longer pairings probably do occur' (Hazlett 1966, p. 32), is mentioned, as is half an hour, and two and a half hours (Hazlett 1972, p. 814); the behaviour of the various species appears to be much the same. Species in which mating is reported only after moulting do not seem to have longer precopulatory associations than other species. The facts are I think too chaotic to include in the test. We might include the single species *Diogenes pugilator* because it is thoroughly studied and its mating habits, if it is considered by itself, seem clear. However, the clarity is soon muddied if we look at the species in the context of the rest of the hermit crabs. I have therefore excluded it from the test.

A similar story must be told for the Paguridae. *Paguritta harmsi* lives on corals in permanent pairs (Patton and Robertson 1980); the well-studied *Pagurus* (formerly *Eupagurus*) *bernhardus* has a precopula of 'several hours to several days' (Hazlett 1968, p. 239), although matings have been seen with both hard and soft females. In *Pagurus hirsutiusculus* 'the male may carry the female about for several days before mating' (MacGinitie 1935, p. 712), and other species may have precopulatory associations of a few minutes to several hours, but the exact time is unclear (Hazlett 1966, pp. 60-70, 74, 79, 85, for four species; 1968, pp. 241-9 for another six). The function of the precopulatory phase is obscure: it may not be a precopula in the sense of this essay. It might be more like a courtship (Hazlett 1970, p. 45).

The other family of the Paguroidea, the Lithodidae, conform to the theory, and can be included in the test. The most studied species is the commercially exploited king crab *Paralithodes camtshatica* (whose mating habits are shown in colour in Idyll 1971). It has a precopula in which the male grips the females first pair of legs; in cages kept in the sea the precopula lasted for 3-7 days (Marukawa 1933, p. 129), which accords with observations of pairs caught in the sea and then kept in tanks, which stayed together for up to three days (Powell and Nickerson 1965, p. 103). All observations agree that the precopula ends with a female moult which is immediately followed by mating (Marukawa 1933, p. 129; Wallace, Pertuit, and Hvatum 1949, p. 22; Gray and Powell 1966; Powell and Nickerson (1965, p. 104, citing 'Kurata MS, 1961') remark that 'ejaculation and oviposition normally are completed within 2 hr after the female moults', and that Kurata found that 'successful spawning did not occur among females which were kept separate from males for the first 5 days after ecdysis'.

Mating must, and does, take place immediately after the female moult; mating is preceded by a precopula: the species conforms to the theory.

Two other species of Lithodidae have been less well studied. MacGinitie (1937, p. 1035) informs us that *Oedignathus inermis* is 'practically always found in pairs', in rock crevices around the low-tide mark. A few specimens of *Lithodes maja* from the North Sea and its environs were observed in the aquarium of Pike and Williamson (1959, p. 565). 'It appears' they write 'that during the period of the moult and until the time of egg laying a male is in attendance on the female, in much the same way as for *Pagurus*.'

Nothing has been published on precopula or its absence in the crabs of the family Galatheidae. Pike (1947) informs us that the female has to moult before mating and egg laying.

Molenock (1975) has investigated the mating habits of four species of Porcellanidae. The relations between moulting, mating, and egg-laying is as confusing as in the hermit crabs. Of the four, *Petrolisthes manimaculis* usually mates, and always oviposits, after a female moult; *P. eriomerus* is similar except that hard females sometimes lay eggs; *P. cinctipes* only mates and oviposits when the female is hard, and in *P. cabrilloi*, both hard and soft females mate and oviposit (Molenock 1975, pp. 8-10, 17). None of the four have a precopula in which the male physically grips the female, but in all four the males are territorial and may defend females for a period before mating. The duration of this period is uncertain; Molenock writes of a 'courtship' which lasts several hours before the moult of the female in *Petrolisthes manimaculis*. *P. cinctipes* males defend a group of from one to four females; but in the other three species the males seem to defend *ad hoc* territories around one female at a time. In all four the female leaves the male soon after mating. A conclusion which is consistent with, but not justified by, these observations is that different species (or different females of any one species) oviposit at different times during the female moult cycle, but at a time which the male can predict, and the male only defends a female just before she is about to oviposit. *Petrolisthes* could thus conform to our theory. The facts, however, are not sufficiently certain for us to use in the test. Two other species of Porcellanidae have been watched by MacGinitie (1935, p. 713; 1937, p. 1035). Both *Pachycheles rudis* and *Pachycheles holosenicus*, she tells us, are 'practically always found in pairs'.

We have reached the final anomuran family for which we have facts: the Hippidae. These extraordinary sand crabs live in the sand and filter food from the sea above. In some, but not all, species tiny neotenous males live attached to the females. Efford (1967) mentions four species in which the males are tiny, and in two of these the males (*E. talpoida* and *E. rathbunae*) definitely live attached to females. In four other species the sexes are about the same size (*E. holthuisi*, in which the males and females are both about 2.5 cm long, and *E. austroafricana*, in which both sexes are about 5 cm long), or the males are not much smaller than the females (in *E. portoricensis* and *E. analoga* the males are about half the length of the females). Let us now consider their mating habits.

Fertilization is external (Menon 1934, p. 500, Wharton 1941), but insemination has been observed at various times during the female moult cycle. In *Emerita asiatica* mating is confined to the period just after the female moult (Menon 1945, p. 500); Subramoniam (1979, p. 198; see also 1977, p. 373) found no intermoult females with attached spermatophores. In *E. analoga* it seems that mating is again confined in time to the period just before oviposition, but in this species eggs are not laid only after the female moult: 'deposition of sperm usually takes place about 12 hours before the eggs are laid, and may or may not occur at the time of moulting' (MacGinitie 1938, pp. 475–6; Efford 1967). *E. emeritus* (a species with relatively tiny males) is the other species for which we have information in the timing of insemination. 'In July [Efford tells us] these small males were found on the underside of, but not attached to, recently moulted females . . . Two or three or more males were attached to the individual females at this time.' In sand crabs, therefore, mating is confined to a short period before oviposition, but this period may or may not be just after a female moult. Do they have a precopula? In *E. analoga* 'males apparently sense the approaching time for egg laying by the female, and they gather round her for as long as two to five days before . . . Females which are about to lay eggs will be accompanied by a considerable number of males, and some two or three of them may even be attached to the female' (MacGinitie 1938, p. 476, likewise Efford 1967). *E. analoga*, at least, conforms to theory, and the observations on the attachment of males to females in the rest of the family are not inconsistent with it.

In conclusion for the Anomura, the Lithodidae and Hippidae we may take to have mating limited in time and a precopula; for the rest of the group the facts are too uncertain to use in our test. None of them appear independently in the test because (outgroup comparison suggests) mating after a precopula during a limited period of the female cycle is primitive in the group. The habits of the Anomura are summarized in Table 3.2 (p. 160).

Brachyura The final reptantian group, those with the shortest tails, are the Brachyura. Our summary will as usual be arranged systematically. Systematics is the first problem. Six tables in Rice (1980) demonstrate the diversity of crab classifications. The arrangement of species into families (or superfamilies: it does not matter which) has, with a few exceptions, been agreed on for a century; so has the limit of the Brachyura as a whole. The arrangement of families within the whole has not. The 'standard' system was invented by Borradaile (1903, 1907) and incorporated (with alterations) into the great treatment of Balss (1957); Kaestner (1968) used more or less Borradaile's system, dividing the Brachyura as follows:

subsection	infrasection
Dromiacea	
Gymnopleura	
Oxystomata	
Brachygnatha	Oxyrhyncha
	Brachyrhyncha

The more primitive sections (Dromiacea, Gymnopleura) are the most difficult. Fortunately we can miss them out: nothing is known about their mating habits. But we cannot miss out the Oxystomata, Oxyrhyncha, and Brachyrhyncha. Carcinologists all agree that these divisions are not phylogenetic; they agree that the similarities which define the Oxystomata and Oxyrhyncha are both convergent; but they have not decided what to replace them with. The phylogeny of the families is just not known.

Guinot (1978) has suggested a classification, which differs from Borradaile's, and is based on the positions of the male and female genital openings. We need to understand the mode of fertilization anyway, so we may as well incidentally learn Guinot's system. Crabs primitively have external fertilization, like that of the other decapods we have been discussing: the female oviducts open on the third thoracic legs, and the male attaches a spermatophore near the opening (Fig. 3.6, left half). All the primitive crabs fertilize in this manner; they comprise Guinot's 'Podotremata' (because they have genital holes in their legs). With the evolution of an internal connexion between the repository of the spermatophore (on the female sternum) and the ovaries, internal fertilization has become possible (Fig. 3.6, right half). This connexion is also the exit for the

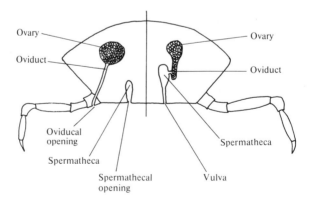

Fig. 3.6. The two main kinds of reproductive organs of female decapods. The more primitive, with external insemination and fertilization is on the left; the derived condition, with internal fertilization, on the right. (Slightly modified from Hartnoll (1968).)

eggs, so now the female genital openings are on the sternum. A similar, but independent, course of evolution has been run by the male genital openings. These too are primitively on the coxae of the (fifth pair of) thoracic legs, and have then moved to the sternum of their segment. Sternal genital openings are found in the females of more species of crabs than are sternal males genital openings, so we may infer that the females changed first in evolution as shown in Scheme 3.8.

Scheme 3.8

The old Oxystomata and Oxyrhyncha are scattered, by Guinot, among the Heterotremata and Thoracicotremata. But, as Guinot explicitly states, the new system is not phylogenetic either. By dismantling the Oxyrhyncha and Oxystomata she dismantled two non-phylogenetic groups. With the 'Thoracicotremata', she has created another one: the male genital openings (she thinks) have moved more than once during evolution. Some families within the Heterotremata contain some species with coxal male genital openings, and others with sternal. The thoracocitrematal families should phylogenetically, be scattered among the families of the Heterotremata, but in a pattern that is completely unknown.

How, then, shall we proceed? The most conservative assumption would be that all (super)families are equally related. The assumption is certainly false. We must fall back on the principle that even an avowedly non-phylogenetic system such as Borradaile's or Guinot's is a better estimate of phylogeny than a scheme in which all families are equally related. We shall steer a conservative course, using such phylogenetic information as we can from Guinot (1978, 1979) and Rice (1980). We will at the end count the frequency of evolution of precopula (and its absence) according to several classifications. We will see what difference the taxonomy makes to our estimate.

I have a few more general remarks before the systematic summary. Modern investigation of the mating of crabs we may date to the observations of the Neapolitan Philippo Cavolini (1787) on *Carcinus maenas*. Cavolini first noticed that mating takes place with a recently moulted female, and is preceded by a precopula. The confinement of mating to the time just after the female moult is not universal in crabs. But we shall come to that. Can we find an anatomical explanation for the species which do mate at the female moult? For the earliest important thoughts on this question we must turn to the great memoir by

Brocchi (1875)[4] on the genital organs of crabs. He wrote (p. 115) that 'in certain cases, in *Cardisoma* [a genus of the Gecarcinidae] for example, the male appendices are relatively enormous, and it seems impossible that there could be a true intromission . . . I believe that two circumstances permit this act: the first is the softness of the female exoskeleton . . . ' (The second? The plasticity of the male organ.) The possibility of such a physical constraint has been repeatedly discussed since then. Williamson (1900, p. 83), writing about *Cancer pagurus*, said that the female genital openings were closed soon after copulation by the contraction of the surrounding muscles. Spalding (1942, p. 419) observed no such muscular closing off of the genital openings in *Carcinus maenas*, but he did provide further evidence of a calcareous constraint. In Fig. 3.7, reproduced from Spalding (1942), we can see that the lower regions of the female genital ducts are continued from the calcareous exoskeleton: the hard region is shaded black. The sperm plug shown in the right hand picture is soft when it is injected but hardens up later. Whether the hard-walled ducts would be impenetrable Spalding does not say but Cheung (1968, p. 117) remarked of Spalding's work that, after a moult, 'the new soft tegumental wall will presumably allow easy passage of the spermatophore from the male'.

The question has since been examined in detail by Hartnoll (1968). The female genital openings of the species which he studied fell into four main

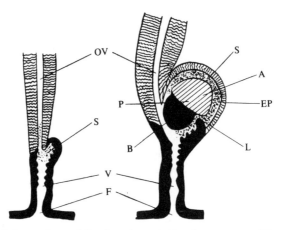

Fig. 3.7. Genital openings of the female crab *Carcinus maenas*. The black regions are calcified exoskeleton, which is shed at a moult. The organs illustrated on the right contain a spermatophore, those on the left do not. (From Spalding (1942).) A, upper part of sperm plug staining red; B, lower part staining blue; EP, stratified epithelium lining upper part of spermatheca; F, external female opening; L, limit of chitin lining spermatheca and vagina; OV, oviduct; P, sperm plug; S, sperms, V, vagina.

---

[4] The title of the memoir bears the author's name 'M. Brocchi', following which Monsieur Brocchi has often been cited as 'Brocchi, M.'; he was in fact christened Paul-Louis-Antoine.

groups. Each group has a different kind of constraint, or lack of constraint, on mating. The four groups are: (1) like that described above, following Spalding (1942), for *Carcinus*: (2) to (4) all have, in contrast to (1), what Hartnoll calls 'concave' genital ducts, which means that one wall of the vagina can be invaginated, by muscles, to close off the duct. The three groups differ according to whether they have an operculum, and if they do, what kind it is. (An operculum is a calcified thickening of the exoskeleton near the vulva (see Fig. 3.6) which can close off the genital opening.) (2) lack an operculum, (3) have a movable operculum, and (4) have an immovable operculum. The operculum of group (3) never inhibits copulation; the immovable kind (4) sometimes does. Crabs of group (4) can only mate when their operculum is decalcified, as it is after a female moult. But females of these species may also undergo a temporary, and local, decalcification of the operculum during the intermoult. Mating can also take place then. The decalcification during the intermoult may be associated with oviposition. We are now going to have to face a difficulty. We have met it before, in the Hippidae (Anomura). The females of some crabs may be receptive for mating during a short period in their intermoult. What difficulty does that raise?

The easiest form of test is to correlate whether or not mating only occurs after a moult with the incidence of precopula. Whether mating females are hard or soft is the kind of fact that naturalists record, so a prediction based on it can be tested. The test assumes that mating is completely unrestricted in species which mate when the female is hard. But this assumption, we now see, is not always correct. In some species the mating female are 'hard', except for their operculum. If the decalcification of the operculum is brief, the theory predicts the evolution of a precopula. It is easy to decide how to score species if we know that they have a temporary decalcification or predictable oviposition. But what when we turn to the species in which we only know that mating females are hard-shelled, in which we do not know the state of the operculum? A temporary decalcification of the operculum is not the kind of thing that naturalists record: it was only discovered in the nineteen-sixties. Perhaps some of the species in which mating females are hard can only mate during a restricted time. How should we enter them in the test? There are two possibilities, one to ignore all species for which we only know that mating females are hard, the other to include them as having unrestricted mating. We shall follow the second course. Some may be scored wrongly, but we will obtain many more facts for the test. The former spoil the test, the latter improve it. Provided the former are not all that common, the overall test will be better off than if we were to exclude all but certain facts.

The anatomical argument itself is hardly certain enough to justify throwing out natural history observations. The crab species which are known only to mate after the female moult are not particularly those in Hartnoll's group (4, which are anatomically constrained to mate only when the operculum is soft). Most of the species which mate when the female is soft seem to come from his groups (1) and (2). Hartnoll does not believe that these species are anatomically incapable of mating at other times. 'In most crabs [he wrote in 1969, p. 165] the structure of the vulva and vagina offers no obstacle to mating during any part of the moult cycle.' Nor is mating confined by the timing of oviposition: in many species there is a long gap between mating and oviposition. Nor do the eggs, once they are being brooded, get in the way of mating. Mating of females which are carrying eggs is rare, but not unknown (Berry and Hartnoll 1970; Hinsch 1968; Raja Bai Naidu 1954; Knudsen 1964; Grigg, private communication to Berry and Hartnoll 1970). The restriction of mating thus seems rather odd. Where can we go from here? We can first notice that Hartnoll's conclusion differs from that of earlier experts. We, as non-experts, can then draw either of two conclusions. Neither is satisfying. We may stand by the most recent and thorough work, that of Hartnoll, and conclude with him (1969, p. 172) that we 'do not know why' mating only takes place after moulting. Or we may declare that the experts disagree, and that it is not for us to decide between them. The point is either moot or mysterious.

However, to test our theory we only need to know whether mating is in fact restricted in the female cycle. So let us now lay aside anatomy, and turn to the natural history of mating. We will, as usual, run through the observations that I have unearthed from the literature on the duration of the sexual association for mating, and the timing of mating in the female moult cycle.

Observations have been published on five of the nine superfamilies of Guinot's (1978) Heterotremata. The five, in the order that we shall consider them, are: Corystoidea, Portunoidea, Xanthoidea, Majoidea, and Leucosoidea. *Corystes cassivelanus* (Corystidae) is our first subject. It was studied by Hartnoll (1968*b*), and is one of the species which mate during a temporary vulval decalcification when the female is hard. Females were not seen to mate after their moult. They only mate during the period of vulval decalcification, which lasts for about 12-23 days. In the laboratory, Hartnoll tells us, males carried females around for 'up to several days'; but he does not tell us whether it is pre- or postcopulatory, or both. Both prediction and result are unclear in this species: let us exclude it from the test.

In the same superfamily Guinot places the Cancridae. Of the Cancridae, Hartnoll (1969, pp. 165-6) states that copulation only takes place just after the female moult. The only contrary remark that I have come across is Chidester's (1911) casual assertion that females of *Cancer irroratus* can mate when hard; but we do not know whether there is a precopula in this species so we do not have to look further into Chidester's remark. The many other cancrid papers, on *Cancer pagurus* (Bell 1853, p. 62; Williamson 1904, p. 101; Edwards 1966, p. 23), on *C. oregonensis* (Knudsen 1964, p. 19), on *C. productus* (MacGinitie 1936, p. 713; Knudsen 1964, p. 21), on *C. gracilis* (Knudsen 1964, p. 22), and on *C. magister* (Cleaver 1949, p. 80; Butler 1960, p. 642; Snow and Nielsen

1966), all agree that mating is confined to a short period after the female moult. The same papers document a precopula (and usually a short post-copulatory embrace as well). Durations of the precopula are as follows: *C. pagurus*, average of 8.2 days (Edwards 1966, p. 23), *C. oregonensis*, 'several days' (Knudsen 1964, p. 19), *C. productus*, 'one to several days' (MacGinitie 1935, p. 713), *C. gracilis*, unspecified duration (Knudsen 1964, p. 22), *C. magister*, several days (Cleaver 1949, p. 80) and at least eight days ($n=1$, Snow and Nielsen 1966). The Cancridae conform to theory.

A precopula, followed by the moult of the female, followed by mating, has also been well documented in the swimming crabs (Portunidae). The blue crab *Callinectes sapidus* (also often called *C. hastatus*) swims in the Atlantic Ocean, off the coast of America from Nova Scotia to Uruguay, and has been introduced to the coasts of Europe. The female blue crab mates only once, just after the moult with which she matures; after this she never moults or mates again. She can spawn more than once, in more than one summer, without re-insemination. The single mating is preceded by a precopula, of 'a day or two' (Hay 1905, p. 405; Churchill 1917-18, p. 118 (these are the main two sources for this species)). B. Harrison, and J. D. Mitchell (both in Rathbun 1895, pp. 371-2) observed pairs of this species. Say (1817, pp. 68-9) was the first to report that the female is protected by a male while moulting, but he did not understand the meaning of the protection.

Another tasty species, *Carcinus maenas*, swims around on both sides of the Atlantic, and also mates only after a female moult (Cavolini 1787; de Lafresnaye 1948; Bell 1853, p. 62; Bethe 1897, p. 444; Williamson 1904; Broekhuysen 1936, p. 259; 1937, p. 159; Spalding 1942, pp. 412-13; Cheung 1966, pp. 107-8, and many handbooks and popular sources of the fishing industry). There are plenty of observations of paired crabs, and Williamson (1904, p. 15) compares the pairing to that of *Cancer pagurus*; but I have found no measurements of the duration of the precopula, and indeed no confirmation that the pairing is precopulatory. It is reasonable to infer that there is precopulatory pairing; but there is postcopulatory as well.

A few facts are available for other members of the family. Of *Ovalipes ocellatus*, Chidester (1911, p. 239) says that 'the method of capture of the female by the male is the same . . . as in the blue crab', a statement from which we might infer a similarity of mating behaviour, or we might not. We know a little about the mating of some species of *Portunus*. Take *Portunus pelagicus* first. According to Broekhuysen (1936, p. 259), it copulated, in one case, when the female was hard: 'from unpublished observations which Dr Verwey kindly put at my disposal, it appears that this female is indeed embraced by the male immediately after the moult but not before five days later [when the female was hard again] the copulation properly took place'. This is but a single observation, and it is so strange that we shall discount it by quoting two observations which contradict it. Chhapgar (1956, pp. 34-5) observed that the female was sometimes still hard when the pair formed; the pair spent two to three days together before copulating, which took place when the female was soft. Fielder and Eales (1972) observed a male sit on the back of a female for four days until she moulted. The female copulated soon after her moult (she moulted in the early morning, during the night, and was copulating by 8.30 a.m. The male stayed with her for less than 24 hours after her moult. I shall dismiss Verwey's observation as aberrant in some way (we are told nothing of the circumstances of observation). *Portunus pelagicus*, by the other reports, conforms to theory. *Portunus sanguinolentus* (Ryan 1967) and *P. trituberculatus* (Oshima 1938) also both mate when the female is soft. *P. puber* evidently also has some kind of pairing and mates after the female moults (Duteutre 1930; see Duteutre 1929 for the meaning of *promenades nuptiales* and *pré-nuptiales*).

The next superfamily is the Xanthoidea. (It is usually only a family, Xanthidae.) We will stick to calling it a family, so the Trapeziinae, Panopeinae, and Menippinae are still subfamilies. Mating females are usually said to be hard in this family, with the single exception of *Menippe mercenaria* (Menippinae), which we will deal with first. These crabs inhabit rock crevices on the shore. There is usually only one crab per crevice, but in August a male can often be seen at the mouth of a crevice which is inhabited by a female. Females thus guarded have either just moulted or are just about to moult (Binford 1913, p. 161; Savage 1971, p. 315). Savage also tells us that the pair are together 'for at least a week', and that in aquaria mating always followed the female moult. The few other observations of mating in xanthids have been of hard females. Knudsen (1960, p. 7; 1964, p. 180) watched the mating of *Paraxanthias taylori* (Xanthiinae) and of *Lophopanopeus bellus* (Panopeinae): the females were hard, and mating was not preceded by a precopula. *Xantho insicus* (Xanthiinae) and *Pilumnus hirtellus* (Pilumninae) do likewise (Bourdon 1962). Thus far we may conclude that there is a precopula and mating after the female moult in the Menippinae, and mating with hard females without a precopula in the other three subfamilies mentioned. Now let us turn to the Trapeziinae.

Trapezid crabs inhabit the branches of pocilloporid corals and often live in monogamous pairs. Crane (1947, p. 83), Garth (1964, p. 142), and Knudsen (1967, p. 52) all describe *Trapezia* or *Tetralia* as living in pairs. In a study of five sympatric Hawaiian species of *Trapezia*, Preston (1973, p. 471) states that 'typically one adult pair per species is found per host. This pair actively excludes all other adults of the same species.' Patton (1974) also found *Trapezia* living in pairs. He thought that the pairs might form as juveniles, and grow up together. The monogamy would then be permanent. Preston (1973) mentioned that some pairs form as juveniles, but argued that most pairs form as adults. Castro (1978) experimentally separated pairs of *T. feruginea* (a species studied by both Preston and Patton) and observed fairly rapid migration of individuals between corals, until they had reformed into pairs. We may suspect, therefore, that pairing is not permanent from the juvenile stage, but individuals of both sexes move around and change (or are forced to change) partners occasionally. Mating has, it seems, not been observed; as we have seen, its confamials variously mate when hard or soft, so we dare not reason from homology. *Cymo andreossyi* (probably a xanthid, but of uncertain affinities: Guinot 1978, p. 274) 'sometimes occurs as pairs, but frequently does not' (Patton 1974, p. 237), on pocilloporid corals.

The next family is the Majidae. As was first established by the French neo-Darwinian Teissier (1935), majids do not moult again after reaching maturity. Both hard and soft females have been seen mating; the soft females are probably mating for the first time. We shall consider eleven species. Matings with hard females are generally not preceded by a precopula; the sexual associations of soft females are less easy to interpret. First, *Hyas coarctatus* (Hartnoll 1969, pp. 168-9). 'The female [Hartnoll informs us] mates soon after moulting, although males were not seen to pay any attention to the pre-moult females.' 'However [he continues], immediately after the female has moulted the male takes up a position over her, and may remain so for several days before mating.' This wait of 'several days' directly contradicts our theory. But, because it is a mere assertion in a secondary review I am going to exclude it from the full test. We have no idea of the circumstances, or sample sizes, of the observation. Hartnoll has also seen hard females mating, 'but in such cases there was no prolonged attendance either before or after mating' (as the theory predicts). The following species have also been seen to mate, when the female was hard, with no more preliminaries than a brief courtship: *Pleistacantha moseleyi* (Berry and Hartnoll 1970), *Macrocheira kaempferi* (Arakawa 1964), *Pugettia producta* (Boolotian, Giese, Farmanfarmaian, and Tucker 1959, p. 217), *P. gracilis* (Knudsen 1964, pp. 24-5), and (probably) *Pisa tetraodon* (Vernet-Cornubert 1958, pp. 6-7).

We can reasonably suppose that, in all the majids mentioned so far, the female can also mate immediately after her final moult; there are no results to contradict this. Watson (1972) observed a pair of *Chionoecetes opilio* in precopula for seven to eight days, after which the female moulted, and they mated. The female laid her eggs within three days of moulting, and the male waited with her for much of the time until she laid. Watson suspects that this species does not mate when hard. *Maja squinado*, like *Hyas coarctatus* discussed above, has been seen mating when females are both hard and soft. At Moorecombe, in South Devon, Carlisle (1957, p. 306) came across a large heap (about a yard in diameter, and half a yard high) made up of 80 or so *Maja squinado*. Females, in the centre of the heap, were casting off their shells and then copulating: 'six or eight pairs could be observed in copula at one time', we are told. But we are told nothing about the duration of pairing. Hartnoll (1965) has also seen *Maja squinado* mating. The females were hard, and there was no precopula. Hartnoll saw nothing resembling the orgies of Devonshire: his crabs paired individually. Schöne (1968, p. 648) reports some kind of guarding in *Maja verrucosa* although he does not state its duration or its relation to the female's moult cycle: 'The pairs seem to copulate once or twice a day. In the meantime the male hovers [sic] over the female, which is crouched between its legs.'

Finally for the Majidae, we will review Hinsch's (1968) interesting observations on *Libinia emarginata*. Mating in *L. emarginata* is not connected with moulting, but it does have a kind of precopula which Hinsch calls 'obstetrical behavior'. Encounters, in aquaria, between males and females 'which have either just released their larvae or are about to release larvae generally result in mating. Females with eggs in the early stages of development were never observed to mate' (p. 275). It appears, therefore, that there is a short and predictable period of the female cycle when she is most receptive to mating. Males can recognize this period and anticipate it. Males fought over females that were about to release larvae, and, 'in sixteen cases, when a female about to release larvae was encountered, the male captured her and placed her beneath him'. The duration of this pairing Hinsch does not tell us; but it looks too like a precopula for us to ignore.

How shall we use the majid crabs in our test? We have species which mate when hard, without a precopula; we have *Libinia emarginata*, which mates only in a short period and has a precopula; we have species which mate when soft, some of which definitely have a precopula while for others we have no proof. By outgroup comparison (and other arguments which we will return to later) the common ancestor of the majids probably had a precopula and mated when the female was soft. *Chionoecetes opilio* retains the ancestral habit. The precopula has been lost at least once, and re-evolved (with mating in the intermoult) at least once in *Libinia*. We will score both of these for the theory. The conclusion, it can hardly need emphasizing, is highly tenuous.

The final Heterotrematan superfamily is the Leucosoidea. A paper by Raja bai Naidu (1954, p. 640), on *Philyra scabriuscula*, is the only one that matters. Before mating, we are told, the male grabs the female; they remain together for from a few minutes to a couple of hours. This is too short a period to count as a precopula. Raja bai Naidu does not mention a moult, and the paired females were carrying eggs: we can infer, as did Hartnoll (1969), that they were hard. It can then be scored as lacking a precopula, and mating when hard.

We can now consider the Heterotremata as a whole. Five families matter to us: Cancridae, Portunidae, Xanthidae, Majidae, and Leucosiidae. The Brachyura probably evolved from a longer-tailed decapodan ancestor, and outgroup comparison would suggest that it had both a precopula and mating confined to the period just after the female moult. (The outgroup comparison is rather far-fetched. The mating habits of the nearest outgroup, the Podotremata, remain a secret to this day.) The most parsimonious conclusion at this stage would be that the Heterotremata have lost precopulatory mate guarding three times, in some of the Xanthidae, some of the Majidae, and the Leucosiidae. The other families retain the primitive condition. The re-evolution of precopulatory mate guarding in the Majidae we have discussed above. (Another possibility would be to suppose that majids primitively lacked a precopula. We could then score one less loss of a precopula, because the Majidae and Leucosiidae could be combined as a single loss. But the number would be made up by the re-evolution of a precopula within the majids, which is not counted on the analysis above. I prefer the analysis above because it uses outgroup comparison more simply.)

The richer phylogenetic information in Rice (1980, Fig. 47) gives a different conclusion. Rice concluded that the Cancridae and Portunidae belong to separate evolutionary lineages, from an ancestral xanthid stock. Just let that 'primitive xanthid stock' have mating habits typical of the modern Xanthidae: let them mate without a precopula when the female is hard, and then we would come to a different score. Precopula would first be lost once from the primitive condition indicated by outgroup comparison. It would then be re-evolved twice: once in the Portunidae and once in the Cancridae. Perhaps a second evolution of mating without a precopula in the Leucosiidae and Majidae could be scored as well. Rice, however, places little confidence in his arrangement (by numerical taxonomy) of families into a phylogeny, so perhaps we should not rush ahead of him. We will draw no more than the most conservative conclusion possible.

Now for the divisions of the Thoracicotremata. Carcinologists have published observations on the mating habits of four of the thoracicotrematan superfamilies (which we will call families): Gecarcinidae, Grapsidae, Ocypodidae, and Hymenosomatidae. We will take them in order. The Gecarcinidae are tropical, terrestrial burrowing crabs. Brocchi (1975, p. 115) instanced *Cardisoma* as a crab in which the moulting of the female facilitated mating; but he did not instance observations that they do in fact mate after the female moult. He may have observed only dead and preserved specimens. Since then, copulation has been watched several times. The females have always been hard (Klaasen 1975, p. 140 reviews earlier work). Weitzmann (in work cited by Klaasen) and Klaasen found that copulating male and female *Gecarcinus lateralis* were both hard; there is no precopula (Klaasen 1975, pp. 145-50). Similarly, *Cardisoma guanhumi* mates without a precopula during the female intermoult (Henning 1975, pp. 479-82). According to Türkay (1970, 1973, 1974), female gecarcinids copulated during a temporary decalcification. A single observation on one pair of another species, *Cardisoma armatum* (Cheesman 1923), suggests a long association. She put a pair together in the laboratory. In July 1922 the male and female dug burrows in opposite corners, then 'in the first week of August' they joined their burrows together by an opening, 'and for the next fortnight they shared the tunnel'. After that the male dug another burrow and the pair were not seen together again. On 18 September the female was 'discovered to be in spawn'. Cheesman's observations are suggestive; but we will not use them. They are of only one pair, and were made in the laboratory. The majority of observations for this family suggest the conclusion that there is no precopula, before a mating with hard females.

The Grapsidae live and mate on the shore, out of the water. With the exception of *Pachygrapsus crassipes*, in which all mating females had moulted within the last 12 hours (Hiatt 1948, p. 199, and see p. 150), all observed mating female grapsids have had hard shells. According to Hiatt, 'pre-nuptial pairing or exhibitionism is lacking' in *P. crassipes*, so this species, as far as the observations go, contradicts our hypothesis. In no other grapsid is there any precopula, but they all mate when hard. All the following have only a courtship before mating, and mate when hard: *Cyclograpsus punctatus* (Broekhuysen 1941, p. 339n; Hartnoll 1959, p. 170), *Pachygrapsus marmoratus* (Vernet-Cornubert 1958a, pp. 113-14), *Hemigrapsus nudus* and *H. oregonensis* (Knudsen 1964, pp. 9 and 13: he does not mention any moult), *Hemigrapsus crenulatus* (Yaldwyn 1966), *Grapsus grapsus* (Kramer 1967, who also does not mention any moult), *Goniopsis cruentata*, *Sesarma ricordi*, and *Aratus pisoni* (Warner 1967, pp. 323 and 333), *Helice crassa* (Beer 1959, p. 201, for absence of precopula, Nye 1977, p. 80, for five out of five mating females hard-shelled), and the remarkable Chinese mitten crab *Eriocheir sinensis*, which has been introduced to the Elbe and Rhine (Peters and Panning 1933, pp. 133 and 136; Hoestlandt 1948, p. 43).

The fiddler crabs, and their relatives such as the 'ghost crabs', which make up the family Ocypodidae, are also inhabitants of the shore. Take the fiddler crabs (*Uca*) first. Except for a speculation which we shall come to, 'all observers agree that in *Uca* females mate when the carapaces are hard' (Crane 1975, p. 503). Copulation was first reported by Pearse (1914), for *Uca pugilator*, and I will not repeat all the references since then; Crane (1975, p. 502) lists them. Likewise, all the observations of courtship provide no evidence of any precopula (Crane 1975). *Uca* pairs do sometimes (but usually do not) stay together for mating for several days in a single burrow, but there is no evidence that this is a precopulatory wait (Crane 1975, p. 503). There is the slightest hint of a precopula, and mating after the moult to maturity, in the following remark by Crane (1975, p. 503). She saw males of two species of *Uca* in New Guinea herding 'slightly immature' (i.e. about to moult to maturity) females. If this herding was a precopula, then the fiddler crabs conform to both predictions of the theory: precopula before the moult to maturity, and no precopula before mating with hard, mature females. For other members of the family, the moult-stage of mating females is not clearly reported. It is probably just assumed that everyone knows that the females are hard-shelled. Hughes (1973, p. 75) implies that one female who mated in the laboratory was hard; the mating was not preceded by a precopula. The Ocypodidae, we may conclude, conform to the theory.

We can finish with a couple of notes on the Hymenosomatidae (which are radically re-grouped in Guinot's classification). In *Hymenosoma orbiculare* there is a precopula of unspecified duration, which ends with the female moult and copulation (Broekhuysen 1955). 'In *Halicarcinus*, the females similarly mate after ecdysis, but the males have not been seen to carry the females during the pre-moult period' (Hartnoll 1969, p. 169, describing a private communication from Lucas). *Halicarcinus*, on this brief note, appears to be an exception to our theory; but I will not count it as such. It is excluded for the same reason as was *Hyas coarctatus* (Majidae) above.

How shall we count the Thoracicotremata? This question is best answered at the same time as the general question of how the Brachyura as a whole are to be counted. We have already noticed that the Thoracicotremata are not monophyletic. So we have two possibilities: we may either treat all the families of both Heterotremata and Thoracicotremata as equally related to each other, or we may assume that, although the dichotomy is not phylogenetic, it is a better estimate of the true phylogeny than assuming all families equally related. Another possible phylogeny is that of the Borradaile-Balss classification. We will therefore work through three methods of counting: all families equally related, Guinot's classification, and Borradaile and Balss's classification. We shall use Guinot's scheme for the test, but the other two are included to see how sensitive the results are. The character states of all families are summarized in Table 3.2 (p. 160).

The first method of counting assumes that all families are equally related. Proceeding with maximum conservativism as usual, we will count what numbers we have to add to those already given for the Heterotremata. If all families are equally related, the thoracicotrematan families can be put anywhere among the Heterotremata to achieve the most conservative count. Thus, we assume that the thoracicotrematan species with mating with hard females and no precopula have evolved from heterotrematan species with the same habit; similarly the Hymenosomatidae would be supposed to evolve from an ancestor which mated, after a precopula, when the female was soft. We would be forced to count the re-evolution of a precopula, when it was not predicted, in the Grapsidae. The result for the Brachyura as a whole is:

	Precopula observed?	
	+	−
Precopula +	1	1
predicted? −	0	3

The second alternative is to treat the Thoracicotremata as a phylogenetic group. It gives but a single difference. Both states are found in the Thoracicotremata, but the ancestor could have had only one of them. So, according to the ancestral state, either precopula, or its absence, will be counted as having evolved once. For the ancestral state we should look to the outgroup: the Heterotremata. We find, to our dismay, both character states; outgroup comparison, as we noticed in Chapter 1, cannot easily be applied. The test will not be affected by which state we take to be primitive: either yields one case in favour of the theory. We could fall back on ingroup comparison within the

Heterotremata. Mating with soft females after a precopula is rather more common: let us take that to be primitive. We then must accept that mating with hard females without a precopula has evolved at least once in the Thoracicotremata:

		Precopula observed?	
		+	−
Precopula predicted?	+	1	1
	−	0	4

The third alternative uses Borradaile and Balss's classification. The dichotomies are shown in Scheme 3.9.

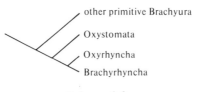

Scheme 3.9

Let us, as before, take the presence of a precopula and mating after the female moult to be primitive (the state of 'other primitive Brachyura'). For the Oxystomata we have the single observation of Raja Bai Naidu on *Philyra* (Leucosiidae). It had no precopula and mated when hard. This is our first evolutionary innovation, and we take it as primitive for the next group (Oxyrhynha and Brachyrhyncha). Now look up the Oxyrhyncha. We see the Majidae (no precopula is primitive: so they have re-evolved precopulas twice, both when predicted) and Hymenosomatidae (which have also re-evolved a precopula, when predicted). All the other families are in the Brachyrhyncha; no precopula is primitive, so it must have been gained at least once. The Grapsidae also contribute their species which has re-evolved mating with soft females, but has not re-evolved a precopula. The result is:

		Precopula observed?	
		+	−
Precopula predicted?	+	4	1
	−	0	1

The results, therefore are very sensitive to the outgroups. In this case the single and vague study of *Philyra* has immense repercussions. If we overlooked that one paper, we would take the presence of a precopula as primitive

CLASSIFICATION OF PERACARIDA   107

in the Brachyrhyncha and (probably) the Oxyrhyncha. The numbers would then be:

		Precopula observed?	
		+	−
Precopula	+	1	1
predicted?	−	0	3

which is very different from the numbers when we include *Philyra* as an outgroup. These differences should prove, if it is necessary, that the actual numbers are highly tenuous, and little trust should be placed in them. But they do also show that the total numbers (although not their distribution in the table) are fairly insensitive to the classification. We are going to use the numbers obtained by using Guinot's classification. Although the exact quantification is almost absurd, we are not grossly biasing the test in favour of the theory by picking on this set rather than any of the others.

Peracarida

The Peracarida is made up of five orders. We know little about the mating habits of the three smaller orders, Mysidacea, Cumacea, and Tanaidacea; about the two larger orders (Amphipoda, Isopoda) we know rather more. The Peracarida form a coherent and distinct superorder of the Crustacea. The phylogenetic relations within it are less certain (Whittington and Rolfe 1963, pp. 81-4). The mysids resemble the primitive decapods (the Euphausiacea: krill); they were once even classified together. But their similarities are probably only shared primitive malacostracan characters, and serve now only to indicate that the mysids are probably the most primitive peracaridans. Below, the sister group relations within the Peracarida implied by Kaestner (1968, III, p. 371) are on the left, and on the right is a slightly modified system. The one on the right abstracts a common feature of three systems discussed by Siewing and in the volume edited by Whittington and Rolfe (1963, pp. 99, 102, and 182): they all have the Tanaidacea as the sister group of the Isopoda. We will use the cladogram on the right of Scheme 3.10. The Pancarida, or Thermosbaenacea, are probably the outgroup

Scheme 3.10

of these five orders. (Whether they should be included in the Peracarida as order Thermosbaenacea or raised into a separate superorder Pancarida does not matter to us.) Females of *Monodella* and *Thermosbaena* have a dorsal brood pouch, unlike the other five orders in which it is ventral. The few papers on the reproduction of the thermosbaenaceans do not tell us whether there is a precopula. A fairly thorough paper by Zilch (1972) on *Thermosbaena mirabilis* does not mention a precopula, which may suggest that they mate without one. But this is unreliable evidence: we must look to a further outgroup, the Hoplocarida, for a reliable outgroup. We will therefore take the absence of precopula as the primitive state in the Peracarida, just as we did for the Eucarida.

The primitive peracaridan mode of fertilization was probably for sperms and eggs to meet in the brood pouch, outside the genital ducts of the female. Such is the practice in mysids and amphipods. The isopods have evolved internal fertilization, with special modifications of the second abdominal legs (pleopods) in the male. In most species mating follows a female moult, after which the eggs are released; it is usually preceded by a precopula. As usual, the main evidence that females are only receptive for a short period after the moult is that females only mate just after their moult: the exact period of receptivity is not known for many species. In the few that it is, it is short: we have already been through the evidence in an earlier section (see Table 3.1).

A few observations on amphipods and isopods may raise doubts. In most species in which mating follows the female moult, it is true, the interval of receptivity is short. It is only about three hours in *Gammarus*, and less than 24 hours in *Asellus*. But in some others it is lengthened: in *Talitrus* (an amphipod which we will come to) it is about three days, in another amphipod *Pontoporeia*, up to four days, in *Caecosphaeroma* and *Tylos*, about a week, and in *Niphargus* up to 17 days.

These raise a similar difficulty to one we met among the crabs. It is a difficulty with evidence. In *Talitrus, Pontoporeia*, and the others, mating follows the female moult. If the only evidence we had to go on was when they usually mate, then we would unhesitatingly count them as species which should have a precopula. But in these few we know that the interval of receptivity is longer. A species with an interval of 17 days probably should not have a precopula; but we can only predict this if the interval is known. What about the species in which the interval is not known? Perhaps the interval of receptivity is more like *Niphargus* than *Gammarus* in some which mate after the female moult: perhaps some may be falsely used in the test.

There are two answers. One is to ignore all the species in which the interval of receptivity is not known; the other is to count all the species which mate after the moult as having a short period of receptivity, hoping that the evidence is generally accurate. As for crabs, and for the same reason, we will choose the latter. Some species may be counted wrongly. But there is evidence that *Gammarus* is the rule, *Niphargus* the exception; and we obtain extra facts for the test. If more of the facts are scored correctly than incorrectly the test will be improved. Because the method is so conservative, it does not matter much whether most of the species are included or excluded. I shall include them in the review anyway. To the review we now turn.

Mysids are either active swimmers in the water column, or inhabit the bottom sand. Those that live in the sand swim up into the water, at night, to mate. Mating has not been seen many times, but no one has noticed a conventional precopula. Blegvad (1922, p. 78) was not rewarded by a single mating despite watching adults of three species of *Mysis* 'for hours'. He concluded that 'probably it lasts only a very short time, and furthermore . . . takes place during the night'. He did observe that 'the females moulted with every new oviposition'. On the mode of fertilization, Van Beneden (1861, p. 23) had stated that it 'takes place inside the female body', but he supplied no further details. Mating has since been described in six species, and from the finer descriptions for three of them (Nouvel 1937; Labat 1954; Kinne 1955), it seems that the sperms are put by the male into the brood pouch, where they fertilize the eggs. A few perhaps are washed up the female genital ducts and fertilize some eggs there. Of the six species, it is definitely stated for five that mating follows immediately after the female moult (*Praunus flexuosus*, Nouvel 1937; Nouvel and Nouvel 1939, p. 8; *Mesodopsis orientalis*, Nair 1939, p. 183; *Heteromysis amoricana*, Nouvel 1940, p. 6; *Paramysis nouveli*, Labat 1954; *Metamysidopsis elongata*, Clutter 1969, p. 147), and for the sixth (*Neomysis vulgaris*) it is implied (Kinne 1955, pp. 166-7, states that its mating resembles that of *Praunus flexuosus*). However, in at least three of these five species, males were only ever put with females that had just moulted: the authors were interested in copulatory behaviour. We have more useful information on the other two. Of *Heteromysis amoricana*, Nouvel (1940, p. 6) says that a male will not mate with a female that has mated more than 12 hours ago, although he does not specify whether he has all females in mind, or just those who have mated with another male during the twelve hour interval. Of *Metamysidopsis elongata*, Clutter and Theilacher (1971, p. 95) state that mating takes place within two or three hours of the moult, and Clutter (1969, p. 147) states that 'the females apparently were receptive to impregnation for a period of only 2-3 min'. Let us accept that mating and oviposition only take place immediately after a female moult. But what of precopulatory behaviour?

As we have already stressed when discussing the Natantia (much of the work on both groups was done or inspired by the Nouvels), observations on mating when the males were only ever put with recently moulted females are of little use to us. This rules out two species, *Praunus flexuosus* and *Paramysis nouveli*. We do not know enough about *Neomysis vulgaris* to use it in the test. Observations on three species are left which, because they provide little evidence of a precopula, do appear to contradict the theory: *Mesodopsis orientalis*, *Heteromysis amoricana*, and *Metamysidopsis elongata*. We will take them in order.

Nair (1939, p. 183) tells us that he noticed no precopulatory behaviour; but he had placed a single male with six females in a laboratory dish. Nouvel (1940, p. 7) did in fact notice that (as our theory predicts) males take an interest in females which are about to moult. Clutter (1969, p. 147) provides insufficient evidence to prove that there is no precopula. There are two courses open to us. One is to exclude the group from the test for want of evidence. It amounts in all to no more than two or three vague sentences of assertion. This is not enough to prove that the mysids lack a precopula, and Nouvel even hints that there is one.

The second course is to accept that the literature as a whole does demonstrate that the mysids lack a precopula. But it also provides a possible explanation, which is consistent with our theory. It lies in the reports (Clutter 1969; and others, e.g. Green 1970) of breeding aggregations. Clutter also suspects that the females release a powerful pheromone at the time of their moult. Now we must return to first theoretical principles. If only females near their moult aggregate, and if they release a pheromone at the moment of their moult, then males may be surrounded by such a density of receptive females that they are not selected to join them in a precopula. But this is to introduce a variable (the effective sex ratio) which is excluded from the simple comparative test. The explanation is consistent with the spirit, but not the most literal letter, of our law. Again, we would do best to exclude the Mysidacea from the test. Only facts which are clearly for or against the theory, and only those that are well-founded, can be admitted to our test. The Mysidacea, therefore, will be excluded.

We now turn to the richer literature of the Amphipoda. The mode of fertilization of amphipods was first thoroughly worked out, in the ubiquitous *Gammarus pulex*, by the Neapolitan Antonio della Valle (1889, 1893, 276ff). It has a 'riding-position' precopula, in which the dorsal surface of the female is below the ventral surface of the male. During the precopula, della Valle observed, the female cuticle is hard, and the female oviducts closed off. Just after the female moult the pair move to a new position, with their dorsal surfaces facing each other. The male introduces his sperms into the brood pouch, while it is still soft. The sperms do not enter the female's genital ducts, so cannot be stored. Re-fertilization is necessary at each moult (confirmed for *G. chevreuxi* by Sexton 1935). Other amphipods for which mating has been closely described have the same method of insemination.

The Amphipoda are customarily divided into four suborders, Gammaridea, Caprellidea, Hyperiidea, and Ingolfiellidea. Bousfield (1978) has recently reconsidered the phylogeny of the group, and he is the first to have applied cladistic techniques to it. In his scheme the Hyperiidea and Caprellidea have their ancestry in different groups within the Gammaridea; only the Ingolfiellidea have a separate ancestry from the Gammaridea. Although not all Bousfield's conclusions have passed into modern classifications (Lincoln 1979), we will make use of his phylogeny. It is probably the best available for the whole group. We will once again compare the results using it with the more conservative results obtained from the less richly divided (and intentionally non-phylogenetic) older classifications. In allocating genera to families, I have followed (where possible) Lincoln (1979) and Bousfield (1973). I have also used them as guides to synonyms. The sister-group relations of the 14 families whose mating habits are summarized below are shown in Scheme 3.11 (following Bousfield 1978). Let us take them in order. The first is the Caprellidae. Lewbel's (1978) paper on *Caprella gorgonia*, a Californian inhabitant of gorgonians, does not contain enough of the right kind of facts for us to use it in our test. He does tell us

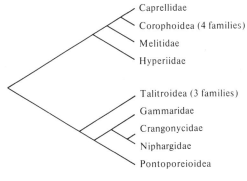

Scheme 3.11

(p. 145) that mating follows the female moult, and that the male protects the female during her moult, and 'carries the female about in her pereiopods for a period ranging from a few minutes to several hours after egg deposition'; but he tells us nothing about the events before the moult.

The four families of the Corophoidea, in the order which we will consider them, are the Ampithoidae, Aoridae, Cheluridae, and Corophiidae. Observations on three species of ampithoids have been published. *Cymadusa filosa* has a precopula (Ginet 1962, p. 81) which is compared by its discoverer to that of *Gammarus chevreuxi*, in which, as we shall see, the precopula culminates with mating after the female moult. *Ampithoe longimana* builds nests, in which a male and a female can sometimes be found together (Holmes 1901, p. 185; 1903, p. 288). The sexual association is probably a precopula, as it is known to be in the congeneric *A. valida* (Borowsky 1978). Borowsky compares the 'tube sharing' of *A. valida* to that of *Microdeutopus gryllotalpa*, which is the only species of our next family, Aoridae, for which we have relevant facts. *Microdeutopus gryllotalpa* builds its tubes among branches of algae, or on a hard substrate, on both sides of the Atlantic Ocean. Langenbeck (1898, pp. 303 and 305-6) first noticed that fertilization follows the female's moult, but she did not observe a precopula (although she suspected that one existed). Holmes (1903, p. 288) was probably mistaking her suspicion for an observation when, reviewing Langenbeck's findings, he stated that *Microdeutopus gryllotalpa* has a precopula in which the male physically holds the female. There is in fact no such precopula; but before the female moult a male may enter the female's tube and stay with her until her moult. Myers (1971, pp. 271-2), observing the species on the British coast, found that males often stayed with females after mating, throughout the incubation period, and then fertilized the next brood: 'not infrequently, pairs remained together for up to 2 months, during which the female produced 11 broods'. Males were sometimes ejected by other, larger males. Borowsky (1980, p. 295) by contrast, watching the species on the North American coast, found that pairing was mainly confined to just before and after the time of the female moult. The male usually left within 24 hours of the moult. 'Although there was much variation in the length of time a given pair shared a tube, in all the pairs observed for two weeks there was always a time when the two animals occupied different tubes.' The two studies were conducted, in laboratory cultures, on different sides of the Atlantic. From our theory we would expect the exact duration of pairing to change with such factors as density and sex ratio, which might well differ between the laboratories. But in both studies, *Microdeutopus gryllotalpa* conforms to the theory.

Next comes the wood-boring *Chelura terebrans* (Cheluridae), the subject of the paper by Kühne and Becker (1964, especially pp. 78-80). Mating follows immediately after the female moult. Azevedo-Gomes (1955, a paper I have failed to track down), they inform us, reported a precopula, and in the form of a *Reiterstellung*, like that of *Gammarus*. But Kühne and Becker were unable to confirm this. The male, however, does visit the tube of the female a day or two before her moult, and remains with her until mating. Oviposition followed a day after the moult. The Cheluridae, therefore, like the Aoridae and Amipthoidae, conform to theory.

The mating habits of *Corophium* are less well known. It lives, sometimes at huge densities, in tubes in intertidal mudflats. The female lays her eggs after a moult, when, we may assume, mating takes place (Sergerstråle 1950, p. 22, on *C. volutator*). Mating has never been seen. They may mate either in the water column at high tide (when some individuals leave their burrows to swim about), or within their burrows. Observations of pairs in the same burrow do exist (Thamdrup 1935; Crawford 1937), which leads some authors (Fish and Miles 1979, p. 365) to conclude that mating takes place in the burrow; but the observations are casual and scanty, which leads other authors (Watkin 1941, p. 89) to conclude that it occurs in the water. A precopula would probably be more likely if they mate in their burrows, because one would expect pairs to have been caught if they have a lengthy pairing in the water. But in the absence of evidence, we will exclude *Corophium* from our test. For other species of the same family there is better evidence. *Erichonius brasiliensis* is another tube-dweller. Zavattari (1920) and Salfi (1939, p. 47) found pairs cohabiting in tubes. Salfi allowed many individuals to make themselves at home in an aquarium; the majority lived alone, but many formed heterosexual pairs. *Siphonoecetes della-vallei* has evolved an extraordinary extension of the tube-dwelling habit (Richter 1978). Each individual builds a tube around itself which it walks about with. (It may use an empty shell as a tube, like a hermit crab, if it can find one.) If a male meets a female he binds her tube to his with threads of silk. He then carries her around as well. A male can easily carry two females with their tubes (one on each side) and may, with difficulty, afix a third to his portable harem. Mating probably follows moulting; but the female remains attached while brooding. The precopula has evolved into a more permanent association. Such scant information as we have for the Corophiidae, therefore, is consistent with the theory; but it is not adequate to stand by itself. It will not appear independently in the test because all its relatives have the same kind of mating: insemination is restricted in time, and is preceded by a precopula.

We turn next to two tropical species of Melitidae. *Melita celericula* (Croker 1971, p. 106) and *M. zeylanica* (Krishan and John 1974, pp. 416-17) both have a precopula. In *M. zeylanica* it has been shown to end with the female moult. The precopula is rather shorter than required by our definition: 2.5-16 hours (mean 11.5), by an indirect estimate, and 2-12 hours, respectively. But these are warm water species: Krishan and John's observations were made at a water temperature of 25-31 °C. For an appropriate criterion we must turn to Fig. 3.2. It does not extend to 25-31 °C, but if one day at 10 °C is the normal criterion, 12 hours would be a conservatively high criterion for 25 °C. *Melita* therefore could be counted as having a precopula or excluded as uncertain. If they are excluded, it is solely for reasons of method: their precopula is obviously analogous to that of other amphipods.

Next on our list is the family Hyperiidae. (It is, as we have seen, normally classified as a separate suborder.) We have a single useful paper, by Sheader (1977, pp. 944-5), on the pelagic *Parathemisto gaudichaudi*. There is no precopulatory carrying of the female by the male. They mate on an attachment site, for which, in the laboratory, a medusa served. 'On the day prior to moulting, the female attached . . . and during this period the male approached and remained attached . . . close to the female.' 'This stage [as Sheader remarks] is probably equivalent to the precopula stage of gammarid amphipods.' On the next day the female moulted, the pair mated, and the female laid her eggs. When mating takes place, 'soon after moulting, the oostegites are not fully extended, making the insertion of the uropods [of the male, for insemination] into the marsupium relatively easy'. Within half an hour the oostegites are fully extended. The Hyperiidae conform to our theory.

We turn now to the other great branch of Bousfield's phylogeny of the amphipods, and come first to three families of the Talitroidea: Talitridae, Hyalidae, and Hyalellidae. The Talitridae are terrestrial, or semi-terrestrial, intertidal 'sand-hoppers'. The family contains five genera in all, of which we shall consider three. The evolution of terrestriality in this family seems to have coincided with the evolution of a shorter precopula, and of a longer period when mating is possible; but the evidence is not of a single accord. Let us start with *Orchestia*. The intertidal *O. gammarellus* is the most studied species. Mating, and egg-laying, follow the female moult (Williamson 1951; Charniaux-Cotton 1957, p. 428; Campbell-Parmentier 1963, p. 477), but there is a longer period (about 72 hours: Campbell-Parmentier 1963, p. 477) than in other amphipods when mating is possible. The precopula in which the male carries the female is very short (0–10 minutes), as recorded by Williamson (1951), but Campbell-Parmentier (1963, p. 467) also saw males staying near females (rather than carrying them) before the moult. She does not specify how long males wait near females. Careful observation is needed to detect these non-contact precopulas. Charniaux-Cotton (1957, p. 428), by contrast, says that there is no pairing [*appariage*] before mating. She also reports seeing mating without a precopula in three other species, of which in two (*O. mediterranea* and *O. platensis*), other authors report a precopula in which the male carries the female (Williamson 1951, p. 53; and Bock 1967, p. 418, respectively). Finally, for this genus, Smallwood (1905, pp. 8, 9, and 19) saw male *O. grillus*, an inhabitant of saltmarshes, carrying females before mating, although 'it apparently does not extend for as long a time with this Orchestia as with Gammarus'; it lasts, she tells us, 'for hours or even days'. Of the other two genera, little can be said about *Talorchestia*. Williamson (1951, pp. 54 and 55) merely remarks that mating in *T. deshayesii* follows the female moult, and that two observed matings of *T. brito* resembled those of *T. deshayesii*. The terrestrial *Talitrus saltator* also mates only after the female moult, but it has no precopula (Verwey 1929, p. 1158; David 1936, p. 342; Williamson 1951, p. 55). For this species as well, we have tantalizing evidence that the interval after the moult available for mating is rather longer than in other amphipods. Successful mating is still possible at least two days after the moult, and eggs were not laid until four days after the moult.

What may we conclude for the Talitridae? They seem to have decreased (*Orchestia*) or lost (*Talitrus*) their precopula, although the possibility of males staying close to, but not carrying, females, introduces a nagging uncertainty into the conclusion. The reduction of the precopula interestingly coincides with a lengthening of the interval after the moult when mating is possible. The change is in exactly the direction that the theory predicts. We may be tempted to include the family in support of the theory, but caution should prevail. We must not pretend that the theory is more precise than it is; although the evolution of the Talitridae is in the predicted direction, we could not really have predicted the loss of the precopula. We do not know whether the lengthening of the period of female receptivity is sufficient, in theory, for precopula to disappear. Although they will be excluded from the test, the reader might still keep them in mind. We have not heard the last of this coincidence.

Three species of *Hyale* (Hyalidae) and *Hyalella azteca* (Hyalellidae) will complete the summary of the Talitroidea. *Hyale nilssoni* (della Valle 1893, p. 283, who calls it *H. prevosti*; Williamson 1951, p. 58), *H. pontica* (della Valle 1893, p. 283), and *H. schmidti* (Gilat 1962, p. 87) all have a precopula in which the male carries the female. In none of them has it been confirmed that mating follows the female moult, although it can be assumed. *Hyalella azteca* (*H. knickerbockeri* and *H. dentata* are both synonyms of *azteca*) has a precopula of 1-7 days (Gaylor 1921, pp. 240 and 242), or of 1-29 days with an average of 'about 4' (Wilder 1940, p. 451), or the duration varies between populations, perhaps because of differences in the intensity of predation (Strong 1972, p. 1108). (Other authors who have recorded precopulas in this species are: Holmes 1903, pp. 288-9; Embody 1911, pp. 8ff; and Jackson 1912, p. 54.) Mating follows the female moult: 'oviposition usually takes place within 12 hours of moulting' (Wilder 1940, p. 451). A pair of *Grandidierella bonnieri* (also Hyalellidae) mated when a male was added to a recently moulted female (Naylor 1956, p. 179).

Observations have been published on 12 species of Gammaridae, ten of them belonging to the highly speciose genus *Gammarus*. We will start with three freshwater species, *G. pulex*, *G. fossarum*, and *G. lacustris*. Mating follows the female moult in *G. pulex*, as we saw at the outset of the amphipodan section. The actual duration of the precopula is about 5-10 days, depending on the temperature and other factors (Heinze 1932, p. 413; Birkhead and Clarkson 1980 give a figure of nine days). *G. fossarum* (Ducruet 1973, p. 1037) and *G. lacustris* (Hynes 1955, p. 366; Berg 1938, pp. 71-2)[5] also have precopulas. In Norway pairs of *G. lacustris* enter precopula in September or October and do not breed until the ice breaks in the following June, which adds up to a precopula of 8-9 months. A precopula has also been reported in the variously marine or brackish water species *G. fasciatus* (Holmes 1903, p. 288; Hynes 1955; p. 373; Embody 1911, pp. 9-10, $t = 4$ days, $n = 1$; Clemens 1950, pp. 23ff, $t \leq 4$ days), which mates after the female moult (Embody 1911, who may be reasoning from analogy rather than observation), *G. locusta* (della Valle 1893, p. 283; Blegvad 1922, pp. 11 and 31,[6] $t = $ 'several days', up to five days; Sexton 1924, p. 343, $t = 1-2$ days) which mates after the female moult (Blegvad 1922; Sexton 1924), *G. chevreuxi* (Sexton 1928, p. 39, $t = $ 'several days . . . sometimes for as many as 8 or 9 days'), which mates only after the female moult (Sexton 1924, 1928, 1935), *G. duebeni* (Le Roux 1933, pp. 29-33; Forsman 1951, p. 228: 'in the experimental dishes it can last for three weeks, under natural conditions [in Scandinavia] at low temperatures probably much longer'; Kinne 1959, p. 192, $t = 4-13$ days depending on tem-

---

[5] Berg wrote of *G. pulex*, but he kept his animals and when Stephenson (1941) re-examined them he found them to belong mainly to *G. lacustris*.

[6] Segerstråle (1947-8, p. 238n) writes of 'the Danish collections [which] include amongst other, the material used by Blegvad for his investigation on the biology of *G. locusta* (1922). Of c.210 specimens from his collection, . . . c.one-third (70 specimens) proved to belong to this species, the rest being *G. zaddachi oceanicus*, and *salinus*. This result is by no means surprising in view of the confusion prevailing . . .'.

perature; Hartnoll and Smith 1978, p. 509, $t = $ 11-12 days), which mates immediately after the female moult: the female lays her eggs within an hour of her moult whether or not she has been inseminated (Le Roux 1933, pp. 29-33 and 38; Kinne 1959, p. 192), and *G. palustris* (van Dolah 1978, p. 198), *G. minus* (Kostalos 1979, p. 116), and *G. lineatus* (Embody 1911, pp. 9-10). A precopula, preceding mating after the female moult, has been recorded in three other genera: *Chaetogammarus marinus* (Vlasbom 1969, p. 321, $t = $ '1 (or less) to 19 days' at 15 °C), *Eulimnogammarus obtusatus* (Sheader and Chia 1970, p. 1082), and *Pallasiella quadrispinosa* (Samter and Weltner 1904, pp. 691-2, $t = $ 3 days, $n = $ 1; Mathiesen 1953, p. 69, $t = $ 6-7 days at 12-14 °C). Thus do the Gammaridae conform to theory.

Less is to be said about the remaining amphipod families. Only one species of Crangonycidae, *Crangonyx gracilis*, has been appropriately studied. Embody (1911, pp. 9-10) informs us that it has a precopula, which end with mating after the female moult. Again, it is difficult to decide whether he is reporting his own observations or his analogical reasoning. The latter may be suspected, because Mackay (1951, p. 8) and Hynes (1955, p. 380) saw no precopula in which the male gripped the female; Hynes even declared that 'it would indeed be impossible because the male is much smaller than his mate'. (Hynes is here inspired by a widespread tradition in amphipodology, that in species with a 'riding-position' precopula the males are bigger than the females, whereas in species with smaller males there is no precopula. We will not test this comparative hypothesis.) *Crangonyx gracilis* does have a precopula, of unspecified duration, in which the male stays near the female before her moult (Hynes 1955, p. 383; see also MacKay 1951, p. 8). The Crangonycidae, therefore, probably conform to theory.

Della Valle (1893, p. 283) reports a precopula in *Niphargus puteanus* (Niphargidae; perhaps a synonym of *N. fontanus* Bate); but the main work on this family is on *Niphargus orcinus virei*, which dwells in underground caves, by Ginet (1967). Copulation is very rare: he saw it only 19 times during six years of continual work. It is brief (p. 246), and not preceded by a precopula (p. 247). Egg-laying, and so (we can assume) mating, follow a female moult, but, 'unlike all other gammarids, this moult takes place many days before egg laying' (p. 247 original italics removed). Laying may be up to 6-17 days after the moult. The lengthening of the period is, according to Ginet (p. 248), the reason why the precopula has been lost. We may recall a similar coincidence in the Talitridae, and, if we keep it in mind, we will meet it again. The great lengthening of the period of receptivity of the female seems to me to warrant including this species in the test in favour of the theory.

The Haustoriidae inhabit sub-tidal and intertidal sand. A precopula has never been seen in this family. As in other sand-dwellers, this could be because there is none, or because males do not physically grip females during the precopula (which is anatomically unlikely: Sameoto 1968, p. 382) so that no pairs would be detected by the crude sampling techniques which are employed (Salvat 1967; Dexter 1971). Mating may occur either in the sand, or when the animals swim in the water column; the latter theory is preferred by the authorities, with little (Segerstråle 1937, p. 90) or no (Sameoto 1968, p. 382) evidence. The only direct evidence is for *Pontoporeia affinis* (Segerstråle pp. 90-1). He saw some pairs, which he inferred from their posture to be in copula rather than precopula. Egg-laying (and so mating) follows the female moult, but here again there is an extended period before oviposition as compared to most gammarids. By an indirect technique, Segerstråle estimated that in three populations the average times from moult to oviposition were 2½, 1¼, and ½-4 days (varying through the season). These should be compared with the one, or few, hours in *Gammarus*. Segerstråle also explains the loss of precopula by the lengthening of the period of receptivity. Like *Talitrus, Pontoporeia* has evolved in the direction predicted by our theory but it is in a rather grey area of prediction. We will therefore exclude it from our test.

Before presenting the conclusions for the Amphipoda, we will just run through three families for which there is a little information but not sufficient to include them in the test. *Leptocheirus pilosus* (Photidae) is a tube-dweller. Goodhard (1939, pp. 321-2) saw two pairs sharing tubes, but the only actual mating that he saw took place outside a tube. Forsman (1956, p. 400) tells us that he has never seen a precopula in this or any other tube-dwelling amphipod. He suspects that mating takes place when the animals leave their tubes, 'perhaps only for the purpose of ecdysis and hasty copulation'. The Ampeliscidae, which also dwell in tubes, have (it is hinted for *Ampelisca macrocephala* by Kanneworff 1965, p. 310) mating after the female moult; but 'the actual process of pairing and fertilization is not known in *Ampelisca* (Mills 1967, p. 322). Mills suspects that they mate in the water-column when they leave their tubes by night. Males, by the way, are the smaller sex. For *Leucothoe incisa* (Leucothoidae, more sand-dwellers), Salvat (1967) mentions neither the presence nor the absence of precopula. If there were a 'riding-position' precopula he probably would have recorded it. *Jassa falcata* (Iscyroceridae) females oviposit after their moult, on the same day (Sexton and Reid 1951, p. 38).

What are we to conclude? We have already seen that the mating habits of the nearest outgroup of the Amphipoda, the Mysidacea, are uncertain. If we go to the next outgroup, we will decide that the primitive state is the absence of a precopula. A precopula has evolved independently at least once in the amphipods. Most amphipods share this habit. Only in the lower half of the phylogeny of Bousfield with which we started do we find precopulas definitely being lost (although they may have in the Corophoidea as well). Here we will find some more numbers for our test. Let us summarize the habits of these families in Scheme 3.12.. A '+' indicates the presence of a precopula and mating after moulting; a '−' no precopula, and mating at any time. (All Amphipoda fit one of the two predicted states of the theory. There are no exceptions.)

PRECOPULA IN AMPHIPODS SUMMARIZED

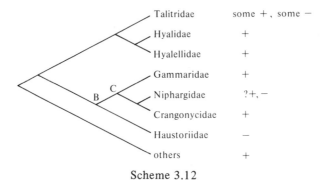

Scheme 3.12

This is the case of a double character reversal mentioned in Chapter 1. Outgroup comparison is not straightforwardly applicable. I have filled in B as '—' and C as '—': Haustoriidae are thus used as the outgroup for C, but 'others' are not used as the outgroup for B. The justification is that 'others' are less related, and more variable relative to Haustoriidae. But in fact it makes little difference to the total count (although it does effect whether the total is made up of +s or —s.). So the common ancestor of group B lacked a precopula. A precopula has probably re-evolved at least twice (Gammaridae and Crangonycidae, the evidence for *Niphargus* is not convinicing). *Niphargus orcinus virei* retains the ancestral state. Precopula has also been lost in the Talitridae. For the three groups without a precopula, the theory dithers, rather than predicts, two, and correctly predicts the third. Only this third, *Niphargus orcinus virei*, can be included in the test. (It is scored independently because the ancestors from which it evolved have been excluded: the theory is not clear for the Hautoriidae.) Although the loss in the Haustoriidae cannot be used in the test it has made an indirect contribution in the cladistic counting. In conclusion, we are going to score three predicted cases of the presence, and one of the absence, of precopula.

The Cumacea is the next peracaridan order. We have information on six species. The cumaceans are all marine, and may live in the bottom sand by day and ascend the water column by night, when pairs in precopula may be found; the female of such pairs typically has one moult to go until maturity (Fage 1951, p. 18). Most of what we know about mating is based only on inference; it comes from the study of *Diastylis rathkei* by Forsman (1938). He never (p. 34) actually saw mating and insemination. The female first develops oostegites at her 'parturial' moult. Observing that paired females had yet to undergo their parturial moult, Zimmer (1926) had supposed that insemination precedes it. Forsman (pp. 5-6) put males with females just before, just at, and 10 hours after, the parturial moult. The females who were with males only before or 10 hours after the moult did not lay eggs; those with males at the moult did lay eggs, which later degenerated. Forsman (p. 6) also observed that the female oviducts were not open to the outside until after the parturial moult, so mating before would be anatomically impossible. Such is most of the evidence that mating takes place at the female moult. Gnewuch and Croker (1973, p. 1016) observed mating once in *Mancocuma stellifera*, and it took place just after the female moult. The scant evidence suggests that the Cumacea conform to the common pattern of mating after moulting. We would predict that mating is preceded by a precopula. But is it?

Pairs, interpreted as precopulatory pairs (Zimmer 1942, p. 119; Fage 1951, p. 18), have been found in *Diastylis rathkei* (this is not mentioned in Forsman 1938, but Fage asserts it in his review), *Lamprops fasciata* (Sars 1900, p. 20 and Pl. XI; he also says that *Lamprops fuscata* has similar habits), *Pseudocuma longicornis* (Foxon 1936, p. 384), and *Mancocuma stellifera* (Gnewuch and Croker 1973, p. 1016 and Fig. 6), and *Iphinoe* (Fage 1951, p. 18). The duration of precopula is not known for any species. The Cumacea, therefore, probably have a predicted precopula. But the information is very vague, so it is probably just as well that they will not appear independently in the test.

Next come the Isopoda. I have not found a modern phylogenetic analysis of the isopods; so we will employ, as our best estimate of phylogeny, the classification of Kaestner (1968, III, with some extra information form Schultz 1979, Fig. 1). The species of interest to us fall into six suborders: Gnathiidea, Flabellifera, Valvifera, Asellota, Oniscoidea, and Epicaridea. Before embarking on the systematic summary, let us say a bit about their methods of fertilization. Unlike the other Peracarida, isopods have internal fertilization. This was (I think) first established by Schöbl, who summarized his findings in 1880. He studied the terrestrial oniscoid *Porcellio scaber*. This species mates in the spring, in late April or early May; the male introduces his sperms into the seminal receptacle of the female (Fig. 3.8a). A few days after mating, the female moults, develops her brood pouch, and lays her eggs into it. The openings of her oviducts are then closed to the outside. Later in the year she may moult, after releasing her larvae, and produce a second brood. Remating is not necessary: she stores enough sperms from the spring mating. The pattern of unrestricted mating before the spring moult is common in the terrestrial Oniscoidea (Friedrich 1883; Vandel 1925, Arcangeli 1948; Legrand 1958; Patanè 1959; Mead 1970, 1977). Mating may (Legrand 1958; Leuken 1968) or may not (Arcangeli 1948; Patanè 1959) take place at later moults as well. Leuken (1968) tells us that mating is restricted to the time of moult, but presents neither observations nor authority. (He may be referring only to moults after the 'parturial' moult at which the oostegites develop. After this moult mating would indeed by physically prevented except at the moult, and even then may be impossible in some species: Arcangeli 1948.) Schöbl's discovery that mating in *Porcellio* precedes the female moult misled certain biologists who studied in *Asellus aquaticus* (Rosenstadt 1888; Leichmann 1890; von Kaulbersz 1913, p. 350) and *Gammarus* (Haempel 1908). In fact, as was first established by Unwin for *Asellus*, it is the other way round: moulting

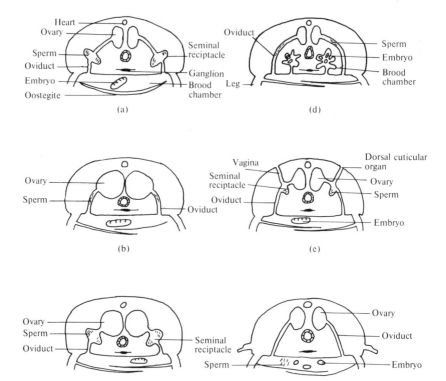

Fig. 3.8. Female reproductive organs of various isopods. (a) Oniscoidea; (b) *Limnoria*; (c) *Asellus*; (d) *Sphaeroma*; (e) *Jaera*; (f) *Epipenaeon*. (a, b, d, e, f from Menzies (1954, after various authors), c from Unwin (1920).)

precedes mating. The method of fertilization in *Asellus* has already been explained in great detail. There are many more methods in other isopods. Figure 3.8, derived from Menzies (1954), surveys some of them. The oniscoid system we have already described (Fig. 3.8a). Some aquatic isopods, such as *Limnoria*, have internal fertilization without a sperm storage organ (b); *Asellus* (c) has a sperm storage organ, although no one has ever demonstrated that a single insemination provides for more than one brood. In *Sphaeroma* (d) the eggs and larvae are no longer brooded in the external pouch, but have an internal pouch formed by an invagination of the ventral surface; the eggs are laid (after a moult) into the brood pouch and are then directly moved into the 'uterus' (Leichmann 1891). In *Jaera* (e) the female genital openings are on the dorsal surface of the fifth thoracic segment, forming a so-called 'dorsal cuticular organ' (Forsman 1944, p. 16). The relation between mating and the female moult in different *Jaera* species varies, as we shall see. Yet another system is found in the parasitic bopyrid *Epipaeneon* (f). Hiraiwa (1935-6, p. 108) inferred that, in this species, fertilization is external. Enough about the details of isopodan methods of fertilization; let us turn to the summary of the incidence of precopula.

The Gnathiidea will not keep us long. The males are armed with enlarged mouthparts which variously function in snipping, crunching, or piercing conspecific males. Each male defends a burrow, which may also contain a number of females. On the timing of mating and fertilization in the female cycle there have been many speculations, but none of them are proved (Monod 1926, pp. 221-3). The female joins a male even before metamorphosing to adulthood (Monod 1926, p. 164), so there must be a long precopulatory sexual association. Many casual remarks record couples, or larger numbers (up to 10 in *Paragnathia*) of females with a male (Monod 1926, pp. 264 and 420). We may conclude that there is a lengthy sexual association, but we do not know enough about it.

We will consider five families of the next suborder, Flabellifera. Their phylogeny, according to Schultz (1979, Fig. 1; see also Brusca 1981, Fig. 4B), is shown in Scheme 3.13. First come the Limnoriidae. These small (*c*. 3 mm

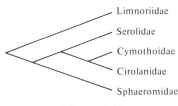

Scheme 3.13

long), white, wood-borers, on all accounts, live in pairs. The male is usually at the outside end of the burrow, and the female at the inside; the young eat out little side-burrows. Pairs in burrows have been reported in *Limnoria lignorum* (Coker 1923; Henderson 1924; Sømme 1941; Eltringham and Hockley 1961), *L. quadripunctata* (Eltringham and Hockley 1961), and in *L. tripunctata* (Menzies 1954). For *L. tripunctata*, Menzies also gives us some idea of how long the sexes stay together for: 'organisms kept in the laboratory have been observed to pair with periods exceeding ten months (one case) and four months (four cases)' (p. 376). Copulation has never been seen, but Menzies thought that re-insemination between each brood would be necessary because of the lack of a sperm-storage organ in the female. *Limnoria* seem to conform to our theory but we do not know enough to be sure.

Next, the Serolidae. Moreira (1973, pp. 110-16) saw and depicted something that looks very like a precopula in *Serolis polaris*. He called it a copulation, but, 'due to the end of the cruise it was not possible to observe the normal separation of the male from the female'. The pairs stayed together for 'more than 4 days' (p. 118). There is not enough information here for us to use.

The Cymothoidae are mainly parasites of fish; many of them live on their hosts in lengthy sexual associations (Brusca 1981, pp. 123-9, gives a fine introduction to their natural history). Copulation has never been seen, so we have to fall back on inference. Szidat (1964, p. 84), writing about *Meinertia gaudichaudi*, said that it must take place when the genital openings first appear, 'in the short time . . . before the development of the marsupial lamellae [oostegites]'. Fain-Maurel (1966, p. 9) similarly informs us that the female is inseminated 'during the moult preceding egg-laying'. Many cymothoids (including *Meinertia gaudichaudi*) are protandric hermaphrodites. This was first shown by Bullar (1876, who only showed hermaphroditism; he was less certain about protandry), and by Mayer (1879), and has since been confirmed in many species; so far as I know, no cymothoid has yet been shown not to be a protandric hermaphrodite.

The two other important features of cymothoid sexual biology are the dimorphism of the sexes, and the part of the host where the parasite lives. Cymothoids may live on the outside of their host, or internally, either in the gill-chamber or in the mouth. Trilles (1969, p. 442) summarily concluded from a study of six species that internal parasites of the gill-chamber do not move from their hosts, whereas cymothoids that live externally or in the mouth do retain, at least in the males, an ability to move between hosts. Some species have extreme sexual dimorphism, with dwarf males, whereas in other species the sexes are of similar size (Szidat 1964). A comparison of the degree of sexual dimorphism with the motility of the males between hosts might be interesting; but such a study has never been made: the significance of the variation in sexual dimorphism remains obscure.

We can infer, from Trilles's generalization, that species which inhabit the gill-chamber of their hosts must live in lengthy sexual associations, but apart from this little can be inferred about how long the sexes spend together. The best evidence that pairing is lengthy is that the sexes are usually found in pairs. Such has been found for *Lironeca convexa* (Menzies, Bowman, and Alverson 1955, pp. 284-8), *L. puhi*, which inhabits gill-chambers, (Bowman 1960, p. 89), *L. vulgaris*, the male of which lives in the gill-chambers, and the female on the tongue, of their hosts (Brusca 1978*b*, pp. 10-12), *Meinertia* 'new sp.', which clings to the tongues of mullet, (MacGinitie 1937, p. 1031), *Phyllodurus abdominalis*, which live on the feet of the thalanassid shrimp *Gebia pugettensis* (Lockington 1878, p. 300), and *Nerocila californica*, which lives attached to the outside of several fish species (Brusca 1978*a*, pl. 1); *Olencira praegustator*, which lives in the mouth cavity of, or externally attached to, its host, does not live in pairs: the males move between hosts, mating when they find females, until they find an unparasitized host, when they develop into females (Kroger and Guthrie 1972, p. 372). This list could probably be lengthened by a thorough search in the taxonomic literature; but a longer list would probably be no more conclusive. Again, this family probably conform to the theory; but the facts do not prove it: the Cymothoidae do not appear independently in the test.

Nor will the Cirolanidae. *Eurydice affinis* and *E. pulchra* have a precopula (Jones 1970, p. 637). Jones himself did not observe mating, although he tells us that Bacesco (1940) did. He suspects that they mate at the female moult. *Eurydice* therefore probably conform to law. There is a hint that *Cirolana* does not. Of *Cirolana harfordi*, Johnson (1976, p. 345) writes that 'there is no precopulatory pairing . . . Mating was not observed, but presumably usually takes place at night when the ripe female moults'. If we trusted a simple and literal reading, this species would have to be counted against the theory. But mating has not been oserved, and the author is probably only reasoning from analogy with other isopods when he presumes, out loud, that it mates when the female moults. Our theory denies that analogy. The species could, furthermore, have some kind of precopula in which the male only stays near the female rather than physically gripping her. (Such, indeed, are the precopulas—if they exist— of the Cymothoidae, and the Limnoriidae.)

The Sphaeromidae variously inhabit marine and intertidal waters, freshwater, as well as hotsprings and underground caves. *Thermosphaeroma thermophilum* has a precopula of up to ten days, ending in mating when the female moults (Shuster 1981, p. 699). Similarly in the intertidal species *Sphaeroma serratum* and *S. hookeri*, there is a precopula of one to two days, although 'sometimes' only a 'few hours', before mating after the female moult (Daguerre de Hureux 1966, p. 34, for *serratum*; Forsman 1956, p. 401, and Jensen 1956, pp. 320 and 334, do not specify a duration for *hookeri*). A female put with a male during her intermoult is not inseminated. Monod's (1930, p. 439) description of three pairs of *Cassidinopsis maculata* '*in copula*' sounds like a precopula. There can be no doubt about how to score all the species which we have discussed so far. Now we come to two more difficult species. In the intertidal *Dynamene bidentata* the male does not physically hold the females, but there is again a long sexual association. A male is usually found, in a rock crevice or a barnacle shell, with 'up to five females' (Naylor and Quenisset 1964, p. 213), or 'a "harem" of females, averaging four in number' (Holdich 1968, p. 150). Holdich further informs us that 'the female is fertilized by the male when in the half moulted condition . . . which may last from one to five days before the oostegites are released' (p. 139). According to Husson and Daum (1953, p. 2346), *Caecosphaeroma burgundum*, a cavernicolous species, also has a long interval (8-9 days) between moulting and oviposition. The male guards the female before her moult, and then afterwards until she lays her eggs. 'The duration of the precopula is very long (up to 45 days)' (Husson and Daum (1953, p. 2346, $n = 15$), while Daum (1954, p. 132) says that it lasts from 8-79 days, with an average of 28 ($n = 13$). These last two species, with their long periods from moulting to oviposition, may remind us of such amphipods as *Talitrus, Niphargus*, and

Pontoporeia. These three amphipods, however, had lost their precopulas. They occupy the grey area between the clear theoretical predictions. Two of the amphipods were excluded from the test: we will exclude these two isopods for the same reason. The remaining sphaeromids we may provisionally take to support the theory. The evidence is provisional only because we do not know that they do not also have a long interval between moulting and oviposition; we only have the qualitative statements of Daguerre de Hureux and of Forsman that females in their intermoult, or before the moult, could not mate.

The Flabellifera make no independent contribution to the test. Most of them retain, or lengthen, the ancestral isopodan precopula. Some may have lost it, but we do not know enough about these to use them. And now for the next group.

The next group is the Idoteidae (Valvifera), for which we have information on seven species. They are all marine. They all have a precopula: *Idotea neglecta* (Kjennerud 1950, p. 25, t = 'often for several days'), *I. emarginata* (Naylor 1955, pp. 274-5), *I. baltica basteri* (Patanè 1962, p. 364, t = 'long duration'), *I. pelagica* (Sheader 1977, p. 663, t = 2-3 days, in laboratory), *I. baltica*, *I. chelipes*, and *I. granulosa* (Salemaa 1979, p. 140), and *Glyptonotus antarcticus* (White 1970, t = up to 190 days). Naylor, Salemaa, White, and Betz (1974, p. 67, for *I. chelipes*) all state that mating takes place immediately after the female moult, but it is not always clear whether their statements are based on independent observation or Aunt Jobisca's theorem. The Idoteidae, to conclude, conform to the theory.

Facts are available for two families from the next suborder, Asellota: Asellidae and Janiridae. The species of Asellidae inhabit fresh water and all those for which definite statements are available have a precopula. Similar evidence confirms that they mate after the female moult. The species are: *Asellus aquaticus* (for duration of precopula: de Geer 1778, VII, p. 508; von Kaulbersz 1913, p. 303; Ridley and Thompson 1979, p. 388; Manning 1980, p. 266; for mating after female moult: Unwin 1920;[7] Maercks 1930, pp. 451ff), *A. meridianus* (Steel 1961; Manning 1980, p. 266). *A. intermedius* (Ellis 1961, p. 83), *A. tomalensis* (Ellis 1971, p. 55), *Mancasellus macrouris* (Markus 1930, p. 229), and *Lirceus fontinalis* (Styron and Burbanck 1967, p. 405). The Asellidae conform to the theory.

The Janiridae conform as well, and contribute an independent test. They have (predictably) lost the precopula in some species while retaining it in others. The species of *Jaera* have, as we have previously seen (Fig. 3.1e), are inseminated through a 'dorsal cuticular organ'. But the exact details of when insemination can occur vary between species. Let us take the *Jaera albifrons* group first. In *Jaera albifrons* the dorsal cuticular organ is open throughout the female moult cycle, so she can be inseminated at any time (Veuille 1978, p. 390, and kindly

---

[7] *A. aquaticus* was split by Racovitza (1919) into *A. meridianus* and *A. aquaticus* while Unwin was studying it; according to Dudich (1925), Unwin's species was the one we now call *A. meridianus*.

confirmed in a letter to me in 1980). They have a precopula but it is so short—
'some seconds to several hours' (Veuille 1980, p. 91)—as not to amount to a
precopula by our definition. (Solignac 1972*a*, p. 1571) tells us that precopula
'can be prolonged for a very long time', but does not quantify; this statement, I
think, refers to *J. (a.) posthirsuta*: Solignac (1972*b*, p. 2236) mentions three other
species of the *albifrons* group with similar habits.) *Jaera albifrons* is an instance of
the predicted loss of precopula. And we shall count it for what it is. Other *Jaera*
species are more conventional. *J. italica*, *J. nordmanni*, and *J. istri* have a precopula, which lasts for about three days (Veuille 1980, p. 91), and insemination
follows soon after the first moult of the reproductive cycle (Veuille 1978,
pp. 386-7, also kindly confirmed in a letter to me in 1980). These other *Jaera*
species fit the theory but will not appear independently in the test.

Next, the suborder Oniscoidea, among whose ranks can be found the many
groups of terrestrial isopods (Crinocheta). The families that will concern us
are classified in Scheme 3.14 (Kaestner 1968, III, with one amendment following Schultz 1979, Fig. 1).

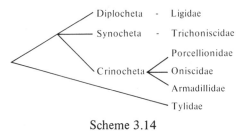

Scheme 3.14

The Ligidae inhabit the shorelines of the ocean, and other moist places,
such as beneath the bark of trees. They have retained the habit of mating only
after the female moult (Nicholls 1931, pp. 670-1, on *Ligia oceanica*), preceded
by a precopula (Nicholls, *ibid*., and, for *L. pallasii*, Miller 1938, p. 118; and
Carefoot 1973, p. 304).

The Tylidae inhabit wet terrestrial habitats, and may contain species which
have both lost, and species which have retained, their precopula. All our knowledge comes from the investigations of Françoise Mead (1964, 1965, 1967), on
two species. *Helleria brevicornis* has a precopula of up to 20 days (Mead 1964).
Copulation follows the female moult: it usually takes place between seven and
fifteen hours after the moult (corresponding to the period before the moulting
of the anterior half of the female), but (in experiments) can take place up to 48
hours (Mead 1965). *Tylos latreilli* (Mead 1967) has no long precopula like
Helleria, and the restrictions on mating seem to have been relaxed. Of 30 matings,
six occurred in the 7-15 hour interval, 13 more before 48 hours, and another
eleven 2-10 days after the moult. In observations on *Helleria*, by contrast, no
matings took place after the moulting of the anterior half. *Tylos latreilli* is one
of those species for which the theory does not make an unambiguous prediction

with the limited facts available. It has certainly evolved in the direction which the theory predicts. But is the relaxation of when mating can take place sufficient to predict the loss of a precopula? We cannot say for sure, so the species is best excluded from the test.

Only inconclusive information is available for the Trichoniscidae. The habit of building a small container (*logette de mue*) to moult in is quite widespread among terrestrial and cavernicolous isopods (Dalens 1967). The female also practises her 'parturial' moult (at which she grows her oostegites and lays her eggs) inside a *logette*. We may therefore be interested by reports of males and females sharing *logettes de mue et de parturition* in *Nesiontoniscus corsicus* (Dalens 1977). Dalens suspects that they copulate inside their *logette*; but he had never actually seen it.

We turn now to the different habits of the terrestrial isopods. We have already been through the relation between mating and moulting in *Porcellio*: the females can be inseminated any time during the great expanse of spring. Precopula has been utterly lost. For the Porcellionidae we have information on four species. Schöbl (1880, p. 127) was, it appears, writing of many species, but he explicitly described mating for *Porcellio scaber*. It mates briefly any time before the female moult. *P. laevis*, we can infer from Patanè (1959, pp. 114 and 124-7), has the same system. *P. dilatatus* has a brief copulation, not necessarily (though sometimes) after a female moult (Legrand 1958). The paper by Arcangeli (1948) on, among others, these three species, is consistent with my summary except that (contrary to Legrand) he denies that copulation can ever come after a moult. *Metaponorthus pruinosus* (Arcangeli 1948; Shimoizumi 1952) and *M. sexfasciatus* (Mead 1970, p. 55) also has a brief copulation some time before the female moult. Mating can in the latter species be extended for 'plusieures hours', but this is not a precopula. (Another porcellionid genus, *Hemilepistus*, contains monogamous species: Schneider 1971; Linsenmair and Linsenmair 1971.)

In the Oniscidae and the Armadillidae the pattern is the same as in most porcellionids: mating precedes the parturial moult, and there is no precopula (Schöbl 1880; Friedrich 1883; Arcangeli 1948).

The final suborder of ispods are the parasitic Epicaridea. It contains two infraorders, the Bopyrina (which are gonochoric), and the Cryptoniscina (which are protandric hermaphrodites). The scant evidence points to lengthy sexual associations in the Bopyridae: 'the male and the female generally live in association for all their life' (Reveberi 1943, p. 112; similarly, e.g. MacGinitie 1937, p. 1035). The male is usually tiny, and lives attached to the female, Bonnier (1900) illustrates such pairs for several species; MacGinitie (1935, p. 704; 1937, p. 1035) documents permanent pairing in *Phyllodurus* and *Argeia*; and Hiraiwa (1935-6, p. 108) in *Epipenaeon japonica*. The timing and mode of fertilization are not known from direct observation, but we can make some inferences. From the fact that the male has no copulatory apparatus and lives permanently attached to the female near her brood chamber, Hiraiwa inferred that fertilization is external. Its timing in the female moult cycle is completely unknown. Nor is it known if (and how) the male remains attached over the female moult. For the Epicaridea I have found only one paper, by Sheader (1977), on *Clypeoniscus hanseni*. It inhabits the brood pouch of *Idotea*. The tiny 'cryptoniscan' male attaches to the female in what Sheader (p. 667) calls a precopula; after insemination, the male detaches from the female but may (one gathers) stay in the same host until the host's next moult when the female departs. A male in the absence of a female changes sex. Although this species evidently has some kind of precopula, its duration, and the timing of insemination in the female moult cycle, are obscure.

The independent contribution of the isopods to our test can easily be summarized. A precopula, or longer association, is found in most species. It is the primitive state so is not counted. It has been lost in the Oniscoidea and in *Jaera albifrons*. These are two independent losses which we will count. It may also have been lost once in the Tylidae (*Tylos*) and once in the Cirolanidae; but we do not have enough information on these two to use them.

What, finally, about the Tanaidacea? We have only one study, by Bückle Ramíres (1965), on *Heterotanias oerstedi*. Most tanaid shrimps live in tubes, and *Heterotanais oerstedi* is no exception. Usually only one individual lives in each tube, but for mating a male leaves his tube and breaks into that of a female. Bückle Ramírez added males, in the laboratory, to tanks containing females. The females had recently moulted (pp. 726-7). The males showed no precopula in the 'riding-position' (p. 733), but this hardly supports his conclusion that there is no precopula at all. We would like to know, before so concluding, what a male does with a female who is about to moult. There would probably be no 'riding-position' precopula in this tube-dwelling form, but, if mating is confined to the period just after the female moult, we would expect the male to stay with her until her moult. Bückle Ramírez saw a courtship (*Paarungspiel*) which lasted about eleven hours, then mating, then the male left. Eggs were laid soon after, but then resorbed: mating had not been successful. These facts, valuable though they are, are insufficient for us.

Our totals for the Peracarida are three predicted losses of precopula and three predicted gains. The entire order contains no certain exceptions to our law.

Crustacea: summary

Let us now summarize the evolution and loss of precopulatory mate guarding in the Malacostraca and in the Crustacea as a whole. The phylogeny on which we are basing our score is shown in Scheme 3.15. The outgroup of the Crustacea, whatever we take it to be, we can assume to lack a precopula: the habit is rare outside the Crustacea. Within the Branchiopoda it remains a minority habit. Most branchiopods lack a precopula, but the habit has evolved in some Anos-

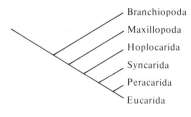

Scheme 3.15

traca. The common ancestor of the Maxillopoda is difficult to determine by outgroup comparison because the Branchiopoda contain both states. Branchiopods with a precopula are a minority and their ancestor lacked a precopula, so we will assume that the ancestor of the Maxillopoda also lacked one. They have acquired the habit twice. Again it is difficult to assign an ancestral state as we move towards the Malacostraca. Let us suppose, for the same reason as we gave for the Maxillopoda, that the Malacostraca also ancestrally lacked one. The Hoplocarida retain this ancestral state. The common ancestor of the Eucarida and Peracarida also lacked a precopula. It may be that both the mysids (primitive Peracarida) and the Euphausiacea (primitive Eucarida) lack one and the habit has been independently evolved in each line. But the evidence does not demand this conclusion, and the most conservative assumption is that the habit evolved once before then. So we now assume that the Peracarida and Eucarida ancestrally had a precopula. The further contribution of the Peracarida we have just summarized at the end of the previous section. The contribution of the Eucarida is compiled from the Brachyura (tabulated at the end of that section) and at least one loss in the Reptantia outside the Brachyura, in the crayfish. The totals for the Crustacea then are:

		Precopula observed?	
		+	−
Precopula predicted?	+	8	1
	−	0	8

Arachnids

Precopulas have evolved in two main groups of arachnids, the spiders and mites. *Limulus* may also have a precopula. I have not read much of the literature on the other groups, but general references and the few detailed papers on particular species that I have read show no sign of any precopulas. We can conclude that, in all these other arachnids, the sexes only pair as adults. As the theory predicts, the sexes do not stay together for long. This is the primitive state in

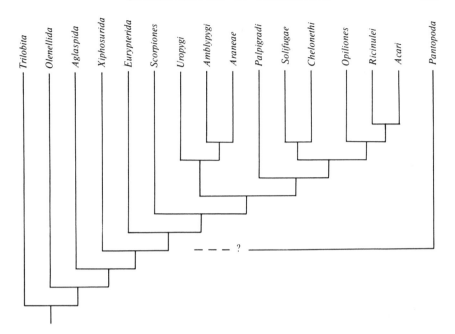

Fig. 3.9. Cladogram of the arachnids. (From Weygoldt and Paulus (1979).)

the Arachnida, so these other groups are not counted independently in the test. I have reproduced the cladogram of Weygoldt and Paulus (1979) in Fig. 3.9. In each of the three groups (spiders, mites, and *Limulus*) precopulas (if they prove to be genuine) have probably evolved independently at least once. We will say no more about the other arachnid groups. We will come back, in turn, to the spiders and mites.

We will deal with *Limulus* straight away. When adult, *Limulus* migrate (at the breeding season) to shallow water. The male, who can smell the female, clasps a mate. There is a precopulatory phase; but unfortunately the literature does not reveal how long it lasts. At high tide the pair go up the shore, the female digs a hole and lays her eggs in it, and the male fertilizes them externally. The precopula therefore probably precedes a short predictable period of receptivity. But we do not know enough to include it in the test.

Now for the spiders and mites. To avoid repetition later, we will discuss here some points which apply to them both. Their life histories, and the position of the precopula within it, are similar in some respects. Besides these similarities, there are also some common problems with the evidence.

First, the timing of mating and mate guarding. Most arachnids, like most insects but unlike most crustaceans, do not moult again after reaching maturity. Maturity is reached some time after the final moult. Most have internal fertilization, and the females possess sperm-storage organs. Mating is usually not possible with immature females because they lack the necessary genital openings, which do not develop until after the last moult. Precopulatory mate guarding may take place when an adult male joins an immature female and stays with her until she moults. He will mate with her immediately she reaches adulthood.

In many spiders and mites, the males reach maturity earlier, after fewer moults, than the females (the males consequently tend to be smaller). Precopulas may then evolve even more easily than if the sexes mature at the same rate. If the generations are separated at all, there is a period when many males have reached adulthood but many females have not. The adult males may seek and stay with immature females. Similarly, if the generations overlap completely, at any one time there will be both adult males and immature females. They may form pairs before mating.

Precopulas, according to the theory that we are testing, are likely to evolve before the female's final moult. Let us spell out the theory in the terms of the arachnids. In most arachnids the female becomes receptive after her final moult: her genital openings (the 'epigyne' of spiders) appear then for the first time; moreover, a moulting female is usually too fragile to resist a male. The male will be selected to wait with an immature female when the time to go until her last moult is shorter than the expected time to find another female and wait until she becomes receptive. Mating (as we have seen) is physically impossible before the moult; but what about afterwards? In many crustaceans, mating becomes physically impossible again soon after the female moult. But arachnids are not like that. Mating is never again physically impossible after the female has reached maturity. Crane (1949, p. 176), for example, saw a male of the salticid *Corythalia fulgipedia* mate with a female just after her moult. She also kept other females isolated after their moult: they were still attractive to males, and able to mate, at least four months later. A similar result could probably be obtained with most spiders and mites. (The mite *Typhlodromus cucumeris* is apparently an exception. Females cease accepting males two to three days after their moult even if they have not mated, in which case they are infertile: El-Badry and Zaher 1961, p. 429.) But if mating does not become physically impossible, the female may still become unreceptive; and, in our theory, psychological unreceptivity has exactly the same consequences as physical unreceptivity. The question is not, For how long after her moult can a female still mate, if she is kept alone? it is For how long, in nature, do females tend to be receptive?

It is easy to phrase the question, but not so easy to answer it. The literature hardly helps at all, although it does implicitly suggest a few relevant variables. In many species, the female develops her eggs over a period of a few weeks to a month (spiders), or a few days (mites) after her final moult. Once she has laid them she (in many species) stops accepting males; even if the female does not become completely unreceptive, she may still become much less so. The female is only receptive after her moult, but before she lays her eggs: this is the first constraint on the period of receptivity. The second operates in species in which the females mate only a limited number of times. In spiders and mites, different species differ in whether a female mates once or several times. Take spiders first. Females of some species simply mate only once. In others females have been seen to mate more than once (Austad 1983). However, an observation of the act of mating does not prove that the mating will contribute to fertilization. We cannot be certain that matings after the first one are not blocked by something (such as male genitalia) left behind by the first male. In some species (such as *Nephila*, Vollrath 1980, and *Phidippus*, Jackson 1978) females definitely mate, efficaciously, more than once. In conclusion for spiders, we do not know enough to specify for how long a female is receptive after her moult: in some species (those in which the females mate only once) the interval must be short, perhaps only a few days; in others (with multiple matings) it is longer, perhaps a month.

In mites the story is the same. In some species females are said to mate only once (immediately after the moult): these will have only a short receptive period. Females of other species may mate many times. For some it has even been shown that multiple mating is necessary for full fecundity. Females who mate more than once lay more eggs than those who are allowed to mate only once (Phytoseiidae, Amano and Chant 1978a; ticks, Feldman-Musham and Borut 1971; *Caloglyphus anomalus*, Acaridae, Pillai and Winston 1969, p. 301). Of other reports of multiple mating by females we should be more suspicious. In *Tetranychus urticae*, for example, males may start to mate with females that have already mated, but they break off the mating much sooner than when mating with a virgin (Overmeer 1972), and only the first mating is genetically effective (Helle 1967). Variation in the time from mating to egg laying (the first constraint mentioned for spiders) may account for some of the differences between reports about the number of times a female will mate. Consider *Typhlodromus occidentalis*. Females studied by Lee and Davis (1968, p. 253) took only on average 1.3 days to start laying, and only mated once; those studied by Laing (1969, p. 980) took about three days, and some mated more than once. For mites, as for spiders, we do not know exactly how long the female is receptive for after her moult. However, the theory still crudely predicts that precopulas are more likely to evolve with immature females, before mating after the last moult, than with adult females. Such is the prediction that we will test.

Before going on to the individual groups, we will look quickly at the problem of negative evidence. In many species of spiders and mites only pairings with adult females have been described. We cannot conclude that these species lack a precopula. To conclude that a precopula is absent we will demand some positive evidence: some passing statement, at least, that (for example) males are not attracted to, or are repelled by, immature females. No such statements exist in the literature about mites, and only very few in that about spiders. Our conservative method will therefore force us to uninformative conclusions.

In the mites we will find only one independent case of precopula. In spiders, too, the numbers will be low. These conclusions probably do not reflect the state of nature: they are imposed on us by a combination of methodic conservativism and factual ignorance. They will be modified by progress. But for our purposes, they are the best conclusions we can come to.

Spiders

The precopula of a spider is usually called 'cohabitation'. The male may simply stay near or on the web of a female (in web-builders), or he may live in a nest right next to her, perhaps building himself a special compartment (in hunters). Whatever the details, cohabitation is what we will be looking out for. We will be searching in the literature for any mention of cohabitation among immature females and adult males. The search is not quite so simple. Almost any lengthy association of a male and female spider is called cohabitation. The word refers to both precopulatory cohabitation, which interests us, and postcopulatory cohabitation, which does not. Our theory predicts only the incidence of precopulatory cohabitation, so it is on this that we must concentrate. It predicts nothing about postcopulatory associations. We must try to recognize descriptions of postcopulatory cohabitation, and then ignore them.

Before starting the systematic summary we have two further tasks: to examine some alternatives to our theory of precopulatory cohabitation, and to make some general comments about the literature on which our summary is based. There are two other explanations of cohabitation. One of them explains it by the fact that adult male spiders do not build webs to catch food. Immature males build webs like any normal spider, but when they reach adulthood they stop spinning webs (except for a small web which is used for 'sperm induction': the male inserts his sperms into the female with his pedipalps—small legs at the front of the head—but his genital openings are on his abdomen; the male transfers the sperms from his genitals to his pedipalps via a sperm web). Adult males of some spcies rely on stealing food from female webs, and they may stay on a particular female's web for a long time just to feed. But we would not expect this kind of cohabitation, cohabitation in order to steal food, to be precopulatory; and furthermore this theory can only explain cohabitation on the female web. In some of the species which we shall be discussing the males do cohabit on the female web, but in others the male and female live together in a special nest, usually of two compartments, and this association cannot be a male adaptation for feeding.

Sexual cannibalism is the basis of the second alternative to our theory. The male, we are told, cohabits with an immature female and then mates with her after her moult (when she is soft and helpless) because the female will then be in no condition to attack the male. This explanation was first (I think) suggested by Menge (1866) and has been repeated often (e.g. Sörensen 1880,

p. 174; Montgomery 1909, p. 560; Nielsen 1932, I, p. 35; Bristowe 1958, p. 119). This theory is of course fairly specific to spiders, and I do not think that there is enough information in the literature to test it. The theory would, we may assume, predict that cohabitation should be found in species in which sexual cannibalism is particularly common.

So much for the other theories. Now let us turn to the literature itself. We have several comments, and may as well start by continuing with the subject of sexual cannibalism. Sexual cannibalism provides the context, in many general works, of facts about cohabitation. Take, for example, the following passage by the future Baron Walckenaer (1837, I, p. 143):

In certain genera, such as the tegenairids, the epeirids [araneids], after the act of generation, the male precipitately departs from the web of the female, who attempts to seize and devour him; but among the *Petitèles*, and the numerous families of the genus *Theridion*, the males cohabit for a long time on the same web with the female before, and after, copulation . . . In certain species of *Epeira*, such as *apoclisa* [later restored as *Araneus cornutus*], the males cohabit with the females, without harming each other.

The student of cohabitation should always consult indexes for entries on sexual cannibalism.

We have dwelt on the importance of distinguishing precopulatory cohabitation with immature females from postcopulatory cohabitation. Other questions, too, must be asked of the literature. Here are some of them. Cohabitation is not the only kind of sexual relation between adult males and subadult females which we may encounter. There are many reports of males' courting immature females (e.g. Montgomery 1910, p. 154; Crane 1949, p. 175). There are even a few, unreliable reports of immature females which have already mated (Montgomery 1903, p. 68). The observations of males courting immature females need not concern us: males not only court immature females, but also males, mature and immature, in many species besides spiders. It probably has no special significance. (It might, of course, be a preliminary to precopulatory cohabitation; but there is no evidence that it is. The suggestion is too speculative to justify transcribing into this summary all the observations of males' courting subadult females.)

Our main problem is that silence about cohabitation cannot be taken as evidence of its absence. When an author does state that immature females do not cohabit, we will record it. But these are only occasional statements. They probably reflect as much the whims of natural history reporting as actual patterns in nature. The reports of mating among adults, unpreceded by a precopula, are anything but occasional. There are hundreds of them, by Peckham and Peckham, Crane, Bristowe, and Locket, Bonnet and Gerhardt, Robinson and Robinson. I do hope that no reader will be ungrateful because I have not transcribed all the names of the species whose mating has been observed. Any observations of mating are, strictly speaking, relevant to our theory. These observations all, as it happens, conform to our theory: the copulation of normal adult spiders is not preceded by a precopula. In the summary I have mentioned only a few examples from each family if pairing has only been seen among adults. I have usually taken these examples from Bristowe (1958), which is an easy source to find. These few examples are not intended as a proof that mating can take place among adults without a precopula. They merely illustrate a principle which it would be tedious, and not worthwhile, to prove fully.

Our main task will be to find out which species have precopulatory cohabitation. The authors whom we have just listed were mainly not interested in precopulatory cohabitation. They have all provided some information on our subject, and I have transcribed it below. But they were mainly interested in taxonomy and the evolution of courtship. They did not attend to matings with newly moulted females, or did not look for them. Most of the observations of Bonnet, Gerhardt, Bristowe, and Locket were made on adults which had been put together in the laboratory, under which circumstances spiders do not cohabit before mating. Cohabitation is particularly boring to the student of courtship because matings with newly moulted females are usually not preceded by courtship: the male does not need to court his helpless partner. We need not be surprised that none of the students of mating in spiders before Jackson (1978, 1982, and many other papers) appreciated the interest of precopulatory cohabitation. Most of the information for spiders is very vague. It is culled from a literature whose purpose lies elsewhere. We learn of males and females cohabiting in pairs, but less often are we told whether the female is immature, less often still whether the cohabitation is precopulatory, ending after copulation, and hardly ever its duration. Many interesting observations have nevertheless been made on spiders. Let us take a look at them.

We shall mention 23 families in all, but we shall only say much about a few of them. We shall, as usual, go through them in systematic order. The first great division (Platnick and Gertsch 1976) in the classification of spiders is between the mygalomorphs and araneomorphs; I have not uncovered any relevant facts about the most primitive group, the Liphistiidae. The classification of these groups is shown in Scheme 3.16. Kaestner (1968) followed the

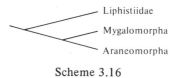

Scheme 3.16

tradition of dividing the Araneomorpha into two suborders: Labidognatha and Cribellatae. But we will follow Levi (1982), who scatters the cribellates among the labidognathan families.

First come the mygalomorphs (which may moult again after reaching adulthood). Not much is known about their mating. The Atypidae build silken tubes and feed on such insects as walk onto it. The male may live with the female in her tube for a long time (Enock 1885; Locket and Millidge 1951, p. 48; Bristowe 1958, pp. 76-7; Clark 1969 for *Atypus affinis*; Enock 1885, p. 391 for *A. piceus*). Enock (by his own account) was completely beside himself when he first found a male in a female tube, and immediately 'ran to the Highgate Post Office, and sent a post-card to the Rev. O. P. Cambridge, apprising him of my success'. But a remark by Bristowe is more noteworthy for us. The males, he tells us, do not enter the tubes of immature females; cohabitation commences only after mating. The pair mate soon after the male enters, although they may stay together all winter afterwards (Clark 1969). In a species of Theraphosidae (*Dugesiella hentzi*, Petrunkevitch 1911, pp. 372-3), as well, immature females will have nothing to do with males, and the brief mating appear not to be preceded by cohabitation. Similarly, in *Cteniza moggridgei* (Buchli 1968, p. 15) only receptive females will admit males, and females only become receptive when they are fully adult. The mygalomorphs, on this scant evidence, lack a precopula. We can conclude that a precopula was originally absent in the group to which the mygalomorphs are the outgroup, the araneomorphs. And to the araneomorphs we now turn.

The araneomorphs can be divided as follows (Kaestner 1968, incorporating the relevant cribellate families from Levi 1982):

Haplogynae	Dysderoidea
	Scytoidea
	Pholcoidea
Entelegynae	Araneoidea
	Lycosoidea
	Clubionoidea
	Thomisoidea
	Salticoidea
	Eresoidea
	Dictynoidea

First, the Haplogynae. They are probably not a monophyletic group (Glatz 1972-3). Little is known about their mating, and they show no sign of cohabitation. The matings described by Bristowe (1958, pp. 95 and 98-9) for the Dysderidae and Oonopidae were between full adults and were not preceded by cohabitation. Similar evidence is available for the Scytoidea (Bristowe 1958, p. 108) and Pholcoidea (Montgomery 1903, p. 116; Bristowe 1958, p. 111). The evidence is even slimmer than for the mygalomorphs: we cannot even say whether immature females allow males near them. If, as is not unlikely, the entire group lacks cohabitation, they would not have appeared independently in the test, because the absence of cohabitation was the primitive state of the group.

We have rather more evidence for the second group, Entelegynae. We will start with six families of the orb-web spiders (Araneoidea). The Nesticidae, Mimetidae, and Tetragnathidae have yielded no evidence of anything more than short matings (Bristowe 1958, pp. 223, 226, and 254-6). In the Linyphiidae long matings have been seen, but they do not seem to be precopulatory, and they are between adults. Emerton, for example, wrote (1878, p. 95): 'In *Linyphia* . . . the male and female live peaceably together for a long time in the same web'. Bristowe (1958, p. 269) likewise remarks, of *Linyphia triangularis*, 'early in August, the male . . . appears . . . in every female's web and for weeks they live peaceably together'. The evidence again points to the absence of a precopula; only full adults form pairs. The same enduring pairs, formed among adults and beginning only just before copulation, are found in the Theridiidae. The ellipsis in the quotation from Emerton a few lines ago concealed only the words 'and *Theridion*': Emerton had seen pairs in this genus too. A pair of *Steatoda bipunctata* put together by Nielsen (1932, I, pp. 198-9) remained together for two weeks until the male died. McCrone and Levi (1964, p. 19) often found an adult male in a retreat with a female (of unspecified stage), separated from her by a silken partition, in *Latrodectus bishopi*. Other reports for this family do not mention cohabitation (Montgomery 1903, p. 104; Locket 1926; Bristowe 1958, pp. 211, 214, 215, and 221). But for the house spider *Archararanea* (formerly *Theridion*) *tepidariorum* precopulatory cohabitation has been reported. In Montgomery (1910, p. 153) we read that 'adult males wait upon the snares of immature females', in Ewing (1918, p. 191) that 'frequently a male was found for several days in the snare of an immature female, waiting for her to reach maturity after which mating would take place', and in Bonnet (1935, p. 354) that if a male chances upon an immature female, 'the male retires to the extremities of the female's web, and waits'.

The remaining family of the six is the Araneidae. Mating without cohabitation has been observed in many species; but we do not know how many of them might also mate after cohabiting if the conditions were right: in other species cohabitation has been confirmed. Let us start with the tropical species, and with one possible case which is all but denied by its discoverer. The species is *Nephila madagascariensis*; the author, Bonnet (1929, pp. 509-11). Bonnet placed some adult males on the webs of immature females. The males initially remained on the webs, suggesting cohabitation to Bonnet; but within a few days all the males had died, probably in the jaws of the much larger females. 'All this [Bonnet concluded] amply demonstrates that the males gain no advantage from their cohabitation with young females and that, on the contrary, they risk succumbing sooner or later to their chelicerae.' We may be tempted to draw a different conclusion. The males had tried to stay with the immature females. They may (albeit with terrible risk) naturally cohabit. The survival of the males probably depends on how much food the female is given, which is an unknown variable in Bonnet's observations. Vollrath (1980, p. 69) saw males of *Nephila clavipes* on the webs of immature females, and Robinson and Robinson (1980, p. 18) state that it is general in the genus. Males cohabit with adults as well as subadults in *N. clavipes*, and Vollrath (1980) measured how long the males spend at any one web site: on average (of some 530 males), a male spends 3.9 days at a site (bigger males stay for longer than smaller males; for comparison, a female stays at a site on average for 9.5 days).[8] These are measurements of how long the sexes stay together; they are not measurements of the duration of precopula. *Nephila*, therefore, shows both pre- and postcopulatory cohabitation. We would predict that males only cohabit before copulation with immature females: we would have to argue that adults only cohabit after copulation; but this has not been proved. Matings have been seen in some species just after the female's final moult. Robinson and Robinson (1973, p. 36, 1976, p. 20) saw copulation then (unpreceded by the courtship that precedes copulation among fully-developed adults) in *Nephila maculata*, and (1978, p. 25) in three more *Nephila* species, and *Argiope aurantia*. *Argiope aurantia* is another species in which males wait upon the webs of immature females (Robinson and Robinson 1980, p. 67). Porter (1906, pp. 341 and 345) saw both *Argiope riparia* and *A. transversa* mate just after female moults. In summary, the evidence from the tropical araneids suggests that precopulatory cohabitation is widespread; we cannot, indeed, assert with any confidence that any species lacks it. We will meet the same problem in the rest of the family.

We turn now to the araneids of the temperate zone. Males and females of the common garden spider *Meta segmentata* may cohabit in pairs for several days (there is a high turnover rate of males: Bristowe 1929, pp. 319-20; 1958, p. 250). Rubenstein (1981) says that 'often for 3-4 weeks before copulation, males can be found in the corner of female webs', but he does not specify whether the males have to wait because the females are immature. The female can only be mated with when there is some food in the web (Blanke 1974), so the male may have to wait for some prey to be caught, rather than for the female to moult. We may conclude that *Meta segmentata* has a lengthy sexual association, but we cannot use it in the test because we do not know when it occurs in relation to copulation. In some other araneids too cohabitation with immature females has not been recorded. In the common *Araneus diadematus*, for example, only pairings of adults been reported (e.g. Gerhardt 1924), and we may be tempted to conclude that if immature females paired in so common a species, it would have been reported. But the arachnological literature hardly supports this conclusion. Another species of *Araneus* provides the best known example of cohabitation among adult males and subadult females. The species is *Araneus cornutus*. Cohabitation is particularly noticeable because the pair cohabit for a long time—all winter—in a special chamber. It is easy to find pairs in the heads of reeds near streams. Walckenaer (1837, II, p. 63) seems to be the earliest record, although Sörensen (1880, p. 173) was the first to record the immaturity of the paired females. Further observations are recorded by Savory (1928, p. 229; 1935, p. 125), Nielsen (1932, I, p. 171), and Locket and

---

[8] I am most grateful to Dr Fritz Vollrath for allowing me to calculate these averages from his original figures.

Millidge (1953, p. 136). The cohabitation of immature females has also been seen in *A. quadratus* (Sörensen 1880, p. 173; Savory 1928, p. 229; Nielsen 1932, I, p. 172), and of females of unspecified stage in '*E. Benigna*' (Staveley 1866, p. 120), and in '*Epeira insularis*' and '*E. trifolium*' (McCook 1890, II, p. 28; these two I presume now have new names). (Montgomery 1910 p. 153 wrote that '. . . adult males wait upon the snares of immature females; and this has been seen by McCook (1890) and me in various epeirids', but McCook did not in fact say whether the females were immature.)

We will conclude that precopulatory cohabitation has evolved at least once in the Araneidae. It has evolved, as the theory predicts, with immature females. Many species have long associations with adult females, but there is no evidence that these are precopulatory.

We may now move on to four of the families of the Lycosoidea: Agelenidae, Lycosidae, Pisauridae, and Oxyopidae. In the Agelenidae, species have been found with short associations among adults for mating, and in others (such as the water-spider *Argyroneta*, Bristowe 1958, p. 205) the male may stay with the female after their first mating. For most species we cannot say whether they lack precopulatory cohabitation with immature females, but the following observation by Campbell (1883, p. 165) suggests its absence in the house spider *Tegenaria guyoni*. He saw 13 matings in all; they did not usually come just after a female moult, but one happened to, although 'up to that time . . . they had taken no notice of each other'. This single remark is inadequate for us to conclude that the Agelenidae have species which lack precopulatory cohabitatin. But there is one species which definitely practises it. The species is *Agelena labyrinthica*. Sörensen (1880, p. 184) found paired males and females of this species, and in one case confirmed that the female was immature. Nielsen (1932, I, p. 110) may have been repeating Sörensen, or may have been recording original observations, when he wrote: 'in the snare, the mouth of whose retreat was divided by a vertical partition into two parts, an immature female and a mature male were sitting one in each part'.

There are no reports of precopulatory cohabitation in the Lycosidae. In *Lycosa pullata*, Bristowe (1958, p. 183) tells us, 'the female often runs about for hours with the male on her back'. All the reports, by Bristowe, Locket, and Montgomery, discuss only pairings among adults. Schoeman (1977, p. 226) measured how long after maturity female *Pardosa crassipalpis* mated. The average delay was 21.8 days (range 10-34). We can conclude that the Lycosidae lack precopulas. The reports of mating among full adults in the Pisauridae (Bristowe 1958, p. 189) and in the Oxyopidae (Bristowe 1958, p. 162) provide no basis for us to conclude whether cohabitation has only not been seen, or is absent. Analogy with the lycosids suggests its absence. Let us turn to the next group, the Clubionoidea.

We shall discuss three families: Gnaphosidae, Clubionidae, and Heteropodidae. For the first, precopulatory cohabitation has been seen in *Drassodes* and *Pros-*

thesima. Emerton (1890, p. 179) first reported it for *Drassodes neglectus*: 'In the early summer [he wrote] a male and a female live together in the nest, the female often being immature'. Montgomery (1910, p. 153) also saw cohabitation, and also saw mating after the female's final moult. Montgomery may also have seen some kind of cohabitation in *Prosthesima acta*: 'A mature male ... was repeatedly observed by me to hold an immature female, and on another occasion to grasp two such at once' (1910, p. 153). Cohabitation of males with immature females, followed by mating after the female's moult, has also been seen in *Drassodes lapidosus* (Bristowe 1929, p. 325; 1958, p. 119; Nielsen 1932, I, p. 78); Bristowe also watched matings among normal adults in his species which were not (it seems) preceded by a precopula. (McKeown 1936, pp. 204-5, also saw male drassids cohabiting with immature females until their moults.)

Cohabitation has also been found in the Clubionidae. Sörensen (1880, p. 174) found male *Cheiracanthium carnifax* paired to immature females. Berland (1927, pp. 18-19) found a pair of *C. siedlitzi*; he put them in a bottle, whence they proceeded to build a shelter of two compartments. The male dwelled in one, the female (who was immature) in the other. After the female moulted, they mated, and the female made as if to leave. 'The female *Clubiona*' McKeown (1936, p. 204) tells us 'constructs a tubular silken nest in which she lives; the courting male erects a small tent close against that of his desired mate. His courting procedure [is to] knock upon the wall of her retreat, sometimes for several days without ceasing, before she finally relents'. Although the stage of the female is not specified, this sounds very like a precopula. Last for this family we may note that Bristowe (1958, p. 131) thought (but without direct evidence) that mating in *Anyphaena accentuata* took place soon after the female's final moult. In the Heteropodidae, precopulatory cohabitation has been observed in the Australian *Isopeda vasta* by Clyne (1971, pp. 244-5). The pair observed by Clyne copulated soon after the female moult, and then stayed together for over a month afterwards. They were, however, being kept together indoors, which might have encouraged, or even forced, them to stay together.

We have little to say about the Thomisoidea. Its single family Thomisidae may lack a precopula. Two reports positively assert that immature females (of *Xysticus nervosus* (Montgomery 1909, p. 562), and of *Dioea dorsata* (Bristowe 1958, p. 123)) drive off adult males. Both authors replaced males with the same females several days after their moults; they then mated.

The mating habits of jumping spiders (Salticidae) have been thoroughly studied. The earliest important paper was by Peckham and Peckham (1889), who saw cohabitation with immature females in *Eris marginata*:[9] 'this is the only species in which we saw males take possession of young females and keep guard

---

[9] The literature on salticids is a field mined by name changes, many of which are very difficult to detect. I have tried to keep the names consistent, and to avoid such nonsense as discussing the same species under different names, by using the list of North American species and their synonyms provided by Richman and Cutler (1978).

over them until they became mature' (p. 50). Two pairs stayed together for about a week (p. 52) before the female moulted and the pair mated. Cohabitation has since then been seen in several more species, although mainly among species of the subfamily Dendryphantinae (see Crane 1949, for classification). Montgomery (1909, p. 559; 1910, p. 153) saw sexual pairs of *Phidippus purpuratus*, although he did not determine whether the females were mature or immature. (He did claim to have found one immature female who had already mated.) The most thorough work on cohabitation in any spider is by Jackson (1978) on *Phidippus johnsoni*. Pairs may form among adults or when the female is immature. Mating among adults is preceded only by courtship, which lasts for about quarter of an hour. The pairs with immature females cohabit in a two-chambered nest, for about seven days ($n=100$, measurements made in laboratory) before the female moults. Snetsinger (1955) reported cohabitation 'for part of a week or more' (p. 13) in two more species of *Phidippus*: *P. audax* and *P. clarus*. He does state (p. 12) that the female mates after her moult, but not whether the cohabiting females were mature or immature.

Cohabitation has been seen in other salticids. Male *Synageles noxiosa* overwinter in the same nests as females who have one moult to go to maturity (Kaston 1948, p. 451). Forster and Forster (1973, p. 132) found two pairs of *Trite planiceps*, with 'a fully mature male closeted in a retreat with a nearly mature female which moulted within two days'. Jackson (1982) found eleven cohabiting pairs of the web-building *Portia fimbriata*. All the females were immature, and so were two of the males. The eleven females made up 41 per cent of the immature females. The duration of cohabitation varied from 2 to 48 days (median 8.5). The males left after the female moulted. Crane (1949, p. 176) caught a male *Corythalia fulgipedia* on the coccoon of a female which was just moulting to maturity; the pair copulated immediately after the moult. Groups of salticids have also been seen, which contain mature and immature individuals of both sexes (Crane 1949, p. 202), or only immature females and mature males (Kaston 1948, p. 479). Jackson (1978, p. 130) suggested that these groups may be analogous to precopulatory pairs. It is not yet known whether mating takes place in these groups.

What may we conclude for the salticids? We can be confident that many species pair both without cohabitation when the female is adult, and with it when the female is immature. In many species cohabitation has never been observed, and their habits remain uncertain. This is a pattern familiar from other families. For the other families it has not usually been possible to conclude that any species definitely lack cohabitation. For the salticids we may be more confident. For salticids, many species have been watched carefully but have never been seen to cohabit. We have quoted the statement by the Peckhams, in which they said that *Eris marginata* was the only species in which they saw cohabitation with immature females. They reported on 15 species in all, and in none of the other 14 has cohabitation subsequently been discovered. Kaston (1948), as we have noticed, recorded one species which overwintered in pairs with immature females. But he listed many other species in which the females overwintered alone, as adults. Crane did see one mating with a female who had just moulted, but she also described it as exceptional for a salticid, and she had herself studied the courtship of 13 species, and reviewed the entire earlier literature. There are not enough observations to support a subfamilial cladistic analysis, but (taken as a whole) they strongly suggest that there are among the salticids species which do not cohabit.

Not much can be said about the former cribellatan families. In two of them (Eresidae (Bristowe 1958, p. 81) and Uloboridae (Bristowe 1958, pp. 83 and 86)) only brief matings among adults have been seen. Cohabitation has been found among the Dictynidae. In *Dictyna arundinacea*, 'the male seems to stay for a month or more in the female's web' (Bristowe 1958, p. 90), and spins 'a rough chamber . . . in which they both live'; but Bristowe does not state whether the female is mature or immature. Jackson (1977) provides us with a more thorough study. He found cohabitation in 10 of 12 species of North American dictynids in which he found at least one adult male (of the two exceptional species, he found only one adult male of one, and only three of the other, so further study would probably reveal cohabitation in these two too). Of the ten species, he found cohabitation with adult females in all ten, and with immature females in eight (in the two exceptions he found only a single pair of each, so further study would probably reveal pairs with immature females). Thus precopulatory cohabitation with immature females certainly exists in the Dictynidae: it can be used, if wanted, in the test. We do not know whether the cohabitation with adult females is pre- or postcopulatory, so it cannot be used in our test.

We have now completed our systematic review of the spiders. What are the main trends? Cohabitation has been found in several families. Typically, in some of the species of the family, the spiders have been found living in pairs. We know more about some of these pairs than others. In some we know that the females are immature, and that the cohabitation is precopulatory; in others, that the females are adult, and the cohabitation postcopulatory. In some of the former, copulation has been seen. In all these cases it took place just after the female moult. The pair then usually separate. We know little about the exact duration of precopulatory cohabitation. In the five species of salticids for which we have at least a hint (and in the two for which we have accurate measurements) it lasts about a week. The overall pattern of mating and pairing among spiders conforms perfectly with the theory: whenever the appropriate observations have been made, precopulatory cohabitation has been found to be with immature females, and cohabitation with adult females to be postcopulatory. All species can pair when the female is fully adult, to mate well after (days or more) her final moult. These matings, which have so often been seen and described, are never preceded by cohabitation.

The general pattern of the literature conforms exactly with the theory. How are we to include the spiders in our test? They cannot make much of a contribution because it is so difficult to be sure that any species lack cohabitation. Our estimates of the frequency of evolution of cohabitation will be hazardous. We will persist with the rule that cohabitation with immature females may be said to be absent if some observer has positively stated that immature females do not pair. If such pairing has merely not been observed we will not conclude that it is absent, unless there are many such observations for a well-studied group. Now we can summarize our knowledge for the spiders, using our modification of Kaestner's (1968) classification (they are also summarized in Table 3.2). Each family in Scheme 3.17 is '+', '−', '?', according to whether it contains species which have been said to cohabit in pairs with immature females (+), said not to (−), or not said to or not to (?). There are two methods of counting from this information. The difference stems from our interpretation of the question marks. We may either assume that cohabitation is really common, in which case it would have evolved early on in the araneomorph line: (a) would have had cohabitation. We would then count cohabitation as having evolved only once (from the common ancestor of the mygalomorphs with the araneomorphs, to the common ancestor of the araneomorphs (a). Then we would count the minimum number of times it must have been lost. That minimum counts some Araneidae, Lycosidae, some Salicidae, and the Thomisidae: a total of four times. The other method is to suppose that cohabitation is rare.

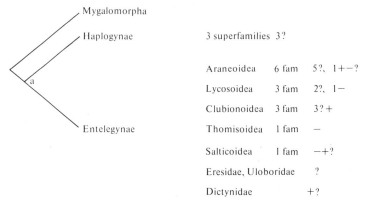

Scheme 3.17

This supposition can be justified: we would expect it to have been found more if it is common: pairs would have been noticed more in field samples. In fact only a minority of observed species have been found to cohabit, which is presumably why Jackson (1977) wrote that 'although it would be difficult to estimate how widespread this phenomenon is, it seems safe to conclude that cohabitation does not occur in the majority of species'. Reason, and the authority of Jackson, combine to support this second method of counting. Now we will suppose that most of the question marks are minus signs, that the common ancestor (a) lacked cohabitation. There are therefore no cases of the loss of cohabitation in spiders. We count only the minimum number of times that it must have evolved. The minimum is made up of some Araneidae, some Salicidae, the Clubionoidea (if we interpret, conservatively, the uncertain species as cohabiters, so the common ancestor of all three families must have done as well), and the Dictynidae (if we suppose their previous separation as Cribellatae to indicate that they are not the sister group of the Clubionoidea.) The total number of evolutionary events is four, one less than the previous method. Let me stress that both techniques almost certainly give a gross underestimate, but it is an underestimate of an unknown quantity. We only admit such evolutionary events that we are forced to: we do not want our test biased by a mass of species from little-studied groups.

Which method of counting shall we use for our overall test? Although the two methods imply almost opposite conclusions about the pattern of evolution in spiders, they do both give almost the same number of evolutionary events. They are similarly conservative. On this count alone neither is to be preferred, so, in a sense, it would not matter which method we used. I prefer the second method for the reasons given above: it seems unlikely that cohabitation would have been overlooked the enormous number of times which the first method implies. It also gives one less entry in the test. But the argument is not essential: we can also appeal to the authority of the expert on the subject, Jackson, which we have quoted. His casual remark may decide us in favour of the second method. Cohabitation has evolved, we conclude, four times in the spiders.

Mites

The other large group of Arachnida to have evolved precopulatory mate guarding is the Acari. The mites practise a much wider variety of ways of life than any other arachnid group. Many are economically important as pests or as agents of biological control of pests. They are small, however, so their biology is not well known. We will have to face once again, and in acute form, the recurrent problems of this chapter: inadequate taxonomy, and the untrustworthiness of negative evidence in little-studied groups. The classification of mites is exceptionally poor for so major a group, with the worst ignorance, as usual, at higher levels. There are (to repeat Kaestner's (1968, II, p. 270n) illustration) more families of Acari than of the largest insect order, Coleoptera, although the Coleoptera contains more than ten times as many described species. The grouping into suborders which (following Kaestner) we shall use is based on the birefringence of the cuticle. The existing classifications are not a stable base on which to stand an analysis, but they are the best we have.

Nor does our knowledge of their reproductive habits amount to much. There are whole large groups for which mating has never been described; and the information which we have is for only about a dozen of the (over 200) families. Precopulas have been described in perhaps eight families, but (as we have seen) we cannot say for sure that any species lack one. To follow the literature we will need to know the names of some of the life stages of a mite. Maturity is preceded by several 'nymph' stages. According to whether there are two or three nymphal stages (it varies between groups), the penultimate stage is called a deutonymph or a tritonymph. At the time of its moult a mite becomes 'quiescent', so adulthood immediately follows a phase termed the 'quiescent tritonymph' (or deutonymph, as appropriate). Males can recognize, and are attracted to, female tritonymphs, by a pheromone (Cone *et al.* 1971a, b; Amano and Chant 1978; adult female ticks attract males pheromonally: Homsher and Sonenshine 1976). Males may guard the quiescent female tritonymphs until the moult. The duration of the precopula is short, from a few hours (our information is vague) to a day or two at most; but the lives of mites are short (the whole tritonymph phase lasts perhaps four or five days), so relative to the life of the animal, the precopula is about as long as those of crustaceans. Mating follows the moult. The form of the mating act itself varies among species. In some the male deposits a spermatophore on the ground, picks it up (usually with his chelicerae), and puts it in the female; in others it is not first deposited on the ground; in others the male has some special organ, a gonopodium or penis, to transfer his sperm. Several forms of sperm transfer may be found even within a single genus (e.g. *Eylais*, Lanciani 1973). But in all species there is insemination, and internal fertilization. The female may remain receptive after her moult; but we have already discussed that. Matings with fully adult females are (as we would expect) not preceded by precopulatory guarding.

The facts with which we must deal are only scattered, unconnected observations. They do not constitute a tradition of research. Measurements of exact periods of receptivity, and so precise testing of our theory, must await the future. (It would not be without commercial interest to know how common is the habit of *Typhlodromus cucumeris* mentioned above: limited period of receptivity independent of mating; unmated females infertile.) But from the scattered observations we can contribute some crude evidence to our test. To this crude evidence we now turn.

The main divisions of the Acari with which we will concern ourselves are (following Kaestner):

Suborder	Infraorder	[Groups]
Parasitiformes	Mesostigmata	Gamasina, Uropodina
	Ixodides	
Trombidiformes	Tarsonemini	
	Prostigmata	Eleutherongona, Parasitengona
Sarcoptiformes	Acaridei	
	Oribatei	

These represent only a part (three of five suborders) of the main groups. Within them, precopulatory mate guarding seems to have been described with reasonable certainty in eight of the families, from three of the infraorders (one from each suborder).

We start with the Gamasina. Precopulatory mate guarding has been seen in some species of the predatory Phytoseiidae, as well as in the Parasitidae, the Dermanyssidae, and *Macrocheles*. Many papers on the life histories of phytoseiids state that the female mates immediately after her moult to adulthood (e.g. Ballard 1954, p. 178 on *Typhlodromus fallacis*; Herbert 1956, p. 702 on *T. tiliae*, El-Badry and Zaher 1961, p. 429 on *T. cucumeris*; Prasad 1967, p. 907 on *Phytoseiulus macropilis*; Lee and Davis 1968, p. 253 on *T. occidentalis*; Zaher and Shehata 1971, p. 396 on *T. pyri*); but do not state whether or not mating is preceded by a precopula. Ballard does tell us, simply, that 'no premating activity on the part of either sex was observed'; but this does not prove the absence of a precopula: Ballard may well have just put males together with recently moulted females, and observed mating without preliminaries. We do have the suggestive remark of El-Badry and Elbenhawy (1968, p. 160) that (in *Amblyseius gossipi*) 'usually mating process took place immediately after the final moult, however males were observed frequently trying to mate with female quiescent deutonymphs'. That 'trying to mate with' is interpretable as 'guarding'. (All Mesostigmata omit the tritonymph stage so the deutonymph is the last stage before adulthood.) For a clear report of precopulatory mate guarding we must turn to the paper of Amano and Chant (1978) on *Amblyseius andersoni* and *Phytoseiulus persimilis*. 'Males of both species [they inform us] were frequently observed waiting near, or sometimes on, female deutonymphs' (p. 199). Females of the two species mate 'as soon as' they reach adulthood. Amano and Chant further remark that the male habit of waiting near female deutonymphs 'is common with many phytoseiid mites'. If we combine this remark with the lack of evidence that any species lacks a precopula, and the apparent universality of mating just after the female moults to maturity, we may provisionally conclude that all phytoseiids have precopulatory mate guarding.

The evidence from a single species, *Parasitus coleoptratorum*, of Parasitidae, points to a similar conclusion. Rapp (1959, p. 309) tells us that in this species mating usually takes place soon after the female moult, although older females can mate too. Males can as a rule be seen gripping the fourth pair of legs of a female with his second pair. The male thus follows the female around for a 'few hours' before the female moults.

The Dermanyssidae may also have a precopula. The snake parasite *Ophionyssus natricis* mates soon after the female moult (Camin 1953). So do other species. The only evidence of a precopula comes from Oliver (1966, pp. 31 and 34), who saw 'adult males riding and embracing female nymphs and adults' in three species: *O. natricis*, *Dermanyssus gallinae*, and *Ornithonyssus bacoti*. The paired nymphs must have been in precopula. But we really need to know how long the adult pairs stay together before we can conclude that these species conform to the theory.

Macrocheles has a precopula. Costa (1967, pp. 320-1) saw a male *M. saceri* leap on a female just as she was about to moult, so there was hardly a precopula at all. But Kinn and Witcosky (1977, p. 141) saw male *M. boudreuxi* with female deutonymphs 'several hours before her last moult'. Copulation took place 'either during or after the female's final moult'.

Precopula may prevail in the Uropodidae as well. Radinovsky (1965, pp. 267-8) states that male *Leiodinychus krameri* may wait with females who have yet to emerge, and Compton and Krantz (1978) inform us that 'female deutonymphs of *C[aminella] peraphora* are attractive to males shortly before female ecdysis'. In *C. peraphora* 'mating proceeds soon after female emergence', and females mate only once.

The ticks (Ixodides) probably all lack precopulatory mate guarding. The literature provides no evidence of its presence; but nor does it provide explicit evidence of its absence: it is silent on the issue. The absence of a precopula can be inferred from some descriptions of mating: we can read about species in which females only become receptive after they have fed (e.g. Arthur 1962, p. 180). and Dr J. H. Oliver has kindly confirmed for me, in a letter, that precopulatory mate guarding is 'not established in several large taxa of acarines'; ticks are one of them. Female ticks of the family Argasidae feed and copulate at the same time, and both activities are necessary between each (of several) ovipositions. The matings reviewed by Oliver (1974) are all among adults, and they are not preceded by a precopula. The sexes stay together for about 10-30 minutes with only a short period of stimulation before mating (Oliver 1974, p. 30). When Feldman-Musham and Borut (1971) put male and female *Hyalomma excaratum* (of unstated stage, so they were probably adults) together, the sexes soon joined up: 'they may remain so for hours before actual copulation . . . but sometimes they copulate after ca. 20 min'. Whether the 'hours' were a precopula can be doubted. Both the circumstances (and even the stage of the female) are so unclear that it is not worth speculating. There does not seem to have been any precopula before the mating of *Ornithodorus moubata* (a troublesome parasite), observed when Nuttall and Merriman (1911) put together 'ready to copulate' females and males. The females were probably adults. The ticks, we can conclude, have no precopula.

We can now move on to the second suborder, the Trombidiformes. Members of the Pyemotidae (Tarsonemini) merit a mention. The extreme incest and female-biased sex ratios of *Pyemotes* are well known. The females, soon after moulting to maturity, mate with their brothers. The act takes place (according to species and circumstance) either inside or just outside the other's body (Cooper 1937; Tawfik and Awadullah 1970, p. 55; Rapp 1972). The male has been waiting continually with his mate since the zygote stage, but this 'precopula' is not clearly a prediction of our theory. The mating system may well have evolved by extension of a once shorter precopula. The force of natural selection identified in our theory may contribute to the evolution of the mating habit of *Pyemotes*. But *Pyemotes* really falls outside the frame of reference of our theory: it neither proves nor disproves: it is best ignored. In this respect it is analogous to such cases as barnacles, and some kinds of permanent monogamy. The theory is relevant, but does not make unequivocal predictions.

The plant-gall mites of the Tarsonemidae have a precopula. The adult male carries the quiescent laraval female behind him, with his hind legs. Such has been seen in *Tarsonemus pallidus* (by Moznette 1917, p. 380, *Hemitarsonemus tepidariorum* (by Cameron 1925, p. 102), and *Hemitarsonemus latus*, the Ceylon tea mite (by Gadd 1946). Only Gadd tells us the time of mating: it was immediately after the female moult. The Tarsonemidae conform to the theory.

We now turn to the other suborder, Prostigmata. The spider mites (Tetranychidae) have been well studied, because of their diet often consists of agricultural crops. According to the reviews (Vitzthum 1940-3, p. 675; Boudreaux 1963, p. 139; Helle and Overmeer 1973, pp. 102-3) precopulatory waiting by males with immature females is certainly usual, and probably universal. No species has been shown to lack a precopula, although plenty have been shown to have one. The spider mites lack a tritonymph stage, so precopulatory pairing is between an adult male and a quiescent female deutonymph. Mating immediately follows the female moult. Such has been recorded for *Panonychus citri* (Quayle 1912, 1938; Ebeling 1959; Beavers and Hampton 1971), *P. ulmi* (Wafa, Zaher, Soliman, and El-Kadi 1967, p. 135), *Tetranychus cinnabarinus* and *T. telarius* (Lehr and Smith 1957, p. 634; Potter *et al.* 1976),[10] *T. marianae* (Moutia 1958, p. 65), *T. urticae* (Ewing 1914; Cone *et al.* 1971, p. 355), and *Eutetranychus sudanicus* (Siddig and Elbadry 1971, p. 807). Most of these authors do not indicate the duration of the precopula, but Lehr and Smith (1957, p. 634) imply a period of about a day, and Cone *et al.* (1971b) of about 1-2 days; Potter *et al.* (1976, Table 2) present a frequency distribution of guarding times which has a mean of 10.34 hours ($n = 776$). Female spider mites may mate when fully adult as well (although such matings may be inconsequential if the female has previously mated: Helle 1967, see above, p. 131). Females of some species are said to mate more than once, others only once; but we should not trust these stated differences much. Even in species in which the females mate only once, the female may (we can assume) mate when fully adult if overlooked by her male conspecifics while she was a quiescent deutonymph. The evidence, in conclusion, is consistent with all spider mates' being able to mate both with and without a precopula, in a pattern which our theory predicts.

A single paper by Spickett (1961, p. 190) on the sheep follicle mite *Demodex follicarum* (Demodicidae) incidentally reconstructs a life history in which the

[10] Potter (1978, p. 218n) informs us that the species investigated by Potter *et al.* (1976) was 'erroneously identified by two tetranychid experts as *Tetranychus urticae* Koch'. It was in fact *T. cinnabarinus*.

female deutonymph moults in her host's follicle, and then waits to be mated. The life history implicitly lacks a precopula, but none of Spickett's observations (he did not observe mating) would be contradicted if it had one.

Precopulatory mate guarding has been seen in the Cheyletidae. Zaher and Soliman (1971, p. 51) state that, in *Cheyletus malaccensis*, 'when the male emerges, it runs searching for a quiescent female deutonymph and settles down beside her. The copulation takes place immediately after the female emerges'. In *Cheletogenes ornatus*, too, mating follows 'immediately after the [female] emerges' (Zaher and Soliman 1971a, p. 87).

In the Tenuipalpidae, mating is said to follow immediately after the female moult in *Cenopalpus lanceolatisetae* (Zaher, Wafa and Yousef 1969a, p. 54), *C. pulcher* (Zaher, Soliman, and El-Safi 1974, p. 369), *Phyllotetranychus aegyptiacus*, and *Raioella indica* (Zaher, Wafa and Yousef 1969b, p. 409). A precopula has not been reported in any of them, unless one re-interprets Moutia's (1958, p. 67) report of copulation in female deutonymphs, 1-2 days before the quiescent phase, in *Raioella indica*.

We will comment on three families from the other division of the Prostigmata, the Parasitengona: the freshwater Hydrachnellae and Eylaidae (the red water mites), and the terrestrial Erythraeidae. The exquisitely detailed description (occupying over 40 pages) of mating in three species of water mites by Böttger (1962) contains no mention of a precopula, but Lanciani (1973) tells us of 'a behavioural adaptation that was observed many times in field and laboratory. Males, alone or in groups, crawl over conspecific females that have not yet left the teleiochrysalis.' One of Böttger's three species was among the 18 species of *Eylais* studied by Lanciani. Precopulas have not been recorded for the Erythraeidae, but Putnam (1966) informs us that adult *Balaustium* aggregate for mating. So there may be no precopula. This, however, is only a suggestion: the evidence is not strong enough for us to use in the test.

We come to the final suborder, Sarcoptiformes. We start with two families of the Acaridei. Copulation with immature females has been thought to be common in this infraorder (review: Vitzthum 1940-3, p. 685). However, the main observation supporting this is just that males are found attached to female nymphs. We may question whether these pairs are actually copulating. Grandjean (1938, p. 289), for example, writing about *Otodectes cynotis*, says that the pairs are not in copulation, but 'a preliminary phase . . . copulation takes place later, with the nascent adult female, when she emerges from her tritonymphal exuvium'. Oudemans (1926) had come to the same conclusion for *Acarus bubulus*. Grandjean thought that 'it is necessary to generalize, and say that, in the Acaridae, copulation never takes place between a male and a tritonymph'. In *Caloglyphus anomalus* (Acaridae) copulation 'takes place a day after the adults have emerged' (Pillai and Winston 1969, p. 301), and males have not been seen carrying tritonymphs. It seems then that at least a few acarids have a precopula and mating after the female moult. Whether we would multiply that few to 'many' depends on whether we would join Grandjean. If, with him, we interpret all those earlier observations of copulation with tritonymphs as precopulas, then there is much evidence of precopulas. If we agree with the original authors, then there is less. The results in our final test are not affected either way, so we will not come to a definite conclusion. Whether few or many species have been shown to have a precopula, the family makes no independent contribution to the test.

Precopulas have also been seen in the other family of Acaridei, the Psoroptidae (mange mites). Gerlach (1857, p. 94) is a possible early report, but for a more definite case we must move on to Downing (1936, pp. 172-3). Of the sheep scab *Psoroptes ovis*, he wrote: 'In cells where there is an equal or greater number of males than females the pairs remained coupled for 48 hours, although during the last 24 hours the female is actually undergoing its moult'. The male drags the female around during this phase. Sweatman (1957, p. 658) gives some more accurate measurements for *Chorioptes bovis*. The male is attached to the 'pubescent' female, for both the active and quiescent phases, by his anal suckers. The precopula lasted an average of 5.35 days ($n = 68$). They copulate immediately after the female moults. Other members of this family have similar precopulas (Evans, Sheals, and MacFarlane 1961, p. 149).

The final group which we will discuss are the oribatids. They serve as a crowning epitome of acarid reproductive biology: it is not known whether they have or do not have a precopula. The only reports, except for a paper by Pauly (1956) which is no help to us, are odd remarks in the taxonomic literature (Taberly 1957). Pauly (1956, p. 282) describes spermatophore transfer, but does not say whether they have a precopula.

What shall we conclude about the Acari? How shall we enter them in our test? The most conservative conclusion, and therefore the one that we shall draw, is that a precopula has evolved only once in all the mites. Only for the ticks do we have acceptable proof that a precopula is absent. Thus we enter them once for the evolution of a precopula, and once for a loss.

These tiny numbers are all that we can extract from the literature. However, they could be very misleading if interpreted wrongly. Dr Oliver informs me that 'precopulatory mate-guarding is exhibited in a minority of species', and I believe him. The numbers are only meant to be the minimum that the literature admits: they are not meant to be biologically reasonable. A more 'reasonable' conclusion might score many cases of the evolution of precopula. However, we have no technique for counting what is 'reasonable': we can only count the smallest numbers that are proved.

ANURA

Among the frogs and toads are the only species outside the arthropods which have precopulas. They typically lay their eggs in a freshwater stream or pond.

At the breeding season, the males, before the females, might migrate to the appropriate pond. Soon after arriving they start, by night, to croak. In species like the common European toad (*Bufo bufo bufo*) the males sit silently in the pond for a few cold nights in April; in others, such as hylids, the male may croak, from a bough overhanging a tropical pond, for many months. The details vary; but the males do generally wait and croak near the oviposition site. They are calling, and waiting, for the females.

Pairs may form between males and receptive, or unreceptive, females. In some species only receptive females pair. A female, if receptive, may approach a calling male; the male (who is the smaller sex) then leaps on the female's back and puts his forelimbs around her in the well-known posture called 'amplexus'. 'The copulation of frogs' as Gilbert White wrote to Thomas Pennant in 1768 'is notorious to every body.' The other kind of pairing is between a male and an unreceptive female. A male may meet a female before she is ready to spawn, on her way to the pond. They may join in amplexus. The male climbs on her back and holds on, despite any attempts by the female to rid herself of her rider. Lengthy precopulas may follow this second kind of pairing.

There are two amplectic postures: 'inguinal' (pelvic) and 'axillary' (pectoral). The pelvic position is used by 'primitive' frogs such as discoglossids and pelobatids; the male grasps the female just in front of her hind legs. The pectoral position is found in more 'advanced' forms, such as ranids and bufonids; the male grasps the female just behind her forelimbs. The difference is distributed almost exactly into phylogenetic categories, but it probably has adaptive consequences. Rabb and Rabb (1963) suggested that pelvic amplexus is less efficient for fertilization (which is external in most anurans): the male and female genital openings are further apart then in a pectoral amplexus, so more eggs may go unfertilized. A comparative hypothesis to explain the difference was suggested by Salthe and Mecham (1974, p. 356): pectoral amplexus may enable the female to move more easily (with the male on her back) when out of water; in a pelvic amplexus the male has to be dragged behind. When underwater a pelvically clasping male can be pulled along, and he may even assist with propulsion. The species with a pectoral amplexus do indeed tend to form pairs when out of water, and the species with pelvic amplexus when underwater. One exception to this *scala naturae*, the midwife toad (*Alyetes obstreticans*), has a pectoral amplexus even though it is a primitive species. It pairs on land. Salthe and Mecham did not formally test their hypothesis, although they did think that it fitted the facts. I would of course argue that a comparative test may not be possible because there is so little evolutionary reversal between the two states.

For our main theme it does not matter what the exact position of amplexus is. What matters is how long the amplexus lasts. Amplexus ends after spawning (except when the sexes are kept together in the laboratory, Wright 1914). The act of spawning may go on for up to a few hours. But how long do they spend together before spawning?

In many species the pair spawn soon (within only minutes or at most an hour or two) after joining up. The theory does predict that if the female is ready to spawn when she meets the male, they should not wait. In others, however, we read of amplexuses that start 'several days', even 'weeks', before spawning. Anurans, it appears, contain both species that possess precopulas, and others that lack one. How can the theory explain that?

The answer, which was given by Wells (1977), is to be found in the length of the breeding season. Let us take a look at the breeding seasons of anurans, and then turn to see how they can explain the incidence of precopulas. We can begin with a quotation from Wells (1977, p. 666):

The reproductive behaviour of anurans can be divided into two basic patterns: prolonged breeding and explosive breeding. Although it is convenient to distinguish these categories, they actually represent two ends of a continuum from single-night breeding in some species to year-round breeding in others. In general, prolonged breeding covers breeding periods of more than a month, whereas explosive breeding refers to breeding periods of a few days to a few weeks.

The length of the breeding season is one of the two main themes in the literature of anuran mating (acoustics is the other). The distinction between short and long seasons dates at least to Boulenger (1896), and has often been written about since (Bragg 1945; Jameson 1955). It is not the easiest of distinctions. There may be many short peaks within what might at first appear to be one long season. There may be different short peaks in different places over the species' geographical range. We will only be concerned with the breeding season at any one place. When we come to our test, we will follow Wells's rough criteria, quoted above, of long and short seasons. But before that, let us spell out how precopulatory mate guarding can be favoured in species with 'explosive' breeding seasons.

A short breeding season is closely analogous to the short period of receptivity (after a moult) of arthropods. There is one small difference. If the period of receptivity is short because of the short breeding season then all the females will be closely synchronized: all the females at the pond breed in the same few days. In the Crustacea, although any one female is receptive for only a day or less, different females become receptive at different times. This difference has no important effect on the theoretical predictions. The mathematics of Grafen and Ridley (1983) actually assumed that the cycles of different females were completely unsynchronized; but this assumption could be relaxed without changing the qualitative, comparative prediction. The theory now runs as follows. If a male frog, of a species with a short breeding season, meets a female making her way to the pond, he can 'predict' that she will spawn soon after her arrival. The conditions for the evolution of guarding are then easy: he should guard if the expected guarding time is less than his expected time to find another female and wait for her to spawn. There is a further force favouring guarding: the sex ratio. It is, at the pond, strongly male-biased: perhaps four, perhaps nine, times as many males are present, which makes for intense male competition (references in Wells 1977, p. 668). Precopulas evolve more easily, and of a longer duration, if the sex ratio is male biased (Grafen and Ridley 1983). So natural selection may well favour a precopula before spawning in species with a short breeding season.

The story is different if the season is long. Now if a male meets an unreceptive female he cannot (unless he has other knowledge besides the breeding season to go on) predict so accurately when she will spawn. She may not breed for months. Nor do these species seem to have strongly male-biased sex ratios and intense fighting between males at the spawning sites. Thus in species with long breeding season the males are less likely to be selected to wait, in amplexus, for days before mating. They wait instead for receptive females to come to them. Then they spawn without delay.

Such is the theory. It predicts a comparative trend by differences in the duration of the breeding season. We may now ask the next question, why do different species have different lengthed breeding seasons? We may . . . but we won't be able to answer it. The obvious answer is that the breeding season is determined by ecological requirements: explosive breeders, for example, usually breed after rainfall in temporary ponds. But the obvious answer has not been properly tested in any species. It is an important question, and Wells even menacingly remarks that 'any future quantitative treatment of the evolution of anuran social systems should include some consideration of these problems' (p. 667). But I have nothing more to say about it, so we can move straight on to the test.

We will test for a relation between the duration of the breeding season and of amplexus. Others have been here before. Wells (1977, p. 672) listed, in a table, these two variables for thirteen species. All thirteen fitted the theory. They may, however, have been chosen more to illustrate the theory than to test it. He mentions a possible exception outside the table, and does not use the table as a proof of the theory. The thirteen came from only three families. We will be able to use rather more facts than Wells. We will also (if we can arrange the anuran facts like those for the arthropods) use them in the overall test. The durations of breeding seasons and amplexuses will have to be split, by arbitrary criteria, into dichotomous categories. The criteria for 'long' and 'short' breeding seasons have already been given. There is a difficulty with the evidence, which is similar to one we have met in crabs, amphipods, and isopods. Seasons which appear to be long may not really be long: they may be made up of many discrete explosions. Boulenger (1896, p. 61), for example, compared the season of the edible frog *Rana esculenta* to that of the natterjack toad *Bufo calamita*: he gave both as 2-3 months. Later research has not altered his facts for the toad, which is still scored as a 'long' season, but has shown the frog to have an explosive season. How do we know that other frogs with long seasons are not erroneously scored? We do not, but we will (like with the crustaceans) trust the facts at their face value, and if all the published observations say that a season is long, then we will score it as long. Sometimes we may be wrong, but we have to risk this sort of error to make the comparative method possible.

For precopulas the usual criterion will do: amplexus must last about a day or more before oviposition to count as a precopula. Amplexus may continue for several hours once spawning has started. Where possible we will use figures for pre-spawning amplexus. Often the authors do not specify what their measurement is of, and then we will treat the figures with less respect. The measurements of the duration of amplexus are actually very poor. It only seems to have been properly measured in two species (*Rana catesbiana* and *R. sylvatica*): most of the observations we will be relying on are casual and unquantitative. The duration of amplexus has been measured in far fewer species than the duration of the breeding season. So we will work through only those species in which amplexus has been measured: in nearly all of them something is known about the breeding season. Many species, whose breeding season is known but whose amplexus is not, will be left out. The habits of the families are, as usually, summarized in Table 3.2 (p. 160).

We need a phylogeny first. Most systems (Inger 1967) agree on a major division between the primitive frogs, and two main lines of advanced frogs and toads (ranoids and bufonoids). The recent cladograms of Lynch (1973, p. 133) and Duellman (1975, Fig. 1) have this overall pattern in common, as shown in Scheme 3.18. Sokol (1977) argues that the pipids are not 'primitive' frogs. He would move the pipid branch somewhere downwards in the cladogram. It will not make any difference to our counting where in the cladogram the pipids are moved to, so we will not have to make a decision about them. Lynch (1973) and Duellman (1975) do differ in how they group the families of bufonoids and ranoids. We will have to choose one of them. It does not much matter which, so let us take Duellman's: it is the more recent. It classifies the families with which we will be dealing as shown in Scheme 3.19. Outgroup comparison suggests that the common ancestor of the Anura had a long season

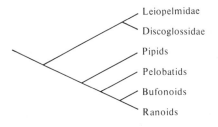

Scheme 3.18

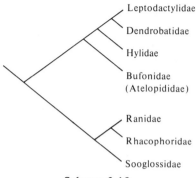

Scheme 3.19

and lacked a precopula. This habit has been retained by most of the species for which we have evidence. Prolonged precopulatory amplexuses only seem to have evolved for sure in three groups (*Bufo, Rana,* and *Atelopus*). But we start with the Discoglossidae.

Alyetes obstreticans, the midwife toad, has a short amplexus. Its brevity is apparent from the classic paper by De L'Isle (1876, pp. 12-13). For a number we must go forwards to Heinzmann (1970, p. 34): 'the clasping lasts for up to 90 minutes'. The season is long, perhaps six months (De L'Isle 1876, pp. 2 and 6; Heinzmann 1970, p. 34). There are many peaks within the total season, but breeding is probably continuous enough to count as a long season. Amplexus is also short (about a minute or two) in *Discoglossus pictus* and *D. sardus* (Knoepffler 1962, p. 69; Weber 1974, p. 43). The season, which lasts from January to September or October in Algeria (Boulenger 1896, p. 136), is again long, and may be made up of many small peaks. Knoepffler saw that a single female lays about four broods a year, at intervals of a couple of months. The three females he watched were not synchronized, and together layed in almost every month of the year. Thus breeding is probably fairly continuous in nature. The discoglossids have long seasons and short amplexuses.

The next family, Pipidae, contains *Xenopus laevis*. It breeds in South Africa for six and a half months (Balinsky 1969, p. 45). Balinsky says that spawning is completed within a night, which agrees with other reports. Most of the time is actual spawning, but there may be a pre-spawning amplexus of an hour or less (Bles 1905, p. 798). The duration of amplexus in another African pipid, *Hymenochirus boettgeri*, is also short. It lasts 3-12 hours depending on when in the day it begins; they normally spawn late at night (Rabb and Rabb 1963). The Rabbs watched their toads in a laboratory in New York. The breeding season is not known. The evidence from the Pipidae suggests that they too retain the ancestral habit.

Not much is known about the Leptodactylidae (which are inhabitants, mainly, of the Southern hemisphere). Three species of *Pseudophryne* have a 'protracted' courtship and then a pre-spawning amplexus of (perhaps) 'several hours' (Woodruff 1976, p. 314); their seasons are all from March to May. This looks like a long season and a short amplexus. *Eleutherodactylus coqui* is similar. Townsend, Stewart, Pough, and Brussard (1981) do not mention the season, but they did collect specimens at a particular early stage of development from January to July. The amplexus lasts about 10 to 15 hours in all (the pre-spawning part is much shorter). Other species also have long seasons, but I have found no more information on the duration of amplexus.

The facts for the dendrobatids are not much better. Most dendrobatids mate without an amplexus. They lay their eggs on land, and fertilize them without the male climbing on the female's back (Silverstone 1976, pp. 7-8). Laying is preceded by a complex courtship, but nothing long enough to count as a precopula. The courtship lasts no more than an hour or two in five species (two *Colostethus*, three *Dendrobates*: Crump 1972, p. 197; Bunnell 1973, p. 283; Wells 1978, p. 151; 1980*a*, p. 196; 1980*b*, p. 203; Limerick 1980, p. 70). Breeding seasons are fairly long: *Colostethus trinitatis* breeds throughout the rainy season, which lasts over a month (Wells 1980*a*, p. 197), and *C. inguinalis* bred in every month except April (Wells 1980*b*, p. 200). The dendrobatids, we can provisionally conclude, have no precopula and a long season.

The Hylidae also have long seasons and short pairings. Thirteen species listed by Wells (1977, Appendix 3) all have seasons of several months. Amplexus, in the six species in which it has been measured, lasts about 5-6 hours or less. *Hyla andersonii* has a season from mid-May to mid-July, and (in one case) an amplexus of less than an hour (Noble and Noble 1923, pp. 423 and 427-8), *H. arborea* a season from mid-April to the end of May, and spawns 'soon after' the beginning of amplexus (Eibl-Eibesfeldt 1956, pp. 383 and 387; Schneider 1973), *H. cinerea* a season for early June to late July, and an amplexus of about five hours (Garton and Brandon 1975, p. 157), *H. versicolor* a season of a bit more than a month, and its amplexus is 'not long', for it spawns on the same night as it pairs (Wright 1914, pp. 46-7), *Pseudacris nigrita* has a calling season from December to March, or even May, and spawns soon after entering amplexus (Martof and Thompson 1958, p. 251), and *Flectonotus pygmaeus*, which may spawn all year, but mainly from February to October, had (in one case) an amplexus of five hours (Duellman and Maness 1980, p. 217). Other hylids may have shorter seasons. Blair (1961) describes the breeding of *Pseudacris clarki* as 'opportunistic'; it breeds in a short interval after rainfall; *P. crucifer* may also breed only in a short period after rain (Gerhardt 1973, p. 84; Rosen and Lemon 1974). In neither of these has the duration of amplexus been measured.

The Bufonidae are more interesting. They contain some species with long amplexuses and short seasons, and others with short amplexuses and long seasons. We also have a phylogeny of the genus *Bufo* (Martin 1972), so an exceptionally informative analysis will be possible. Let us take the species with short seasons first. The common European toad *Bufo bufo bufo* may breed in a single day of a year (Boulenger 1896, p. 64), but the season is usually a bit longer: a week (Huesser 1961, p. 8), or 1-10 days (Smith 1969, p. 100). Pairs may form when the toads are migrating to the pond, and can stay together for 'many days' (Eibl-Eibesfeldt 1950, p. 218); 'if the weather is cold' it may last 'a week or two' (Smith 1969, p. 104). Pairs may also form at the pond, in which case amplexus might only last a few hours. The story is the same in Asia. *Bufo bufo asiaticus* and *B. raddei* have seasons of only two or three weeks and amplexus may last up to ten days (Liu 1930, p. 49; 1931, p. 58). And America has comparable species, although we know less about their amplexuses. In *Bufo americanus* the laying season is about four days (Miller 1909, p. 648; Oldham 1966, p. 71; Paton and Paton, personal communication to Wells 1977, p. 672), and males may join females on their way to the pond (Miller 1909, p. 654). This observation suggests a long amplexus; Paton and Paton told Wells that it lasts 12-24 hours. *Bufo cognatus* spawns explosively for about three days after rain (Bragg 1936). The duration of its amplexus can only be inferred. Bragg (1936) found hundreds of pairs, already together, on his first visit at 9 p.m.; when he looked next morning no eggs had yet been laid, but at nine that evening there were several fresh strings of spawn. The pairs must have been together for at least 24 hours before spawning.

What of the species with long seasons? The natterjack toad *B. calamita*, for example, breeds for about 2-4 months (Boulenger 1896, p. 64; Wells 1977, p. 672; Smith 1969, p. 114). How long is its amplexus? Days or weeks, like the explosive breeders? No, spawning in all takes only 'a few hours' (Smith 1969, p. 116). *Bufo regularis* breeds, in South Africa, from late August to mid-January, but its amplexus (although not exactly measured) is less than 12 hours (Balinsky 1969, p. 45).

The duration of the breeding season, but not that of amplexus, is known in other species. We will be able to use some of them in the cladistic analysis. Martin's (1972) cladogram, reduced to show only the species that matter, is shown in Scheme 3.20 (where S and L stand for short and long). I have included a number of species for which only the season is known, to help with the counting. Wells (1977) gives the references for these species. Now for the counting. The ancestor of *Bufo* we take, by outgroup comparison, to have had a short amplexus and long season. The first line to consider leads to *regularis*. It is most parsimonious to deduce that *regularis* retains the ancestral habit, even though it has many relatives with short seasons (the one drawn in, *marinus* is just one of many, as can be confirmed by looking together at the full phylogeny and Wells's Appendix 5). *Bufo americanus* and *B. cognatus* have both independently evolved long amplexuses. From here on the duration of amplexus has reversed so much between long and short that outgroup comparison becomes difficult to apply. Between *cognatus* and the other species for which we know the duration of amplexus there are forms with both long and with short seasons. The ancestral season of *calamita, raddei,* and *bufo* is difficult to work out. We will suppose that it was long. We have two (feeble) reasons. One is that the

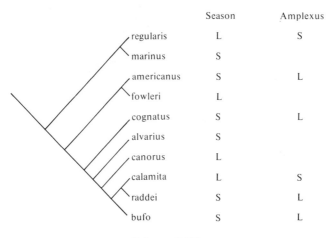

Scheme 3.20

nearest outgroup has a long season; the other that long seasons have generally been the ancestral state in *Bufo*. Thus we count *calamita* as retaining the ancestral, state, and *bufo* and *raddei* as independently evolving long amplexuses. That makes, in all, four cases of the evolution of a long amplexus in *Bufo*.

We will look at the Atelopididae next. Their systematic position is uncertain; they are sometimes included as a part of the Bufonidae. They have a long amplexus. A pair of *Atelopus cruciger* watched by Sexton (1958) stayed together for at least 19 days; a pair of *Atelopus varius* kept by Starrett (1967) spawned at least 19 days after forming. Pairs of *A. oxyrhynchus* (Dole and Durant 1974) spawn after a similar period, but one pair stayed together for 125 days. Pairs of *A. oxyrhynchus* may form long before migrating to the stream where they spawn. Others may form on the day, and others after arrival, but they undoubtedly count as having a long amplexus. And the season? It looks long, Dole and Durant found pairs in the stream from May to July. Two to three months is too long to count as an 'explosion'. *Atelopus* thus appears to contradict the theory. But it may not. Males may guard females only when it maximizes their expected rate of spawning. Our comparison of long and short seasons is a relatively crude test of the underlying theory. But we have settled on a crude method and must see it through: *Atelopus* must be counted as an exception.

Next, the ranoid families. We have the most facts for the Ranidae itself. Like the Bufonidae, it contains some species with long amplexuses and short seasons, and others with short amplexuses and long seasons. Again, we will take the former first. There are six species. First, the common European grassfrog *Rana temporaria*. It may complete its spawning in as little as one day (Boulenger 1896, p. 64), but Smith (1969, p. 129; also Geisselmann, Flindt, and Hemmer 1971) says 7-10 days, and Wells (1977, p. 672) 12-14 days; Gilbert White said that they could be seen paired for a month in the Spring. The amplexus is long. Geisselmann *et al.* (1971, p. 532, quoting Kopsch) say that it lasts days or even weeks; they give one instance of eleven days. Smith (1969, p. 128, following Savage) writes that 'under normal breeding conditions not more than 24 hours are spent in amplexus', and Wells summarizes with '1-12 days'. The season of the second species, *R. pretiosa*, is ten days (Licht 1969, p. 1294), or three weeks (Schaub and Larsen 1978, p. 412); its amplexus 'may go on for days' (Turner 1958, p. 97, one pair stayed in amplexus, in a jar, for six and a half days). Third, the edible frog *R. esculenta*. It probably has a short season. Van Gelders and Hoedemaekers (1971) found that it breeds in short bursts of two to three days, separated by intervals of about ten days. Heusser (1961, pp. 13-14) says that the population bred synchronously, all in a 'few' nights, and Smith (1969) that spawning in a pond was 'more or less simultaneous.' If the season is longer in some places it may be made up of more shorter bursts. The amplexus? Long again: one, two, or more days (Smith 1969, p. 148). Next comes the North American treefrog *R. sylvatica*. Its season is 4-6, or 14 days (Wright 1914, p. 91), or 5-9 days (Banta 1914, p. 177), ten days, six days (Herreid and Kinney 1967, p. 582), or three days (Meeks and Nagel 1973, p. 189): it is explosive (Howard 1980, Fig. 1). Amplexus lasts a 'day or two' (Banta 1914, p. 177), or 1-4 days, although 'most' pairs caught in the field spawned on the same night (Wright 1914, p. 89) or 23.1 hours ($n = 27$, range 1.6-85 h; Howard 1980, p. 708). Fifth, *R. palustris*. The season is about two weeks long (Wright 1914, pp. 62 and 64), and Wright singles it out for the long length of its amplexus. One exceptional pair in his laboratory stayed together for two weeks; they more often stay together for 2-5 days (Wright 1914, pp. 11 and 63). *Rana aurora* may be a sixth. Licht (1969, also Storm 1960) found that it has a short season, of about 15 days. He also suspected that males may mount unreceptive females making their way to the pond; but he did not actually measure the duration of amplexus.

When we turn to species with long breeding seasons, we find, once again, shorter amplexuses. Take *Rana catesbiana* first. It breeds for a month or two (Ryan 1980; see also Blair 1961; Wright 1914, p. 81) although mating can be (at least in one pond studied by Howard 1978, p. 855) concentrated in short bursts of only two or three days. (The concentration in bursts may be normal or it may just be the result of a single year's sample of only about 20 matings.) The complete amplexus lasts nearly 50 minutes (Howard 1978, p. 859, $n = 32$). This species could be omitted from the counting because the season is difficult to classify. The numbers in the final count would not be altered. Two other *Rana* species also seem to have long seasons and short amplexuses: *R. clamitans* breed for two to three months and has an amplexus of 'a few hours' (Wells 1977, p. 672; Wright 1914, pp. 15 and 72), and the African *R. angolensis* breeds from (at least) April to November and pairs spawn on the same night as they form (Balinsky 1969, pp. 47-8). Two species of *Pyxicephalus* have similar habits. *P. delandii* and *P. adspersus* breed from mid-October to mid-

February; (in the former) pairs spawn on the same night as they form, and (in the latter) 'the egg laying started almost immediately after the beginning of amplexus' (Balinsky and Balinsky 1954, p. 55; Balinsky 1969, p. 47).

Three other species of *Rana* are not so easy to classify. *Rana pipiens* probably has a short season and a long amplexus. Wright (1914, p. 54) said that most of the spawning took place in three weeks out of a total of one and a half months; Merrell (1968, p. 275) described the season as 'short'. Amplexus, according to Wright, does not normally last more than a day, but sometimes goes on for two to three days, and one (probably aberrant) pair stuck out five weeks in the laboratory until the male died. The facts for *R. agilis* and *R. dalmatina* are also ambiguous. In both of these, we can read, the season is longer, and the pairing shorter, than in *R. temporaria* (look at Boulenger 1896, p. 65 for one, and Geisselmann *et al.* 1971, p. 532, quoting both Kopsch and Mertens, for the other). This is a trend in the theoretically predicted direction, but is not stated precisely enough to use in the test.

I have not found a phylogeny of *Rana*: so we can draw only the most conservative conclusions for the family as a whole. The common ancestor presumably had a long season and short pairing. Its habit has been retained in *Pyxicephalus*, and some species of *Rana*. A long amplexus and short season must have evolved at least once in the other species of *Rana*.

We will discuss two more families, for completeness only. The information on both is inconclusive. They are the Rhacophoridae and the Sooglossidae. The systematic position of the latter is uncertain (Nussbaum 1980). We will take the rhacophorids first. *Rhacophorus* mainly appear to have long breeding seasons (Liu 1950, pp. 365, 377, and 382). I have found a couple of casual remarks on the duration of amplexus. *R. schlegelii* probably has an amplexus of less than 24 hours, and a season of about one month (Ikeda 1897, pp. 114-15). Of *R. taipeianus*, Liang and Wang (1978, p. 191) merely assert that 'amplexus never exceeds three days', without giving any further hint of the normal duration. The overall season lasts about three months. At any one place, I suspect, it can be much shorter: Liang and Wang found eggs at one site only in early December. A case of explosive breeding with long amplexus cannot be ruled out; but nor can it be ruled in: we will exclude it from the test. Nussbaum (1980, p. 3) found a pair of *Neomantis thomasseti* (Sooglossidae) which stayed in amplexus for a further two and a half days. The amplectic clasp was pelvic, which perhaps suggests that they should be classified among the 'primitive' frogs rather than the ranoids.

What is the total contribution of the Anura? Nearly all the observations fit the theory. Most species retain the (probable) ancestral habit of mating quickly within a long season. So most of the species cannot be counted as independent trials. The frogs and toads contribute, in all, six trials: a long precopula has evolved five times when predicted and once, in *Atelopus*, when not.

Table 3.2 Summary of families

	Mating time	Sexual association
TARDIGRADES	moult	precopula
CRUSTACEA		
BRANCHIOPODA		
Anostraca	moult	precopula
Notostraca	unconfined	no precopula
Conchostraca	unconfined	no precopula
Cladocera	unconfined	no precopula
MAXILLOPODA		
COPEPODA		
Calanoida	unconfined	no precopula
Cyclopoida	unconfined	no precopula
	moult	precopula
Harpacticoida		
Ectinosomidae	moult?	precopula
Harpacticidae	moult	precopula
Tachidiidae	unconfined	?
Tisbidae	unconfined?	precopula?
Peltidiidae	moult	precopula
Thalestridae	moult	precopula
Canthocamptidae	moult?	precopula
Cletodidae	moult?	precopula
Laophontidae	moult?	precopula
MALACOSTRACA		
EUCARIDA		
NATANTIA		
Penaeidea		?
Caridea		
Pandaloideae		?
Alpheoideae	moult	pair
Crangonoideae		? and pair
Palaemonoideae	moult	? and pair
Stenopodidea		pair
REPTANTIA		
Palinura		
Palinuridae		
Stridentes	unconfined	no precopula?
Silentes	moult	precopula?
Astacura		
Nephropidae	moult	precopula
Astacidae	unconfined	no precopula
Callianassidae	?	pair
Anomura		
Diogenidae	variable	variable
Paguridae	variable	variable
Lithodidae	moult	precopula
Galatheidae	moult	?
Porcellanidae	variable	precopula
Hippidae	oviposition	precopula

SUMMARY OF PRECOPULA

	Mating time	Sexual association
Brachyura		
Corystidae	short?	precopula?
Cancridae	moult	precopula
Portunidae	moult	precopula
Xanthidae		
Menippinae	moult	precopula
Xanthiinae		
Pilumninae	hard	no precopula
Panopeinae		
Trapeziinae	?	pairs
	hard	no precopula
Majidae	?hard	precopula
	oviposition	precopula
Leucosiidae	hard	no precopula
Gecarcinidae	hard	no precopula
Grapsidae		
one species	moult	no precopula
the rest	hard	no precopula
Ocypodidae	hard	no precopula
Hymenosomatidae	moult	precopula
PERACARIDA		
Mysidacea	moult	no precopula?
Amphipoda		
Caprellidae	moult	?
Ampithoidae	moult?	precopula
Aoridae	moult	precopula
Cheluridae	moult	precopula
Corophiidae	moult?	? + pair
Melitidae	moult	precopula
Hyperiidae	moult	precopula
Talitridae	(moult)	(precopula)
Hyalidae	?	precopula
Hyalellidae	?	precopula
Gammaridae	moult	precopula
Crangonycidae	moult	precopula
Niphargidae	<17 days after moult	no precopula
Haustoriidae	<4 days after moult	no precopula
Isopoda		
Gnathiidea	?	harem
Flabellifera		
Limnoriidae	moult	pairs
Serolidae	?	precopula?
Cymothoidae	?	pairs
Cirolanidae		precopula
Sphaeromidae	moult	precopula
Valvifera		
Idoteidae	moult	precopula
Asellota		
Asellidae	moult	precopula
Janiridae	moult	precopula
	unconfined	no precopula
Oniscoidea		
Ligidae	moult	precopula

Table 3.2 (*cont.*)

	Mating time	Sexual association
PERACARIDA (*cont.*)		
Trichoniscidae		no precopula?
Porcellionidae	unconfined	no precopula
Oniscidae	unconfined	no precopula
Armadillidae	unconfined	no precopula
Tylidae	moult	precopula
	unconfined?	no precopula
Epicaridea		pairs
ARACHNIDA		
ARANEAE		
Mygalomorpha	unconfined	no precopula
Haplogynae	unconfined	no precopula
Entelegynae		
Nesticidae	unconfined	no precopula
Mimetidae	unconfined	no precopula
Tetragnathidae	unconfined	no precopula
Linyphiidae	unconfined	no precopula
Theridiidae	unconfined	no precopula
	moult	precopula
Araneidae	unconfined	no precopula
	moult	precopula
Lycosoidea	unconfined	no precopula
Cluionoidea	unconfined	no precopula
	moult	precopula
Thomisoidea	unconfined	no precopula
Salticidae	unconfined	no precopula
	moult	precopula
Dictynidae	unconfined	no precopula
	moult	precopula
Eresidae	unconfined	no precopula
Uloboridae	unconfined	no precopula
ACARI (Assume all can pair with no precopula when adult: only those known to have or not to have precopulas are listed)		
Phytoseiidae	moult	precopula
Parasitidae	moult	precopula
Dermanyssidae	moult	precopula
Uropodidae	moult	precopula
Ixodidea	unconfined	no precopula
Tarsenomidae	moult	precopula
Tetranychidae	moult	precopula
Cheyletidae	moult	precopula
Demodicidae	moult	precopula
Tenuipalpidae	moult?	precopula?
Hydrachnellae	moult	precopula
Eylaeidae	moult	precopula
Acaridae	moult	precopula
Psoroptidae	moult	precopula
Oribatidei	?	?
ANURA	(season)	(amplexus)
Discoglossidae	6–9 months	*c.* mins

	Mating time	Sexual association
Pipidae	6½ months	c. mins
Leptodactylidae	2-6 months	few hours
Dendrobatidae	months	c. hour
Hylidae	months	<5-6 hours
Bufonidae	few days	few days
	2-3 months	few hours
Atelopididae	2-3 months	days-weeks
Ranidae	few days	few days
	2-6 months	few hours
Rhacophoridae	c. month	<day

SOME GENERALIZATIONS

We have finished with the detailed habits of individual species. We can now step back and look at some of the main trends. We can also look at some of the main general difficulties of the test. Let us first quickly re-cap the theory.

The theory states that precopulatory mate guarding should evolve when a male can expect to spawn more quickly by waiting with a female until she spawns than by searching among the rest of the females in the population. This theory cannot be tested directly with any published facts. But it does (and not all that indirectly) predict a comparative trend. It relates the incidence of precopulas to the reproductive cycle of the female. If the female becomes receptive for mating at a predictable time, if (in her reproductive cycle) there are sudden changes between receptivity and unreceptivity, precopulatory mate guarding is likely to evolve. Predictable changes in receptivity are common in species in which the reproductive cycle of the female is connected to her moult cycle, or in which the breeding season is short and predictable. If the female is continuously receptive, or her periods of receptivity are unpredictable, then precopulatory mate guarding is less likely to evolve. The comparative prediction allows only an indirect test of the theory. It requires many 'other things to be equal'. We have already met many of those other things, and we shall return to one of them, the sex ratio, later.

Do the facts support the theory? In Table 3.3 I have listed all the independent contributions of the various taxa. The totals are added up at the bottom: of the 20 times that precopulas have evolved, 19 are correctly predicted; of the 11 times that they have been lost, 10 are correctly predicted. The distribution has a probability, by Fisher's exact test, of 0.000 003: the facts do support the theory.

The method of independent trials is, for reasons which have been emphasized enough in Chapter 1, very conservative. The numbers in Table 3.3 are low: 31 in all. But this number conceals the fact that nearly all the literature on the subject fits the theory. Table 3.4 may illustrate this. In it I have listed the

Table 3.3 Independent trials by groups

Animals	Predicted		Not predicted	
	Precopula	None	Precopula	None
Tardigrades	1	0	0	0
Branchiopoda	1	0	0	0
Copepoda	2	0	0	0
Malacostraca	1	0	0	0
Brachyura	1	4	0	1
Other Reptantia	0	1	0	0
Peracarida	3	3	0	0
Araneae	4	0	0	0
Acari	1	1	0	0
Anura	5	0	1	0
All others	0	1	0	0
Totals	19	10	1	1

number of species discussed in the sections on Arthropods and Anura which fit into one of the four categories of the test. (Most species have only one paper written about them, so the numbers can also be roughly read as numbers of studies.) I will not justify the categorization of each species in Table 3.4; the table is only intended to indicate the intensity of documentation. The trials listed in Table 3.3, by contrast, are intended as a formal test, so I have justified every number in it. The number of species listed in Table 3.4 as predictably lacking a precopula is almost meaningless: it is only the number mentioned in the systematic review, but the review missed out most such species (insects,

Table 3.4 Numbers of species fitting and not fitting the theory

Animal	Predicted		Not predicted	
	Precopula	None	Precopula	None
Tardigrada	2	0	0	0
Branchiopoda	1	4	0	0
Copepoda	20	16	0	0
Natantia	3	0	0	0
Brachyura	13	42	0	1
Other Reptantia	6	9	0	0
Peracarida	56	6	0	0
(Araneae	40	46	0	0)
(Acari	47	56	0	0)
Anura	11	23	1	0
All others	0	(100s)	0	0
Totals	199	202+	1	1

Parenthetic entries are uncertain for reasons explained in the text.

molluscs, fish, etc.). A true estimate of the number of species in this category would be (I know not how many) hundreds higher. The other three figures in Table 3.4 are probably fairly complete (for the literature, not for nature). A comparison of Tables 3.3 and 3.4 does reveal one striking similiarity: the number of exceptions in exactly the same (two) in each case, although there are many more species supporting the theory in Table 3.4 than trials in Table 3.3. Thus the method of independent trial has not excluded from the test any exceptions to the theory (although in principle it could have). Because the method has made it difficult to prove the trend, we can be even more confident that the trend is true.

The whole case against the theory rests on those two exceptions. So far these exceptions have been treated with the gentle evenhandedness that is essential in a comparative test. But now that the test is over, we can treat them a little more roughly. Let us see whether they can stand up under detailed examination. What were they again? There was that grapsid crab, *Pachygrapsus crassipes*. All the mating females, Hiatt tells us, had moulted within the previous 12 hours, but 'pre-nuptial pairing or exhibitionism is lacking' (Hiatt 1948, p. 199). But how many was 'all'? We are not told, but the number was clearly small, perhaps only four. Furthermore, it takes careful observation to prove the absence of a precopula. Many other crustacean groups have a precopula in which the male only stays near the female before her moult, rather than physically gripping her. Who dare assert that *Pachygrapsus* lacks one of these? The more likely solution to this exception, however, is that it does not normally mate after a female moult: none of its confamilials do, and they all lack a precopula. Then there was a species of frog, *Atelopus oxyrhynchus*. It may have a long precopula, but a long breeding season. I have already pointed out that this may not be an exception to the underlying theory: males may only enter precopulas when it is advantageous. The duration of amplexus was only measured in a few pairs, and they were singled out for their durability. The breeding season was only casually described. We have seen what complexities can underly an apparently continuous season of breeding. I would not be surprised if further observation brought both these species into line.

The case against the theory could also call on a couple more crabs: *Hyas coarctatus* and *Halicarcinus*. Both were excluded from the test because we know so little about them. They both appear only in Hartnoll's (1969) review. That majid crab, *Hyas coarctatus*, could be credited with an unpredicted precopula. And the hymenosomatid *Halicarcinus* could be credited with lacking one when it ought to have one. But what is the evidence? A single private communication cited in a secondary review for each. We have no idea of the circumstances of observation, or the sample size. This is hardly evidence enough to frighten our theory. Further observations would probably find them out.

The exceptions can thus easily be crumbled away by a little criticism. It might even be asked why they were counted against the theory in the first place. In case it is, I will give the answer. The answer is that if we applied this level of criticism systematically we would precipitate a general disaster. Few of the facts are above criticism: we could probably eliminate nearly all of them. To produce a reasonable test from the available information we have to take the facts more or less at face value. We may look behind our authors' conclusions, and see whether they are really supported by their facts: sometimes they are not; sometimes they support a different conclusion: then we can follow their facts, and not their conclusions. But if we are to produce a comparative test, as distinct from a more or less tendencious compilation, the facts themselves have to be accepted. We are trying to test a theory, not to argue in favour of it. We have to use published facts, even when we do not trust them, in order for the whole method to be possible.

However, once the test has been completed, and the theory found to be supported, there is a reason for looking again, rather less kindly, at the exceptions. If the theory is supported but does not explain everything, it is interesting to know whether there are any more important factors besides the theory under test. In this case, the exceptions are so few, and so poor, that it is likely that there are no other important factors. Nor do the exceptions share any feature which might also matter. The point can be made backhandedly by anticipating the conclusion of the next chapter. The theory under test in Chapter 4 will be supported by the facts; but it has many exceptions, exceptions which do stand up to criticisms, exceptions which do suggest other important factors. The fate of the theory of Chapter 4 is much commoner in comparative biology than that of the theory of this chapter. Our theory of precopulatory mate guarding has been astonishingly successful. Comparative hypotheses nearly always have plenty of good exceptions, and simple hypotheses nearly always end up having to admit 'other factors'. Yet it appears that differences in reproductive cycles may be the only factor controlling the incidence of precopulatory mate guarding.

The test of the theory is now complete. I want to discuss two more questions before concluding. First, did the facts fit easily into the test's arbitrary dichotomy? Second, are any other factors (besides the continuous receptivity of females) associated with the loss of precopulatory mate guarding?

The test was dichotomous; the underlying variables were not. We used a couple of arbitrary criteria to divide up the continuum. A precopula, to count as such, had to last about 24 hours or more. The criterion should be relative to the total reproductive cycle. so, more strictly it was 24 hours if the total cycle is about a month or more. We realized at the outset that there might be a grey area between a long courtship and a short precopula. But how often in fact have we met such species? Hardly ever: precopulas are nearly always obvious. The simple criterion, of 24 hours, is usually enough without worrying about the total cycle. The mites were the main group for which we had to take the length of the moult cycle into account. The tropical amphipod *Melita* was another case, but we used an objective method (the graph in Fig. 3.2, p. 60) to classify it. In a few species, such as a couple of frogs (*Rana*), the precopula was of ambiguous duration. Interestingly, in these two, the other variable (the breeding season) was also in the grey area. Thus both species fit the comparative trend, but they fit it in a region where they cannot be used in the test.

The other variable, the reproductive cycle, was more difficult. We have dealt in several sections (Peracarida, Brachyura, Anura) with particular difficulties. The variable in the theory is the predictability of periods of female receptivity to mating: the variable in the test is the kind of intervals when females are receptive: the actual facts we finished up using were often even further removed from the theory. The best information for most crustaceans is simply whether or not they mate after a moult. In some species of amphipods and isopods the interval of receptivity has been lengthened compared with related species. In most of these cases the precopula has been lost as well. Again we have a trend which the theory predicts, but we have had to exclude the species (except one, *Niphargus*, in which the lengthening is extreme) from the test. Like the two *Rana* species mentioned above, these observations have been excluded solely because they do not come down one side or the other of the arbitrary criteria.

We will finish by looking at some other possible generalizations about the loss of precopulatory mate guarding: sex ratio, living in the open water and interstitial sand, terrestriality, sexual dimorphism. Take the sex ratio first. It can modify the theoretical predictions. Precopulas (other things being equal) are more likely to evolve if the sex ratio is more male-biased. The sex ratio can alter the evolutionarily stable durations of precopula: precopulas are longer as there are more males per female. An earlier sample from the literature (Grafen and Ridley 1983) suggested that species with permanent sexual associations tend to have equal, or male-biased sex ratios, while species with only temporary precopulas have female-biased sex ratios. The sex ratio may help explain why some species have permanent, and others temporary, sexual associations. It is only one factor of several: Menzies (1954), for example, suggested that isopods with permanent monogamy tend to lack sperm-storage organs, while those with temporary precopulas possess them. The references brought together in Grafen and Ridley (1983) were insufficient for a formal test: they were only meant to be suggestive. Much more information on the sex ratio exists in the literature, but I have not included it in this summary. First, the facts are not good enough to support a formal test: second, here we are not trying to explain differences between the quantitative durations of the precopulas of different species; but only its qualitative incidence. We have therefore assumed that sex ratios are fairly similar between species: variation in sex ratio is relatively unimportant relative to variation in the reproductive cycles of females. The sex ratio does lurk in the background as a factor which might, if sufficiently extreme, become of over-riding importance. Let us look at one case where it may matter. It is hypothetical. It is the case of species which form breeding aggregations. The mysids are an example. Their mating habits are so surrounded by uncertainties that they were excluded from the test. Female mysids may only aggregate when they are very close to the time when they can mate. The males would then experience a hugely female-biased sex ratio, and so would not be selected to enter into precopulas. In terms of the original model, there is another way of explaining this effect. The original model assumed that males meet females at random throughout their moult cycles. If females only come out to breed when they are about to moult, this assumption is violated. The predictions are altered: precopulas are not expected.

Species, like the amphipod *Corophium*, which live in tubes in the bottom sand, may only briefly leave their tubes when they are ready to mate (Forsman 1956). Then (as in mysids) we would not predict a precopula. But some tube-dwelling forms do definitely have precopulas, and we can make another generalization about these. In their precopulas the male does not physically grip the female; there is no precopulatory 'riding position' like in *Gammarus*. As we have noticed before, this habit may explain why precopulas are not often found in these species: they will not show up in crude samples.

Another generalization, which has often been mentioned for crabs (Bliss 1968; Schöne 1968; Hartnoll 1969), is that terrestrial species lack precopulas. Terrestrial crabs mate, after only a courtship, when the female is hard; aquatic species mate, after a long precopula, when the female is soft. This generalization holds for nearly all the crabs which have lost their precopulas (in Table 3.2); it is true of all of the independent losses contributed by the crabs to Table 3.3. What about the other four losses? Crayfish 'No: they are aquatic. Isopods and amphipods? *Niphargus* is cavernicolous, and we will come back to that; but the other two losses are also in terrestrial species: *Talitrus* and all the terrestrial isopods. Why this convergence among crabs, amphipods, and isopods? The obvious answer is water loss. Moulting is a dangerous time for terrestrial crustaceans: they are liable to dry out. *Talitrus* does still mate after a moult, but it is not as fully terrestrial as the isopods. It is only a small taxon among many aquatic relatives: perhaps it has only recently taken to the land: perhaps its mating is becoming uncoupled from its moulting.

There may be a trend towards the loss of precopulas in caves. *Niphargus*, as we have seen has lost its precopula. A cavernicolous isopod *Caecosphaeroma burgundum* may also have relaxed the interval when mating is possible after the female moult; but it has retained a (very long) precopula. I have been struck by the absence of any observation of pairs in general papers on the yearly cycle of other cavernicolous forms. In a species like *Asellus aquaticus*, pairs show up in graphs of the frequency of different states of individuals through the year In a cavernicolous *Asellus*, however (Henry 1964), no pairs were reported. But we are straying into a kind of evidence, negative evidence, which I have kept out of this review. It is too unstable to build an argument on.

The last factor which may be associated with the loss of precopulas is sexual dimorphism. As a rough rule for amphipods, precopulas are absent in species in which females are the larger sex. I do not trust this association. But I have not compiled all the facts bearing on it. For a start it is probably associated more with the loss of the 'riding position' kind of precopula than with the real loss of precopula. The male is smaller in *Crangonyx* and he does not sit on the female's back before mating; but that does not stop him from staying near her. Secondly, the association (if it exists at all) is confined to amphipods. Of the groups with precopulas, the males are the smaller sex in some spiders, some mites, nearly all frogs, some crabs, some copepods, and some isopods. Indeed many of the species show that fascinating phenomenon discussed by Ghiselin (1974), male 'dwarfs'. Tiny males form precopulas with females in some parasitic copepods, parasitic isopods, spiders, and *Emerita*. We can at least conclude that if there is any relation at all between precopulatory mate guarding and sexual dimorphism, then it is not at all simple.

The main comparative analysis of this chapter did not aim to discover what factors drive the evolution of the diversity of female reproductive cycles. We cannot yet say what effects cavernicolous, terrestrial, and interstitial habits exert. We aimed to answer the prior question of whether differences in the reproductive cycles of females explained the incidence of precopula. The answer to that question we can confidently pronounce: On the evidence available, it certainly does.

4 On being the right sized mates

INTRODUCTION

Snapping shrimps owe their name to the pops which they let off by snapping their claws together; they can be heard in shallow waters all around the tropics. They live as adults in monogamous pairs. *Alpheus armatus*, for example, whose mating habits have been watched by Knowlton (1980), lives in pairs on a particular species of anemone in the shallow waters of Discovery Bay, in Jamaica. Knowlton collected up dozens of those pairs, and measured them. She then displayed the sizes of the pairs on a graph, with male size along one axis, an female size up the other. Each pair is a single point. The graph of points for pairs of snapping shrimps always shows a correlation of the sizes of mates: bigger males pair with bigger females, smaller males with smaller females. Such a graph could (in principle) show any of three patterns: homogamy, random mating, and heterogamy. Snapping shrimps are homogamous. Homogamy means 'like mates with like'; it is the prior synonym of assortative mating.[1] Heterogamy is its opposite; it means that unlike forms pair, big males with small females, small males with big females. In fact no examples of heterogamy for size seem to exist. This chapter will be concerned with the comparative incidence of homogamy and random mating. We are after a law to explain which species mate homogamously, and which ones randomly.

Homogamy has been investigated in all the great traditions of research on mating habits. Let us start by going through these traditions, in chronological order. We need to know the theoretical interests of the biologists whose work we will be drawing on. Then we will know when their observations fit in with their ideas, and when they do not. We need, in addition, to know the main alternatives to the theory which we will be testing. A knowledge of history is also valuable to the comparative biologists simply as a research technique: it helps in the actual work of reading old papers.

Homogamy was only discussed for one species, man, before the twentieth century. Even for man almost no facts were collected before Pearson's research at the turn of the century. The absence of facts had not prevented the development of an almost proverbial belief that in humans 'opposites attract one another'. Darwin was aware of this. In 1837 he wrote in his B notebook (p. 6) 'In man it has been said, there is an instinct for opposites to like each other' (de Beer 1960, p. 42). Human heterogamy was later to form a minor part of the opposition to Darwin's theory of evolution. Heterogamy would tend to

---

[1] 'Assortative mating' is now the more usual term. I prefer homogamy, which has priority, is shorter, is etymologically preferable, and more populist: it is in more dictionaries than is 'assortative mating'. Their antonyms are 'dissassortative mating' and 'heterogamy'. This use of homogamy should not be confused with its botanical meanings.

preserve the type, and prevent evolution. Thus Murray (1860, p. 277) wrote that 'it is a trite to a proverb, that tall men marry little women ... a man of genius marries a fool', a habit which Murray explained as 'the effort of nature to preserve the typical medium of the race'. The same thought was expressed by the vast intellect of Jeeves, to explain the otherwise mysterious attractions of Bertie for all those female enthusiasts of Kant and Schopenhauer. The source of this proverbial belief is not certainly known; but one possibility can be ruled out. It did not originate in observation: humans mate homogamously (or perhaps randomly) for both stature and intelligence. As Darwin wrote of Murray's paper to Lyell 'it includes speculations . . . without a single fact in support'.[2] Darwin again stressed the lack of evidence of selective mating in humans in *The Descent of Man*. It has never been the case, he wrote, that 'certain male and female individuals [have] intentionally been picked out and matched, excepting the well-known case of the Prussian grenadiers' (1894 edn, p. 29). The Prussian grenadiers were renowned for their great stature, which (it was believed) was enhanced by selective breeding. The selection was personally supervised by King Friederich Wilhelm. For the King, Dr Johnson tells us in his biography, 'to review this towering regiment was his daily pleasure; and to perpetuate it was so much his care, that when he met a tall woman he immediately commanded one of his Titanian retinue to marry her, that they might propagate procerity, and produce heirs to the father's habiliments'.[3]

Homogamy interested Darwin because of its possible importance in speciation. He introduced the subject in the *Origin* as part of his argument that geographical isolation is not crucial to speciation. He summarized the evidence for homogamy for several traits, though size was not among them.[4] It also mattered to Darwin because of his difficulty with 'blending' inheritance. Variation is removed rapidly by blending, but homogamy would slow down the removal. If heredity 'blends', Darwin's critics (such as Fleeming Jenkin) pointed out, then rare advantageous mutants would be swamped out of existence by interbreeding with the majority type long before natural selection could increase their frequency. We need not be surprised, then, when we see that it was the biometrical geneticists that did the first important work on homogamy. They generally held blending theories of heredity, thought natural selection the cause of evolution, and homogamy necessary for speciation. They were also the vanguard of statistics. Karl Pearson, the inventor of the correlation coefficient, measured

[2] Darwin to Lyell, ?4 January 1860, in F. Darwin (ed.), 1887, Vol. II, p. 262. W. H. Harvey, a botanist, raised the same objection in reviews of the *Origin* (1860, 1861). Their are more remarks on homogamy in Darwin's correspondence. See, for example, F. Darwin and Seward (eds.), 1903, Vol. I, pp. 202, 272, 308-9, and 333; Vol. II, p. 232.

[3] In *The Works of Samuel Johnson* (Oxford, 1825) Vol. VI, p. 436. The biography was actually of Frederick the Great.

[4] C. Darwin, *The Origin of Species*, ed. J. Burrow, Penguin, Harmondsworth, 1969, pp. 149-52. There was a longer passage in the big species book (Stauffer, ed., 1975, pp. 257-9), which was largely repeated in Darwin (1868, pp. 102-4).

the heights of husbands and wives (in humans). He found low but positive correlation coefficients (Pearson 1899, Pearson and Lee 1902). A number of other biometrical studies of homogamy appeared at the same time. The two largest, on *Paramecium*, were by the American biometrician Pearl (1905, Pearl became a Mendelian in about 1910) and Jennings (1911). But when Mayr (1947) came to review the evidence for homogamy in a critique of the theory of sympatric speciation he did not find much. In this early literature there is little sense of diversity, of the idea that some species may be homogamous and others not. Alpatov (1925), Spett (1929), and Mayr all wrote as if homogamy were either true (of all species) or not.

The next phase of research was inspired by a particular idea about the mechanics of mating: the 'lock-and-key theory of copulation'. It too has a close connexion with the theory of speciation. It was originally the theory that species are isolated by the mechanical impossibility of fitting their genitals together (Dufour 1844; Jordan 1896; Krauss 1968). It was later extended and applied in many directions. Jordan (1896, p. 520) was to apply it to variation within species. From there an explanation of homogamy was not far off. Pearl (1905) and Tower (1906) explained homogamy in *Paramecium* and Colorado potato beetles respectively by the mechanics of mating. This theory was to be used most extensively by Crozier (1918 and elsewhere) and Pomerat (1933 and elsewhere). Their theory had two requirements: genital size must vary allometrically with body size, and there had to be quite a tight genital fit at copulation. Then big males would be mechanically forced to mate with big females. Pomerat developed this idea into a true comparative hypothesis. He reasoned that species with hard exoskeletons would be more constrained mechanically than species with soft outsides. Similarly, species with external fertilization would be unconstrained. He collected new evidence for toads (soft, external fertilization), *Limulus* (hard, external fertilization), and a beetle (hard, internal fertilization), as well as reviewing the earlier work. It all fitted his theory (Willoughby and Pomerat 1932).

Pomerat's is a different, and competing, hypothesis from the one that we will be mainly concerned with. But we have not heard the last of it. At the end, we will see how well it stands up to all the latest comparative evidence. We will then have something more to say about it.

The next important idea comes from Lorenzian ethology. In his classic paper on the social behaviour of birds, *Der Kumpan in der Umwelt des Vogels* (1937), Lorenz argued that in species with what he called 'labyrinth-like' mating habits, the male of the pair must be dominant to (and therefore usually larger than) the female. In other species, with 'cichlid-like' habits, the female of the pair could not be much smaller than the male because of certain behavioural requirements at pairing. We will come back to those 'requirements' when we reach the cichlids below. Lorenz directly inspired several studies of the dominance relations necessary for pairing in cichlids (e.g. Oehlert 1958). Later Barlow (1970) did further more systematic experiments on the success of pairing in relation to the relative size of the male and female. Similar kinds of observations have been made on other speices. They are summarized in the appropriate sections of the main summary; I have called them 'Lorenzian' observations.

The latest and most prolific phase of research dates mainly from the midnineteen seventies. It has been stimulated in one way or another by the theory of sexual selection. We have no need to go into all the various ideas: none of them are exactly the same as the theory we shall be concerned with, but most are closely related to it. Let us now state the hypothesis which the chapter will test.

HYPOTHESIS: THE SEXUAL SELECTION OF HOMOGAMY

The incidence of male choice is the crucial variable of the theory. Female choice is heard about more often in discussions of sexual selection than is male choice. Female choice, indeed, is probably much commoner in nature than is male choice. Males are not usually selected to be choosy about their mates because they invest so little in each mating. If a male puts no more than his sperms into a mating, he will be selected to mate with as many females as possible. A mutant choosy male, who chose not to mate with certain kinds of female, would be selected against. But what about a species in which the males invest substantially in each mating? Now our reasoning does not apply. If some females are more profitable mates than others, males may now be selected to choose to mate with them. Larger females, for example, may lay more eggs than smaller females. Then a male who mates with large females might be favoured. The conditions for the selection of male choice are in principle easy to specify, if not to measure. The male can expect to fertilize a particular number of eggs by a female of a certain size. If, during the time he would spend with that female, he could expect (were he not to stay with her) to meet, sufficiently quickly, a more fecund female, the male would be selected to choose not to mate with the smaller female. Natural selection acts to maximize the rate at which a male fertilizes eggs. If the sexual association lasts a long time, males will be selected to prefer to mate with large females.

So far we have two necessary conditions for the evolution of homogamy: a long enough mating, and a trend towards increasing fecundity in larger females. But they are not sufficient. One more condition must be met: larger males must be at an advantage in the competition for mates. With only the two previous conditions, all the males are choosing larger females, but they are all also equally good at effecting their choice. If small males are as good at holding on to females as are large males, the result is random mating. If larger males are better at holding on to females, the largest males will monopolize the largest females, so only the smaller females will be left for the smaller males. Homogamy results.

Other causes of homogamy are possible. We have seen one, the theory of mechanical constraints, in the last section. That is not the only alternative. There are others from the theory of sexual selection. Female choice is the favourite. Homogamy could result if different sized females preferred different sized males. I, however, have left female choice out of the main theory. And I can tell you why. I have two reasons. A comparative test is only possible if a comparable kind of evidence is available for many species. What kind of evidence do we have for female choice? For a start, we do not have any direct evidence. The same is also true of male choice. In the absence of direct evidence we have to fall back on indirect evidence. For male choice there is indirect evidence: if larger females lay more eggs then males will be selected to mate with them. Fertilizing more eggs, we can assert, will definitely be selected for. The relation between size and fecundity has been measured in many species. But where is there an equivalent indirect case for female choice? Males can offer nothing comparable to eggs. In some species bigger males may have better genes, or defend better oviposition sites, or be better parents. But all these add up to only a plausible case: they are less convincing than eggs. Anyway they are not available for enough species to support a comparative test. (A third, weaker reason for ignoring female choice is that it will usually support rather than contradict the trend predicted from male chice.)

The theory needs to be modified for hermaphrodites. In a hermaphrodite with reciprocal fertilization and in which larger individuals lay more eggs, a long duration of mating is not necessary for the natural selection of homogamy. An individual invests substantially (all its eggs) in mating, so there will be selection to be choosy about its mate. It should choose as large a mate as possible.

The theory will apply, *mutatis mutandis*, to many other characters as well as size. It might apply, for example, to claw length, if claw length is correlated with fighting ability and fecundity.

To summarize, the theory is that homogamy for size will be found in species that have the following three properties: (i) larger females lay more eggs; (ii) larger males have an advantage in competition for mates; and (iii) the duration of mating is long. This theory is not original to me, although I do not think that it has been stated as a comparative hypothesis before. Many of the recent papers which will be reviewed later contain similar arguments.

METHODS

We shall be examining four variables: homogamy for size, the duration of mating, and the relation between size and fitness in males and females. Homogamy has not been studied as much as the other three variables. We will work through all the species for which something is known about homogamy for size, but we will consider the other three variables only for these species. So we will not be considering all that is known about mating duration, and the relation between size and fecundity, or success in fighting. It is not as easy to compare the four variables in different species as it might at first seem. Nearly all the facts in the literature were collected for purposes far removed from ours. (So when, in the systematic review, I bluntly dismiss some measurements I do not intend any disrespect for the research. 'This research is no use' only means 'it cannot be used to test the theory'.) The measurements which we will review were collected by many different investigators over about three-quarters of a century, and were presented in many, frustratingly different, forms. Here, for each of the four variables in turn, are some comments on how we will try to make them commensurable. The comments implicitly reveal the conditions for including a study in the final test of the theory.

The relation of female size to fecundity is the easiest of the four. The relation has been measured in many of the species in the summary, or at least of close relatives. One possible difficulty is that although larger females may lay more eggs, they may not lay more at any one mating. A larger females may mate with more males instead. When we know something relevant I have mentioned it, but usually we do not, so I have just ignored it.

A trend towards increasing fecundity with larger size is so common that we can assume that it holds true even in species for which it has not been proved. At the end we can do two tests. The first will include only the species in which the relation is definitely known. The second will also include the species in which the other variables are known, but fecundity does not happen to have been measured. The extra species, we will assume, have a positive regression of fecundity on female size. The second test will include more species; but will be slightly less certain.

The relation of male size to success in mate competition. We will use two main kinds of evidence. The first is direct observation. Let two (or more) males, one larger than the other, be put together with a female. It can be seen whether the larger males tend to win the female. If they do, we will count it as evidence that larger males are more successful. The other kind of evidence comes from population statistics. Let the sizes of many males, both paired and unpaired, be measured in a natural population. If the average size of the paired males is greater than the average size of the unpaired males, then we have evidence that the larger males are at an advantage. The evidence, however, contains a slight ambiguity: it is only valid if the different sized males mate for the same amount of time. If, for example, larger males mated for longer than smaller males then the average size of the mating males (in a sample taken at any one time) would be higher than the average size of the non-mating males, even if the larger males were not mating with any more females. Pairs with larger males would be more likely to be sampled. But do different sized males mate for different times? Not much is known. In a couple of species, the amphipod *Gammarus* (Ward, in preparation) and the isopod *Asellus* (Ridley and Thompson 1979) it has been shown that larger males pair for longer; in the king crab

Paralithodes camtschatica 'small males were observed by divers to be grasping their female partners 56 per cent of the time as compared to 80 per cent for large males' (Powell, Shafford, and Jones 1973, p. 85). In the beetle *Brentus anchorago* (Johnson 1982), and the frog *Rana catesbiana* there is no relation (Howard 1978). Such is all we know, for the species under review; we do not know if the effects are large enough to bias the population statistics. We do not even know why the effect exists: one theory suggests the opposite trend (Grafen and Ridley 1983). We do not know enough about the ambiguity to take it into account. We will therefore ignore it.

A higher fitness in larger males is, like the analogous trend in females, very common. Once again, therefore, we can do two tests, one which includes only the species we are certain about, and another including the species in which the relation between male size and fitness does not happen to have been measured. When the relation is not known we can assume that it is positive. I will actually stand my conclusions on the second, more inclusive but less certain, test. The test, therefore, boils down (for the most part) to a correlation between mating duration and homogamy. I have two comments on this. One is that the test only follows nature. If, in most species, larger individuals have higher fitness, then that is how they must appear in the test. The test may have four potentially variable characteristics, but that does not mean that all four will necessarily vary much in real species. The second comment is that the relations between size and fitness sometimes are not positive, and they have then contributed crucially to the test (sometimes to support the theory, sometimes not). So although mating duration is the main variable, the other two do matter. And because they sometimes do matter, I may sometimes have scored species wrongly for or against the theory in the second, looser test. These errors are likely to be rare. Positive relations between size and fitness are so common among species in which they have been measured that it is unreasonable to suppose that they are not common as well in species in which they have not been measured. But why bother with the uncertain species? Why assume a relation when it has not been proved? The justification is that we want to include as much evidence as we reasonably can in the final test. More evidence makes the test more convincing.

Duration of mating. Mating here means the total sexual association: all the time the sexes are together for reproduction. This may include precopulatory or postcopulatory mate guarding, courtship, and parental care as well as copulation. The theoretically important variable is the amount of time for which a male cannot be seeking or mating with another female. The measurement is easy to understand: it is just time. But we have another difficulty. The theory (as we are using it) is unfortunately vague. It does not specify exactly what is a 'long' mating (perhaps leading to a prediction of homogamy), and what is 'short'. The underlying theory is capable of predicting, if enough facts are fed in, whether there should be homogamy. But we do not have enough facts. We have, at best, only rough measurements. If we are to proceed with a comparative test we will be forced to use predictions that are uncertain. We can say that if mating takes a long time, such as more than a week, then (if the other conditions are satisfied) homogamy is more likely than if it takes only a minute or two. Fortunately, in many species mating is either unambiguously long (such as with permanent monogamy), or short (if the sexes meet only for a short courtship and mating and then separate). For a 'long' mating we will use the same criterion as we used for a precopula in Chapter 3: a long mating is one that lasts more than a day. We will meet some ambiguous cases and we will take them individually as they arise.

Homogamy for size we will try to quantify as a product-moment correlation coefficient (r). Sometimes this statistic was calculated in the original paper; sometimes I have calculated it from the original measurements; sometimes I have calculated it from an incomplete set of measurements, in which case my procedure and assumptions are stated in the appropriate section. I have not stated when I have simply repeated a correlation coefficient, and when I have calculated it. Sometimes I have not been able to calculate a correlation coefficient. Then I have settled for the best statistic possible. I have also included any qualitative reports of homogamy that I have come across. These have been included for completeness only: they are too vague to use in the final test. Similarly, I have included any 'Lorenzian' experiments on the success of pair formation among different sized animals. These cannot be used in the main test; but the theory does predict which species should show Lorenzian difficulties in pair formation, and which should not. This prediction we can test.

The theory applies to homogamy within a single population at any one time. The correlation coefficient for the sizes of pairs should therefore be for a single sample, at one time and one place. No animal should be sampled more than once. Some of the correlation coefficients in the literature have been obtained after combining samples from more than one time or place. These statistics may be useful for some purposes but they are not for ours. Our theory is for mate choice, so the test must avoid any 'correlations' due to differences between populations in time or space. When an author has explicitly published measurements for more than one time or place I have calculated an appropriate statistic. The danger is that some of the published statistics may conceal a combination of several samples beneath a facade of statistical propriety. Invisible lumping cannot (by definition) be identified. It will confound our test in proportion to its frequency.

What tests will we do on the correlation coefficients? There are two. The main one is a 2 × 2 table for 'predicted homogamy/random mating' and 'observed homogamy/random mating'. In this test we will ignore the quantitative value of the correlation coefficient and concentrate on its statistical 'significance'. If $p < 0.05$ it counts as homogamy (I have found no cases of heterogamy); if $p > 0.05$ it counts as random mating. We are immediately faced with a difficulty. Randomness, as usual, is the null hypothesis. It is therefore easier to 'confirm' random mating than homogamy, because to count as homogamy the significance has to pass below 5 per cent. Any probability greater than 5 per cent is going to count as random mating. The difficulty is greatest at small sample sizes: a very large correlation coefficient is needed to reach statistical significance. A first step against the difficulty is not to count non-significant correlations as random mating if their sample size is small. If the sample size is below 20 we will not count a non-significant correlation as random mating; we will, of course, count a significant correlation as homogamy. The figure of 20 is not important: the figures in the final test would not be altered much if it were 50.

The first test has the merit of not comparing the exact values of the correlation coefficients. It might naïvely be supposed that the theory predicts a higher correlation in some species (those with longer durations of mating, etc.) than in others. We might be tempted to correlate the exact values of the correlation coefficients with those of the other three variables. This kind of test does appear more precise than mine, but it is less sensible. There are two reasons. One is that the theory has not been worked out thoroughly enough for us to know how to combine the three variables to predict the exact value of the correlation coefficient which describes homogamy. We may suppose, for instance, that (as in 'optimal foraging theory') there is a threshold duration of mating below which males should not discriminate at all, and above which they should prefer larger females. A second reason is that the relative values of the correlation coefficients do not mean much. They have all been calculated for size. 'Size', however, is measured most conveniently and consistently in different ways in different taxa. In some size is measured by the length of a limb, in others by the longest length from back to front, in others by bill length, and so on. The comparison of correlation coefficients will be confounded by the different ways of measuring size in different species. Even if size were identically measured in all species (which would be impossible), they would still not be exactly comparable. Size happens to be an easy dimension for biologists to measure; but this is no guarantee that it is a dimension which matters to the animals. To compare species quantitatively we would need to know how the animals themselves size up their mates. The theory could indeed make more precise predictions than that some species show homogamy but others do not, but to test them we would need to know much more.

Does not the second objection apply to the test I have proposed? If it is sauce for the quantitative goose it must also be sauce for the qualitative gander. Yes... but in a cruder test it is less pungent. The test I have proposed is possible because size (as measured by a biologist) is correlated with nearly all other aspects of an organism: whatever dimension the animals use is probably correlated with the measurement. The correlation between the organism's and the biologist's dimension will vary between species in unknown ways. If the actual correlations for homogamy were compared between species these mysterious differences could throw up artefacts. The measurement of size is crude. I have tried to match the crudeness of measurement with a crudeness of test.

It is enough that this second difficulty should have been pointed out. Just how destructive it is may remain open to argument. I shall argue about it no more here. The first difficulty is clearly insuperable until we are rescued by some heroic mathematician. Any test which crosses it will be worthless. There is another possible test, suggested to me by Alan Grafen, which is continuous in the dimension of homogamy but dichotomous in the predicting variables. We can rank the correlation coefficients and then observe, through a Mann-Whitney U test, any difference between the correlations of the species we predict to be homogamous and those we do not. Such will be the second test.

We now have a complete set of theory and methods. We are ready to face the facts.

SYSTEMATIC SUMMARY

This section summarizes all published facts about homogamy for size. For each species in which it is known whether mating is random or homogamous I have also summarized the facts about the relations of size and fitness in males and females, and the duration of mating. I have included much more information in the text than it is possible to use in the final test. I have, for instance, included qualitative statements that are not precise enough to use. Table 4.1 summarizes all the quantitative measurements of homogamy (and the other variables). I have not repeatedly referred to this table in the text.

The species are ordered roughly according to the *scala naturae*. We start with plants and protozoans; these are really included only for completeness and curiosity: their homogamic habits are not (plants) or probably not (protozoans) comparable with the animals that come after them and make up most of the chapter.

Plants

There is one study of homogamy for size in plants. It is by Levin and Kerster (1973), and it is on 'assortative pollination for stature in *Lythrum salicaria*' (the purple loosestrife, family Lythraceae). *Lythrum* owes its homogamy to the habits of its pollinating bees. Individual bees tend to fly at a fairly constant height, so they bump into and pollinate plants of similar heights. Levin and Kerster found a high correlation among the heights of mating *Lythrum* ($r = 0.86$).

Protozoa

Sexual selection or mechanical constraints may cause homogamy in protozoans. But it is more likely that, like plants, protozoans are completely incomparable

Table 4.1 Summary of species and variables

Species	N	r	p	Female fecundity	Male fight	Male pop. stat.	Mating duration	Homogamy obs.	Homogamy pred.
Olivella biplicata	197		S			(+)	long	+	+
Chomodoris zebra	148	0.49	S	+				+	+
Magicicada septendecim	428	0.02	NS				hours	−	−
Magicicada cassini	340	0.05	NS				hours	−	−
Pyrrhocoris apterus	167	0.03	NS				several days	−	+
Plecia nearctica	20–25	0.22–0.35	NS	+	(+)	+	56 h	+	+
Drosophila melanogaster	212–568	0.11–0.21	S	+	+	0	½ h	−	−
Drosophila subobscura	678	0.006	NS		+		8 min	−	−
Scatophaga stercoraria	239	0.08	NS	+		+	50 min	−	−
Popillia japonica	126	0.19	S				2 h	+	
Anisoplia segetum	56	0.19	NS					+	
Chauliognathus pennsylvanicus	221	0.31	S			+	'prolonged'	+	+
Leptinotarsa	v. high	high	S				min–12 h	+	
Coptocephala unifasciata	156	0.02	NS					−	
Tetraopes tetrophathalmus	9–31	−0.25–+0.58	NS	+		0	days	?	−
Brentus anchorago	?	?	S?			+,	45 min		
Phryganidia californica	58, 37	0.01, 0.18	NS			0, +	'several hours'	−	
Pseudocalarus	19	0.33	S	+				+	
Alpheus armatus	8–15	0.79	S	+				+	+
Alpheus heterochaelis	98, 52	0.73, 0.68	S		+		month	+	+
Alpheus lottini	21	0.76	S					+	+
Paguritta harmsi	28	0.63	S				long	+	+
Paralithodes camtschatica	12–35	−0.46–+0.29	NS	+		+	days	−	+
Carcinus maenas	12	−0.53	NS				days		+
Trapezia cymodoce	20	0.79	S	+	+		long	+	+
Trapezia ferruginea	15	0.59	S	+	+		long	+	+
Uca rapax	11	0.65	S	+	(+)		h–days	+	+?
Gammarus locusta	60	0.91	S	+		+	days	+	+
Gammarus fasciatus	71	0.69	S	+			days	+	+
Gammarus pulex	18–79	0.25–0.58	S	+	+	+	days	+	+
Gammarus palustris	175	0.92	S	+			days	+	+

SUMMARY OF EVIDENCE

Species	N	Value	Sig					Duration		
Thermosphaeroma thermophilum	54	0.27–0.57	S	+		+		days	+	+
Asellus aquaticus	25–123	0.27–0.57	S	+		+		days	+	+
Jaera albifrons	32	0.09	NS	+		0		s–h	–	–
Jaera istri	103	0.09	NS	+		0		3 d. +	–	–
Jaera italica	83	0.32	S	+		+		3 d. +	+	+
Jaera nordmanni	35	0.59	S	+		+		3 d. +	+	+
Limulus polyphemus	100	0.1	NS						–	
Cichlasoma maculicauda	117	0.6	S	+	(+)			long	+	+
Onchorhynchus nerka	?	0.16	S	+	+			5 days	+	+
Cyprinodon	?	low?	NS	+	+			short	–	
Scaphiopus couchi	13–28	0.1–0.29	NS			+				
Scaphiopus bombifrons	10	0.27	NS			+				
Scaphiopus multiplicatus	9	0.53	NS			+				
Eleutherodactylus altamazonicus	9	–0.07	NS	+						
Eleutherodactylus vaiabilis	10	0.09	NS	0						
Eleutherodactylus croceoinguinis	9	0.08	NS	0						
Hyla versicolor	6–9	0.14–0.76	NS			0		hours		
Hyla marmorata	9	0.54	NS			+, +				
Hyla crucifer	8	0.14	NS			0, +				
Hyla cinerea	66	0.18	NS			0		few hours	–	
Hyla garbei	19	0.01	NS	0						
Triprion patasatus	16	0.55	S			0				
Rana sylvatica	35–99	0.09–0.44	S	+	+	+		24 h	+	+
Rana catesbiana	55	0.34	S	+	+	+		50 min	+	+
Bufo typhonius	64	0.13	NS			0			–	
Bufo americanus	7–63	–0.01–+0.78	S	+		+		day	+	+
Bufo quercicus	13–25	–0.23–+0.32	NS	0		+			–	–
Bufo w. woodhousei	17–80	0.07–0.33	NS			+			–	–
Bufo w. fowleri	?	?	NS			+				
Bufo bufo	16, 41	0.57, 0.24	S	+		+		days	+	+
Chen caerulescens	48	0.21	(S)	+		+		long	(+)	+
Larus argentatus	7	0.21	NS			+		long		+
Larus marinus	10	0.5	NS			+		long		+
Geospiza fortis	61	c. 0.3	S			+		long	+	+
Homo sapiens	many	–0.26–+0.63	?							

[see p. 182 for footnotes to table]

with the animals to come. All the protozoans that are discussed in this section are ciliates.

How are we to quantify homogamy in a species in which there are no sexes? In dioecious species the data are 'ordered' in the sense that there are males whose size can be plotted on one axis, and females that can be plotted on the other. In protozoans, and all hermaphrodites, the data is unordered. When calculating the correlation coefficient, which member of the pair should be put on which axis? A number of methods have been used, and we are going to look at them first, before the facts. The first study was by Pearl (1907; Pearl 1905 was an abstract) who entered each pair twice in his frequency tables. (The table has size on both axes. The frequencies of mating combinations are written in the table.) This method, which Pearl described as 'obviously . . . the correct thing to do' (p. 250), does not bias the estimate of the correlation coefficient, but it does spuriously inflate the degrees of freedom ('significance' tests did not exist then). Jennings (1911), by contrast, believed that each pair should be entered only once, the larger individual always being put on one axis, the smaller on the other. Jennings' method obviously distorts the coefficient; it could even discover wholly spurious correlations. But he used this method in his research, which we will come to shortly. Enriques (1908) used another method. The statistic that he invented got round the problem of unorderedness by using only the difference in sizes of the two individuals of a pair. If there is homogamy the average difference between the sizes of the members of the pair is smaller than if pairing is random. Enriques calculated the average difference of the members of real pairs, and the average difference if they had paired random. His statistic was the ratio of the latter over the former: it is one if there is random mating and greater than one under homogamy. Unfortunately the sampling distribution of Enriques's statistic is not known, so the probability ('significance') of a ratio cannot be calculated. Were it not for this problem I would have applied his test to all the protozoan data.

Now for the facts. *Paramecium* is the most studied genus. Pearl measured the size (in three aspects) of paired *Paramecium caudatum*. For five separate cultures the correlations for length were the highest of the three measures (r ranged from 0.43 to 0.79, n from 12 to 200, the probability cannot be calculated

---

[*footnotes to Table 4.1*]

N is the sample size. If more than one sample has been measured I have indicated the range of sample sizes.

r is the correlation coefficient. (Mainly product-moment. Some non-parametric.)

p is the statistical significance. S stands for significant, N for not.

Female fecundity. + means bigger females are more fecund, 0 that they are not.

Male fight. + means bigger males beat smaller males in fights for females.

Male pop. stat. + means that the mating males are larger than the average male size in the population, 0 that they are not.

Homogamy obs. means observed (for size), pred. means predicted. − means no homogamy, + means homogamy present.

for the reason given earlier). Now we can move on to Jennings. His correlation coefficients (which also suggest homogamy) are biased, as we have seen. Jennings had found multiple races of *Paramecium*, and the races differ in average size. His results therefore might have reflected preferential mating by race rather than by size. He then (Jennings 1911a) started some cultures from a single *Paramecium* to be sure that there was no racial mix. He again obtained positive correlations, although they were smaller than Pearl's. (He ran 11 such cultures, and the r ranged from 0.193 to 0.507, while the n ranged from 28 to 336.) The correlations in his original cultures which may have contained more than one race ranged from 0.27 to 0.51 in six large samples (Jennings 1911a, p. 9). The fact that the correlations were higher in the ordinary cultures suggests that some inter-racial pairing may be inflating the correlation. Jennings and Lashley (1913) published their original measurements so it is easy to calculate correlation coefficients by a less distorting method than theirs. They put the larger individual on one axis and the smaller on the other, and calculated $r = 0.388$ ($n = 92$). I recalculated r by using a table of random numbers to choose the larger (if the random number was even) or the smaller (if odd) member first. I found $r = 0.39$ ($p < 0.01$). But even this is not a good estimate of the true correlation because Jennings and Lashley's culture was (it seems) one of those ordinary cultures which may contain a racial mix. The coefficient may be inflated by an unknown amount. I am therefore not satisfied by the correlations so far: Pearl's and Jennings and Lashley's are spoiled by racial mixing: Jenning's (1911) by his method of calculating the correlation coefficient. The fact that all the measurements point to homogamy is fairly persuasive, however.

Both Pearl and Jennings demonstrated, by various techniques, that the homogamy was not caused by the contortions of the paired *Paramecium*, or by the exchange of material between them, or by local size differentiation within the culture. The explanation that they favoured was a mechanical constraint on the size of mating individuals, but they provided no evidence. What about the theory of sexual selection? We need to know three variables: the duration of pairing, the relation between size and success in competition to pair with other sexually receptive individuals, and the relation between size and some equivalent of fecundity (such as rate of division). There is some information on the first. According to MacKinnon and Hawes (1961, p. 294) 'the time taken for conjugation varies with temperature and other factors, but in general *P. aurelia* and *P. caudatum* remain joined for 12 to 15 hours'. This figure in itself falls in the ambiguous zone in the theory, but that does not matter much because the other two variables are sufficiently confusing to rule the species out of the test. I have no idea whether it is advantageous for a *Paramecium* to pair with a larger than a smaller individual (perhaps larger ones reproduce faster?). Conjugating *Paramecium* are smaller than non-conjugating individuals in the same culture (e.g. Pearl 1907; Jennings 1911a, and papers cited by them), but not because of sexual selection for small size. *Paramecium* shrinks while differentiating into the conjugatory form, so the non-conjugating forms are likely to be larger. We would need to know whether, of the potentially conjugating individuals, the actually paired ones are larger. Absolutely nothing is known about that, so we will not meet *Paramecium* in our final test.

Enriques (1908) studied homogamy in *Chilodonella uncinata*. He quantified homogamy by the ratio which we have already discussed. He took nine samples, six from a conjugation 'epidemic', and three more later. He was uncertain by what factor to correct for the bending of the conjugating individuals. So he calculated two ratios, one with a correction factor of two, the other of three. If we write the two ratios in that order, and put the small samples in brackets, his results for the six early samples were: 1.08, 0.89 ($n = 47$), (1.5, 1.88, $n = 11$), (1, 1.02, $n = 8$), 1.06, 1.16 ($n = 60$), (0.93, 1.07, $n = 13$), 0.97, 1.12 ($n = 97$). It is difficult to assess these ratios, but the fact that they are not even consistently bigger than one, let alone much more than one, suggests that there was no homogamy. There was a stronger hint of homogamy in the later samples: 1.08, 1.22 ($n = 85$), 1.32, 1.42 ($n = 66$), 1.33, 1.5 ($n = 44$). He also calculated a correlation coefficient for the sixth and eighth sample. His method? He alternately took the larger and smaller individual first (1908, p. 254). Sample six gave a correlation of zero, and sample eight of 0.4 ($n = 66$, $p < 0.01$). This difference also bears out Enriques's observation that homogamy is shown in late cultures but not in early ones.

We cannot assess Enriques's facts statistically. Perhaps there is homogamy, perhaps there is not. Perhaps there is a trend with time from the absence to the presence of homogamy. Anyway, none of these possibilities are explicable. The only statement on the duration of pairing which I have come across is by MacDougall (1925, p. 373), who writes of 'the close of conjugation, 34–30 hours [*sc.* after the beginning].'

Collin (1909) stated that *Anoplophyra branchiarum*, a blood parasite of *Gammarus pulex*, has homogamy for size. He did not actually measure any pairs, but documents it verbally and qualitatively thus: 'the two conjugating individuals of a pair are nearly always [*très generalement*] of about the same size' (p. 352), and 'the proportion of couples of unequal size (I mean obviously unequal, appreciable at first sight without a micrometer) is certainly not above 5 or 6%' (p. 376). He suggested a mechanism (p. 378) but gave no support for it so it need not detain us.

Finally, Watters (1912) measured the sizes of 279 pairs of *Blepharisma undulans*. She presented her measurements (p. 199) as the number of pairs in which the two were equal in size, differed by ½ a unit, 1 unit, ... up to 5½ units. Only 24 of the 279 pairs differed by 3 or more units. *Blepharisma* is clearly homogamous for size; but again this would be difficult to prove statistically from the facts. Nor can it be explained. The conjugants were smaller than the non-conjugants, but this is presumably no more relevant than it was for *Paramecium*.

In all, there is impressive evidence for homogamy in protozoans. Its significance, however, is a mystery.

Molluscs

Crozier (1918, 1920) measured the total length (among other aspects) of paired *Chomodoris* (Opisthobranchia, Chomodoridae). He collected copulating couples 'during the month of April 5 to May 5, 1917' (1918, p. 249) from a single locality. The lumping of the pairs to calculate a correlation coefficient was probably justifiable, because 'there is no detectable tendency for individuals of any given size to copulate at a different time of day, or season, than those of any other size' (1918, p. 256).[5] *Chomodoris* are simultaneous hermaphrodites. Crozier unfortunately used Jennings's (1911) method to calculate r: he put the larger individual on one axis and the smaller on the other. However, he did publish his original measurements (1918, p. 257) so it is possible to recalculate the coefficient. Once again, I used a table of random numbers to choose the larger or smaller individual first. By this randomizing method, $r = 0.49$ ($n = 148$, $p < 0.01$), which is lower than the 0.61 by Jennings's method. Either way there is significant homogamy.

The theory predicts homogamy in hermaphrodites, provided that larger individuals (in their female capacity) lay more eggs. Crozier (1918, p. 282) proved that they do. He explained homogamy by a mechanical constraint, but that will not stop us from counting it in favour of our theory.

Homogamy has also been studied in the prosobranch *Olivella biplicata*, an intertidal snail. The evidence, for our purposes, is confused by the snail's habit of segregation by size: larger *Olivella biplicata* tend to be found higher up the beach. They tend to mate locally, so homogamy is caused by the geographical segregation (Edwards 1968, Fig. 2). Many intertidal species show similar segregation by size, for reasons which are not known (Edwards 1969), so this kind of homogamy may be common. But it does not count as homogamy under our theory: we are interested in homogamy at any one place where the distribution of sizes is constant. We must therefore turn to another sample by Edwards (1968): his Fig. 1. This sample was taken from a beach where there is no segregation by size. Thus, 'the lengths of 210 nearest neighbors of 97 pairs were found to be independent of the courting females' sizes' (p. 299). Homogamy here? Yes: 197 pairs showed a significant regression. From a glance at his Fig. 1 I would guess that the correlation coefficient is about 0.2 (between about 0.15 and 0.4); but neither Edwards nor I have calculated it. The significance of the regression is enough for our test: *Olivella biplicata* is homogamous.

Olivella's pairing 'may persist for a long time' (p. 298), and Edwards instances cases of 31 hours and three days. We can provisionally count it as a long pairing,

[5] Jones (1928, p. 122) must have misread Crozier. He wrote of it that the large–large pairs formed 'at different seasons of the year' from the small–small pairs.

provisional because the facts are sparse. The males of courting pairs were slightly larger than the non-courting males, although the difference did not reach statistical significance ($p = 0.1$). The relation between female size and fecundity is not known. For our looser test we will assume that larger females lay more eggs (as many snails are known to), and count *Olivella* in favour of the theory. Edwards (1968, p. 300) explains homogamy by a mechanical constraint. Small males, we are told, do not have long enough penises to serve large females: 'since in mating a male must extend his penis most of the length of the female's shell and into the mantle cavity, probably small males simply cannot accommodate larger females'. Although Edwards casts aspersions on the virility of small males, he provides no evidence that they have any difficulty in fertilizing a large female once they have paired with her. I would suggest that the small males' difficulty is in obtaining females, in the face of competition from larger males, to begin with.

Insects

Hemiptera

Cicadidae

Dybas and Lloyd (1962) measured the wing lengths of pairs of two cicada species. Neither *Magicicada septendecim* ($r = 0.02$, $n = 428$, n.s.) nor *M. cassini* ($r = 0.05$, $n = 340$, n.s.) showed homogamy. Dr Dybas, in a letter, has informed me that the frequency distributions of sizes of pairs contain more than one sample. They are probably made up of two main samples; but all the pairs were collected within about two weeks from a single population. My calculation of the correlation coefficient therefore assumes that the frequency distribution of sizes remained constant through that fortnight.

I have found nothing on the relation of size and fitness in males and females. There is some information on the duration of copulation. It is tantalizing but, in the end, too vague to allow this species to be used in the test. Dybas and Lloyd state that 'once copulation has begun, the genitalia typically remain locked together for more than an hour' (1962, p. 453). White (1973, p. 575) makes a similar statement: 'intra-specific matings were ... of long duration—physical contact in three of the five *septendecim* control matings lasted longer than three hours'. *Magicicada* appears to lie somewhere in the hypothetical grey area: we cannot clearly predict how they should choose their mates. Furthermore, the statements are vague and the sample sizes small. Let us exclude these two from the test.

Pyrrhocoridae

Alpatov (1925) measured the body and antennal lengths of 167 pairs of *Pyrrhocoris apterus.* mating was not homogamous for either trait. The sample, he says, was 'collected over the course of some days in July', so we are assuming a constant size-frequency distribution over this period. I have found no information on the relation between size and fitness in males and females. But I have found one casual comment on the duration of copulation: Seidel (1924, p. 432) says that it lasts for 'several days'. On this limited information, the theory predicts homogamy. And it is wrong.

Diptera

Bibionidae

Thornhill (1976, 1980) measured the sizes of mating pairs of *Plecia nearctica* in Florida (1976) and in Mexico (1980). In Mexico he measured the lengths of their foretibiae and thoraxes (the two measures were highly correlated, $r = 0.9$), and in Florida just their foretibiae. He measured samples of pairs from two separate swarms in Mexico ($r = 0.22$ and 0.35, both $n = 20$) and from one swarm in Florida ($n = 25$, $r = 0.33$; R. Thornhill, private communcation). Each correlation coefficient, taken by itself, is statistically insignificant. However, they are all positive. Their insignificance may reflect the low sample sizes. I tried combining the three tests by Fisher's method (Sokal and Rohlf 1969, pp. 622-3), but the combined probability is still 0.1. I suspect that a larger sample size would reveal homogamy (for example, if the real correlation were 0.3, a sample size of 50 would be needed to prove it statistically); but we will not use my suspicion as evidence.

Thornhill also demonstrated that mating males are larger, on average, than non-mating males (1976, 1980), and that 'it appeared' when one male took over a female from another male 'that the successful males were always much larger than the original copulating male' (1976, p. 845). Larger females lay more eggs (Thornhill 1976, p. 846). Mating takes a long time: copulation lasts, on average, 56 hours. Adult males live for only (on average) 105 hours after the start of their first mating, and females 86 hours (Thornhill 1976); Hetrick (1970) observes that 'males live for 2 or 3 days; females may live for a week or longer'.

Plecia nearctica satisfies all three conditions which predict homogamy for size. But it does not show it. We shall score it against the theory. Two reasons why it does not fit can be given. One we have already seen: perhaps it is really homogamous, but the sample sizes are too small to prove it. The other, preferred by Thornhill, is that there is exceptionally intense competition among males in this species. Thornhill saw as many as eight or nine males struggling for one female, and it was usual to see two or three males doing so. In conditions of extreme male competition the prediction of the theory changes. Even if all three conditions are satisfied there should not be homogamy for size. If it is sufficiently unlikely that a male will obtain a female at all, it pays him to hold on to any female, whatever her size.

Drosophilidae

Dobzhansky, while considering whether species might be isolated by the mechanical impossibility of inter-specific matings, wrote that 'variations of body size within a species do not hinder copulation. In *Drosophila* giant and dwarf mutations and large- and small-bodied flies produced from well-fed and from starved larvae, cross easily and give offspring' (1970, p. 326). 'Cross easily', however, is not quantitative. There are two sets of measurements for us to consider. One is for *Drosophila melanogaster*; the other for *D. subobscura*. They contradict each other.

The earlier of the two were made by Parsons (1965), with *Drosophila melanogaster*. He discovered a slight but statistically significant homogamy for size. He actually measured the sternopleural bristle number of mating flies, but he had previously shown that this is strongly correlated with size (Parsons 1961). He used flies in a laboratory mating chamber. He simply sucked out and measured the pairs as they formed. As pairs were sucked out, the number of flies remaining in the chamber decreased. Once the number was down to about half the original, Parsons stopped the experiment and measured the sizes of all the remaining unpaired flies. The average size of the unpaired males did not differ from that of the paired males: larger males had no advantage in mating. He ran three experiments in all, in separate mating chambers; all three gave similar results.

His experimental design can be criticized. The pairs were not all sampled from the same population: as the pairs formed, and were removed, the total numbers decreased and (we can assume) the size composition changed. If the flies do choose a mate of similar size, the estimate of the correlation would be biased. The most complementarily sized pairs would form first; as the experiment proceeds the remaining, unpaired flies would be of sizes that were decreasingly inclined to pair. Thus the correlation among the earlier pairs would be higher than that among the later pairs: the overall correlation would decrease as the sample increased. Parsons calculated the correlation coefficient after about half the flies of the original population had mated. Whether he over-, under-, or correctly estimated the natural correlation depends on how natural it is for half of a sample of 1000 or so flies to pair up. I would guess that if the flies had paired in the wild, their mates would have been drawn from a much larger population. If this guess is correct, Parsons has under-estimated the true correlation. But this is only a guess. The amount (and even the direction) of the bias is unknown. So when we use Parsons' estimate, we should remember that it is rough.

Larger females are more fecund (Robertson 1957). But, in Parsons' vials (as we have seen), the larger males had no advantage. This is an exceptional result. Previous (Ewing 1961, 1964), and subsequent (Partridge and Farquhar 1983), experiments have demonstrated that larger males have a higher mating success.

(Ewing (1961, p. 97) noticed that smaller males were relatively less unsuccessful when competing for small females than when competing for large females: another hint of homogamy.) But the hypothesis cannot be saved by Ewing and Partridge and Farquhar. Whether or not larger males are more successful, mating is definitely quick. There is a brief courtship (of a few minutes, less than a quarter of an hour) followed by copulation, which averaged 18 minutes 14 seconds in one study (Spieth 1952). The observation on the duration of mating is decisive. The theory predicts that *Drosophila* should not have homogamy. But it is wrong. *Drosophila* is homogamous, in Parson's observations at least. Parsons' flies must be counted against the theory.

The second estimate is for *Drosophila subobscura*. It was made by Monclus and Prevosti (1971). They obtained a non-significant correlation of 0.0061 of 678 pairs. The circumstances in which the mating pairs formed are not clear, but the flies were obtained from a laboratory culture. In their experiment, like in most experiments on *D. melanogaster*, the larger males had a faster mating speed (so did the larger females).

Mating is quick (Spieth 1952), and (we can assume) larger females lay more eggs. Again the quick mating is decisive: the theory predicts random mating; the facts support it: we will count *D. subobscura* in favour of the theory.

It is improbable that the one species of *Drosophila* is homogamous and the other not. But I have entered them separately in the test. I do not mean to imply what is improbable. A more sensible interpretation is that they are two independent trials of the hypothesis, so both must be included. We might, instead, have counted the evidence only once, or not at all, in the test: we might have interpreted them as two sets of evidence bearing on some single 'real' habit of *Drosophila*. Then three lines of argument would be open to us: we could trust Parsons but not Monclus and Prevosti; Monclus and Prevosti but not Parsons (both papers contain defects); or we could argue that the results are so uncertain that we must exclude *Drosophila* from the test. The course I have taken, of including one species for, and the other against, has much the same consequence as the third alternative.

Scatophagidae

The dungfly *Scatophaga stercoraria* mates on fresh dungpats. Once a male has intercepted a female the pair proceeds, male on top of the female, to the dung. There they copulate and the male guards the female until she lays her eggs (Parker 1978). Borgia (1981) measured the wing length of 239 pairs and found no evidence of homogamy.

Larger female dungflies lay more eggs (Parker 1970b; Borgia 1981), and 'larger males have been shown consistently to be over-represented in samples of amplexing female flies' (Borgia 1981, p. 74). (In part because the larger males occupy better sites for intercepting females.) Mating is quick, so the absence of homogamy is predicted. Copulation itself lasts about 35 minutes (or in some cases 96 minutes, Borgia 1981, p. 78), and postcopulatory guarding a further 15 minutes (Parker 1970b, pp. 1322-3). There is also very intense competition among males: the sex ratio is about four, and up to eight, males per female (Parker 1970a; Borgia 1981), which may also select against homogamy. Anyway, dungflies fit the theory.

Coelopidae

The Coelopidae live on seaweed. Butlin, Read and Day (1982) provide some evidence of homogamy for size, but it is too ambiguous to use; and nor is the other crucial variable, mating duration, clearly specified. Butlin *et al.* set up experiments with a single female and two male flies. Of 257 (p. 56, or was it 237?, p. 57) cases in which a single male fathered the offspring (the father was determined genetically), in 200 the larger male had been successful. They also noticed that larger males were relatively more successful with larger females than with smaller females, and suggested that this may result in homogamy. But we do not know whether the effect was large enough to produce a significant correlation of mate sizes. Of the other three variables, larger males are more successful as we have seen, the relation between female size and fecundity does not seem to have been measured, and the duration of mating has not been directly specified. If mating was long they probably would have remarked on it, and they do say (p. 60) that courtship is brief to the point of non-existence. But the true habits of *Coelopa frigida* remain concealed in the sea mists.

Coleoptera

Scarabaeidae

Bateson and Brindley (1890), in their paper on male dimorphisms, reported the observations, privately communicated to them, of Baron von Hügel on the mating of the horned beetle *Xylotrupes gideon*. The Baron had seen the males seize the females, and then carry them, on their horns, 'with evident satisfaction' (p. 590). He also 'noticed that large males were often attached to small females and the reverse, but there appeared to be no regularity in this'. Such a statement is not precise enough for us.

Pomerat (1932) provides quantitative measurements of the sizes of mating pairs of the Japanese beetle *Popillia japonica*. This beetle was imported from Japan into America earlier this century and soon became a pest. Pomerat measured pairs in America. He measured eight anatomical aspects of every beetle in 126 pairs. The most direct measure of size, total ventral length, had a correlation coefficient of 0.19 ($0.05 > p > 0.01$). The correlations between other parts were mainly smaller: *Popillia japonica* shows only slight homogamy for size, if any at all.

The general review on this species by Fleming (1972) does not mention any measurements on the relation between size and fecundity. Nor is there any information to tell us whether larger males have a mating advantage. If these two variables are unknown, the other is ambiguous. Fleming says that 'coitus may be brief or prolonged for several hours' (1972, p. 43), Barrows and Gordh (1978, p. 344) that (in the laboratory) copulation itself lasts, on average, 132 minutes ($n = 33$), and is followed by a period of postcopulatory guarding which goes on for 'up to 2 hours'. This information, in the round, it not adequate to predict whether *Popillia* should be homogamous; and the facts about homogamy are on the border line. Let us exclude it from the test.

The relation between the sizes of mating pairs of *Anisoplia segetum* is also known, but none of the other three variables are, so this species cannot be used. Spett (1929) measured the breadth of the clypeus and the length of the right hind femur in 56 pairs, and found (non-significant) correlations of 0.19 and 0.25 respectively. It is not homogamous, but that is all we know.

Cantharidae

Mason (1972) took eleven samples (with n of 14 to 37) of mating pairs of the common American soldier beetle *Chauliognathus pennsylvanicus* in New York State. He measured the width of their prothoraxes, and obtained correlations ranging from -0.22 to $+0.48$. Some of these were statistically significant, but no more than would be expected by chance. McCauley and Wade (1978) measured the dry weights of a further 221 pairs, in Illinois, and obtained a (significant) correlation of 0.31. Which result should we trust? McCauley and Wade point out that Mason's sample sizes were too small to detect a correlation coefficient of 0.31 (which requires $n > 40$). We will therefore use McCauley and Wade's figure. (There could, of course, be variation between populations.)

Mason and McCauley and Wade both proved that mating males are, on average, larger than non-mating males. The relation between size and fecundity, however, has not been measured. The duration of mating? Mason (1980, p. 179) calls it 'prolonged'. This does not add up to a conclusive case, but it looks like a species on which homogamy for size is both predicted and observed.

Melandryidae

Champion (1907) is a communication of the work of Mr J. Edwards on *Osphya*. (The position of *Osphya* in this family is uncertain: Crowson 1965.) After describing the copulatory behaviour of *Osphya*, Mr Edwards continued, 'the effect of these circumstances is to secure the pairing of individuals of suitable size, for the small males were quite unable to hold the large females whilst the small females escaped with ease from the embrace of the normal males' (p. xxv).

Chrysomelidae

Tower (1906), in a paper which is most famous for an alleged discovery of a case of Lamarckian inheritance, included some observations on mating and homogamy in the Colorado potato beetle *Leptinotarsa*. Tower divided the beetles into ten size classes, 'in which 1 represents the smallest size condition and 10 the largest. The unit used in the measurement ... was the perpendicular distance between parallel lines passing through the apex of the elytron and the median posterior side of the pronotum' (p. 238). He then gave his measurements of the mating pairs; but unfortunatley he gave only the percentages of each female size class mating with each male size class. We are not told the total sample size, and it is not possible to calculate the correlation coefficient. However, it is clear from his text that the sample size was very large. For example, he studied homogamy for colour in four other species, with sample sizes of 500, 300, 200, and 100 pairs. Then if we assume that the lowest percentage in the *Leptinotarsa* table is a single individual, we obtain a minimum estimate of n as 5000: only 4 per cent of the class 1 males mated with class 3 females, 0.5 per cent of the males were class 1, so there were $25 \times 200 = 5000$ pairs. It is obvious from inspection of the table that *Leptinotarsa* shows strong homogamy.

(Various correlation coefficients, ostensibly calculated from Tower's table, can be found in the literature. Alpatov (1925) simply asserted that the correlation was 0.87; Spett (1929) repeated this figure. Willoughby and Pomerat (1932, p. 224) tell us that 'on the basis of reasonable assumptions' they calculated a 'product-moment correlation coefficient of .89'. That they were reasonable is all we are told about the assumptions. I have not been able to conjure either 0.87 or 0.89 out of Tower's table. The coefficient obtained by mistakenly supposing that the entries in the table are numbers, not percentages, of pairs, is 0.97.)

Tower gives no facts about the relation between size and fitness in males or females. There is plenty of variation among females in fecundity (p. 237, Table 104), but no indication of whether it is correlated with size. He hints at normalizing selection for size: 'the number of unmated individuals in any population is far greater among the extremes than among the more mediocre [closer to the average] individuals' (p. 239). This statement is too vague for us. (We have a further reason to suspect it. Tower understands neither the meaning nor the consequences of homogamy. He thinks (e.g. p. 241) that homogamy has a normalizing effect. In fact, by increasing the proportions of extreme types, relative to random mating, it does the opposite.)

Males live, on average, for about 21 days, females for 29 (Tower, Table 103). Courtship is short, and 'copulation lasts from two to three minutes up to ten or twelve hours, the first coitus, however, being usually far longer than those which occur subsequently' (p. 234). This large variation is unfortunate. Two minutes is clearly a short copulation. But twelve hours? It is a long time, perhaps long enough to select for discriminatory mating. Tower does not tell us the relative frequencies of long and short matings, so we cannot calculate the average. One possibility is that the first long copulation is 'natural' and the shorter later ones the artefacts of laboratory sexuality, or even not real copulations at all. Once again, this species cannot be included in the test for want of clear facts.

Tower himself favours a 'genital fit' (lock-and-key) explanation for homogamy: 'the attempt of a large male to mate with a small female invariably results in failure, owing to his inability to force an entrance into the small genital passages of the female. On the other hand, large females and small males do not seem to be able to mate with any degree of success on account of the lack of adaptation in the size of the reproductive organs' (p. 240). It soon becomes clear that this was all assumption, not observation. 'I have attempted to obtain statistical data on this point, but the difficulty of dissecting out the organs and the distortion that inevitably results are so great that the error introduced is probably greater than the variation' (p. 240).

Spett (1929) studied homogamy in another chrysomelid, *Coptocephala unifasciata*. He measured the lengths of the right wing covers ($r = 0.002$, n.s.) and the right fore tibiae ($r = 0.02$, n.s.) of 156 pairs. I have not uncovered any facts about the other relevant variables.

Cerambycidae

The observations of Michelsen (1957, 1958, 1964) suggest that *Rhagium* may have non-random mating for size imposed by the exigencies of their copulatory posture. In *Rhagium bifasciatum* 'the genital organs are never protruded more than 1 mm and therefore the abdominal tips of the two sexes are close together during mating', whereas in *R. mordax* 'the genital organs are protruded several mm during the mating and the abdominal tips of the two sexes may move rather independently' (1964, p. 163). During copulation the male must 'lick' the female on her scutellum if mating is to be successful. In *R. bifasciatum* 'mating is successful only if the male is slightly smaller than the female. This is due to the fact that larger males will "lick" in front of the effective place'; but in *R. mordax*, with their more relaxed genital connexion, 'males may "lick" in the right place, almost irrespective of their size' (1964, p. 163). (These observations fit in with the pattern of sexual dimorphism. *R. bifasciatum* males (13-18 mm) are smaller than the females (16-21 mm): *R. mordax* is not dimorphic for size.) *R. bifasciatum* may have non-random pairing; *R. mordax* may not: but Michelsen did not measure any natural pairs. We have nothing precise on the three variables by which I would try to explain non-random pairing. Michelsen does say that in a study of 16 species of Lepturinae (including the two *Rhagium* species), in most species mating lasts from 10 to 40 minutes with an average of 20 minutes' (1964, p. 160).

Mason (1964) measured the lengths of the left elytra of paired males and females of the North American beetle *Tetraopes tetrophthalmus*. He took 17 samples (with *n* ranging from 9 to 21 pairs), most from geographically separate areas. The correlation coefficients varied from -0.25 to $+0.58$; some were significant, but no more than would be expected by chance. Thus

T. tetrophthalmus does not show homogamy for size. Eanes, Gaffney, Koehn, and Simon (1977, p. 51) inform us that pairs of this beetle 'remain *in copulo* for extended periods (days)'. So copulation is lengthy, but what of the other two variables? The relation between female size and fecundity is not known; but the relation between male size and fitness is. There is normalizing selection for male size; in Mason's samples (and it has since been confirmed in larger samples by Scheiring 1977, and (probably) McCauley 1979), the average size of mating males was equal to that of non-mating males. Homogamy is not predicted, therefore, and nor is it shown.

Brentidae

A paper by Johnson (1982) on *Brentus anchorago* explains homogamy for size by both male and female choice. His discoveries are therefore relevant to this study, but (as we shall see) they are not clear enough to include in the test. Take first his evidence for homogamy. He divided his beetles up into three size classes and, using some 336 pairs, calculated a χ^2 for the association of mate sizes. It was significant ($p < 0.001$). However, what I have just called 'pairs' Johnson called 'copulations'. He studied the beetles from July to September. It seems likely that the pairs may have been measured over several days or weeks, and that the same individual could have been counted more than once. Females drill holes in wood and lay only a single egg at a time; they may copulate whenever they lay an egg, so a single female could copulate many times. Thus we may not want to include Johnson's brentid as a definite case of homogamy by the criteria which we are using.

What of the other three variables? Females do only lay a single egg at a time, but Johnson demonstrated that larger females lay larger eggs. Thus it is probably better for a male to mate with a larger female. He also proved that larger males win more fights over females, and that (although no statistical test is given, from his Fig. 5 it is obvious enough) the average size of mating males is larger than that of non-mating males. And how long do the sexes spend together? It is not clear: from the description on pages 253-4 it appears to be less than 45 minutes. Whether this is long enough to be counted as predicting homogamy in our test may be doubted. Johnson (p. 259), however, described the 'male mating investment' as 'sizeable'. He does not actually give its size, but uses it to explain an apparent male preference for mating with larger females. We have said enough to exclude this species from the final test: the evidence on homogamy is indecisive; while my impression and Johnson's (more authoritative) verbal testimony on the male investment in mating point in opposite directions.

We will exclude the species, but we have not finished with it yet. I mentioned above an apparent male preference for mating with larger females. We are of course explaining, in the almost complete absence of direct evidence, the entire incidence of homogamy by such male preferences. We may therefore be interested by some real evidence. What is Johnson's? He gives two kinds. One is that the average size of mating females is larger than that of non-mating females, which is consistent with but does not prove the male preference. The second is that copulations with larger females are more likely to be interrupted by another male than are copulations with smaller females. This does indeed suggest male choice. We thus have better evidence than usual on male behaviour, but there remain too many ambiguities in the rest of the evidence to use it in the test.

Lepidoptera

Dioptidae

Mason (1969) measured the sizes of mating pairs of the California oak moth *Phryganidia californica*. He took samples on two successive days in 1965, at the same site, and measured the wing length of the pairs. There was no homogamy on 8 June ($r = 0.011$, $n = 58$, n.s.) or on 9 June ($r = 0.188$, $n = 37$, n.s.).

Copulation, Mason tells us, 'lasts for substantial periods, at least as long as several hours in cases' (p. 55). He compared the mean size of mating males with that of non-mating males and on one day found the mating males to be significantly larger, but on the other they were not. The evidence on copulatory duration is vague (but it looks short enough to predict random mating) and that on the relation of male size and fitness is contradictory, so we will leave this moth out of the final test.

Hymenoptera

Formicidae

As well as the normal ('macrogyna') form of the female of the common species of ant *Myrmica rubra*, a smaller form ('microgyna') can sometimes be found. The frequency distribution of female sizes is then bimodal, but for the male it is continuous. The two forms may be subspecies. Brian and Brian (1955) determined whether smaller males tend to mate with microgyna females, and larger males with the macrogyna form. They measured the sizes of 189 copulating pairs and calculated a correlation coefficient: it was statistically significant ($r = 0.42$, $p < 0.001$). The pairs, however, were collected from three different sites and lumped in the overall calculation. We will come back to that.

I have not uncovered any information on the relation between female size and fecundity. (The correct relation would be between size when copulating and later productivity of the nest.) Nor have I uncovered the exact duration of copulation. They copulate during a 'nuptial flight', and copulation is (I suspect) not long enough to predict homogamy. Males should not discriminate among females, and nor (it seems) do they: 'males are very catholic in their choice of objects for approach' (Brian and Brian 1955, p. 282). Furthermore, the mating males were no larger, on average, than the non-mating males in Brian and Brian's sample (p. 282). The theory would predict no homogamy. But what is the evidence that they are?

The correlation given above was for a sample lumped from several sites. This alone would be enough for us to rule it out. But we can go further. Brian and Brian actually explain the correlation by spatial, and temporal, variation in size. They did not actually prove that mating at any one time and place is random, but the fact that they explain the overall homogamy by spatial variation is enough for us to conclude that there is no good evidence of real homogamy. We will not, however, go further and include it as a case of predicted random mating. We will simply exclude it from the test.

Crustacea

Copepoda

Hart and McLaren (1978) set up several pots, each containing a single male and several females of a *Pseudocalanus* species (near *P. minutus*, Calanidae). In each pot they put either a large or a small male, together with a range of sizes of females (a similar range in each pot). They watched which females were inseminated and so they were able to draw a graph of the sizes (cephalothorax length) of the mating pairs. They calculated a non-parametric correlation coefficient, Kendall's τ ($\tau = 0.33$, $n = 19$, $p < 0.05$; in fact of the 19 pairs, '16 were mated by different males and 3 by a single male' (p. 26), so there were only 17 independent males). Of some further experiments on *Pseudocalanus*, McLaren and Corkett (1978, p. 355) wrote that homogamy for size 'could only be shown in the present experiment with the inclusion of the unusually large pair'. Thus they have shown homogamy, although only just, in two sets of experiments.

Larger female *Pseudocalanus* are known to lay more eggs (Corkett and McLaren 1978, p. 87), but nothing is known about the other two variables. Copulation in other Calanidae is quick (p. 73-4). (From here on, I refer back to Chapter 3 for the duration of mating. Isolated page references, for mating durations, refer to Chapter 3.)

Malacostraca

Homogamy has been studied in four different decapodan groups, and some peracaridans. We will take the decapods first.

Natania

Stenopodidae The cleaner shrimp *Stenopus hispidus* lives in pairs (p. 84), and there is some 'Lorenzian' evidence that pairing may be non-random. Johnson (1969) placed a male with a female in an aquarium and found that if the female was larger than the male, they fought; but if the male was the same size or larger than the female, he courted her. (Johnson did 15 trials in each combination.)

Alpheidae The sizes of paired snapping shrimps (*Alpheus*) have been measured in several species. Knowlton (1980) sampled pairs of *Alpheus armatus* from two sites; she took three separate samples from one site and two from the other. She combined the different samples of each site, and calculated two correlation coefficients; the results suggested homogamy at one site but not at the other. This and other differences between the two sites provided one of the main themes of Knowlton's study, but the dichotomy is inappropriate here. The two sites do not differ in any of the factors under test as causes of homogamy, so we will treat the five samples equally and combine them, by Fisher's method, in an overall test. Nancy Knowlton has kindly sent me her original measurements for the five sites which enables a re-analysis. She measured the claw length and the rostrum-telson length of each animals. The two are highly correlated (Knowlton 1980, Fig. 4), so I have only re-analysed one of them, the rostrum-telson length. The five samples have r ranging from 0.002 to 0.79, with n from 8 to 15. Individually only two of them are significant, but when they are all taken together they show significant homogamy ($0.01 < p < 0.05$). That is the result that interests us, so now we can move on to some other authors. Homogamy for size in *Alpheus heterochaelis* has been demonstrated twice (Schein 1975, Fig. 4, $r = 0.73$, $n = 98$, $p < 0.01$; Nolan and Salmon 1970, $r = 0.68$, $n = 52$, $p < 0.01$), and in *Alpheus lottini* once (Patton 1974, Table 3, $r = 0.76$, $p < 0.01$). All the evidence points to homogamy in snapping shrimps.

Sexual pairs of *Alpheus* cohabit in shelters for a long time (p. 82). Bigger males beat smaller males in fights for shelters in *Alpheus heterochaelis* at least (Nolan and Salmon 1970, p. 307). Larger females lay more eggs (Knowlton 1980, p. 162). *Alpheus* provides a well-documented case of observed and predicted homogamy.

Reptantia

Astacura

Nephropidae For the American lobster *Homarus americanus* there are some qualitative, and Lorenzian hints that pairing is non-random. The success of mating in the aquarium depend on the relative sizes of the sexes (Templeman 1936, p. 9; Hughes and Matthiessen 1962, p. 419; Aiken and Waddy 1980, p. 226). Mating in artificial pairs is most likely if the male and female are about the same size, and becomes decreasingly likely as the male is smaller or larger than the female. Lobsters therefore are probably homogamous.

Lobsters have a lengthy pairing (p. 88); larger females lay more eggs (Aiken and Waddy 1980, Fig. 9); and, 'two males placed in the same tank with a [receptive] female . . . will invariably compete . . . the larger . . . male eventually driving the other one off' (Hughes and Matthiessen 1962, p. 419). We even have a hint of male discrimination against small females: 'males that are too large for a given female show little interest in mating with her' (Aiken and Waddy 1980, p. 226). The homogamy of lobsters, if it really exists, fits the theory.

Anomura

Paguridae Twenty-eight pairs of the monogamous hermit crab *Paguritta harmsi* were homogamous (Patton and Robertson 1980, $r = 0.63, p < 0.01$). These crabs live in pairs in pairs in the niches of corals. Patton and Robertson thought that homogamy was caused by the pairs forming as juveniles and growing up together. They may be correct, but they provided no evidence for their theory. Patton has put forward the same theory of homogamous pairing in *Alpheus* and *Trapezia* in which (as we have seen) it is probably false.

Lithodidae The king crab *Paralithodes camtschatica* is fished commercially off Alaska, Japan, and elsewhere. The Alaskan population has been well studied. Powell and Nickerson (1965) and Powell *et al.* (1973) measured the carapace lengths of 106 and 35 pairs respectively. Powell and Nickerson (1965, p. 110) state that 'no relationship between size of partners was evident'. They provide no statistical analysis, but they published their original measurements, which makes an analysis possible. First, what is the correlation of all the pairs are lumped? It is positive, slight, and significant ($r = 0.25, p < 0.01$). This correlation, we soon find, is entirely caused by variation in the average size of the crabs in the different samples. An analysis of variance of male sizes reveals significant differences between samples ($F_{6,97} = 23, p < 0.001$), and in fact there is a slight negative correlation within each sample. The next step is to find out whether these negative correlations, when taken together, are significant. Powell and Nickerson's data divide naturally into five samples, with n varying from 12 to 29, and r from -0.14 to -0.46. The negative correlations are severally insignificant, and nor do they reach significance if combined by Fisher's method ($\chi^2_{10} = 7.3$, n.s.). Powell and Nickerson's measurements thus provide no evidence for heterogamy. Powell *et al.* (1973) also found no deviation from random mating ($r = 0.29$, n.s.); but they do not specify whether their sample ($n = 35$) was all collected at one time.

The king crab has a lengthy precopula (p. 91). Larger males have a mating advantage: only the larger males are found naturally paired, but the smaller males are capable of mating (Powell *et al.* 1973). Larger females lay more eggs (McMullen 1969, p. 2739, $r = 0.9, n = 54$). The theory therefore unambiguously predicts homogamy: the facts unambiguously deny it. The king crab will be scored against the theory. But are there any possible explanations of the antinomy? Yes, we have two. First: the evidence on the duration of mating is not good: the duration may be short enough that homogamy is not selected for. Second: the environment of the species is unnaturally populated with fishermen. They aim for the large males, and have caused a noticeable decrease in the average size of males in fished populations (Nickerson, Ossiander, and Powell 1966). The sex ratio is female biased: there are twice as many females in all, and four times as many in the large size classes (Gray and Powell 1966). Perhaps the artificially imposed shortage of large males has made a naturally homogamous population disobey the theory.

Brachyura

Chidester (1911, p. 240) hints at a mechanical constraint on the sizes of mating pairs of *Cancer irroratus* (Cancridae); he wrote: 'in the case of a large male and a small female, the adjustments of their abdomens [at copulation] is very difficult'. Long precopulas are the rule in this family. They are also the rule in the Portunidae (p. 99), of which *Carcinus maenas* is a member. Duteurtre (1930) gave measurements of the cephalothoraxes of twelve pairs (too small a sample to use in the test), which show a non-significant negative correlation ($r = -0.53$). For *Trapezia* (Xanthidae), Patton (1974) published measurements of 20 pairs of *Trapezia cymodoce* and 15 pairs of *T. ferruginea* form *areolata* (which may not be separate species). They are homogamous ($r = 0.79, p < 0.01$ and $r = 0.59, 0.01 < p < 0.05$, respectively). Preston (1973, p. 471) also asserts that five species of *Trapezia* are homogamous, although it would be laborious to demonstrate it from the statistics that he published. We may conclude that *Trapezia* are homogamous. They have lengthy pairing (p. 100), and larger individuals of both sexes win territorial fights against smaller individuals of the same sex (Castro 1978). Preston (1973, p. 471) cites his own PhD thesis (1971, which I have not seen) to prove that larger females lay more eggs than smaller ones. *Trapezia* is another case of predicted and observed homogamy for size. The only other brachyuran which will concern us is the fiddler crab *Uca rapax* (Ocypodidae). Greenspan (1980) measured the breadths of the carapaces of eleven pairs: they showed homogamy (Spearman rank correlation $= 0.65, p < 0.05$). Greenspan also showed that larger females lay more eggs, and (probably) that larger males have a mating advantage. More females go to the burrows of large males than to small males (p. 389). The duration of mating is uncertain, but may be long enough to predict homogamy. Females enter the burrows of males, and may stay down them for hours, or even days (p. 388).

Peracarida

Amphipoda

Gammaridae

Homogamy for size has been relatively intensely studied in gammarids. First, Snyder and Crozier (1922) measured the lengths of pairs of *Gammarus locusta*; they re-published the same measurements in Crozier and Snyder (1923). They measured 'the total length, from anterior margin of cephalothorax to posterior margin of last abdominal segment . . . measured along the curved dorsal outline' (Crozier and Snyder 1923, p. 99). They found strong homogamy ($r = 0.91$, $n = 60, p < 0.001$). (There is, as first noticed by Willoughby and Pomerat (1932, p. 225), an inconsistency between Crozier and Snyder's two papers. Snyder and Crozier (1922) give $n = 62$ and, in the legend to a graph of the average size of mate of each of 15 male size classes, $r = 0.8$; Crozier and Snyder (1923) variously state $n = 61$ or 62, and $r = 0.91$. I recalculated the correlation coefficient from the points (of which there are 60) on the 1923 graph, obtaining $r = 0.91$.)

Crozier and Snyder (1923) also measured the lengths of 71 pairs of *Gammarus fasciatus* ($r = 0.69$, $p < 0.01$). Clemens (1950, p. 26) remarked of the same species that 'there may be some selective coupling in *G. fasciatus*'. More recently, van Dolah (1978, p. 199) measured the body lengths of 175 pairs of *Gammarus palustris*, and found significant homogamy ($r = 0.92, p < 0.01$). Hughes (1979) and Birkhead and Clarkson (1980) have measured *Gammarus pulex*. Hughes (1979) took twelve samples. Eleven of them had from 67 to 79 pairs, the twelfth only 18. The correlation in the small sample was not significant; of the other eleven, nine were significant. (Hughes expressed his results as regressions. It is not possible to calculate the correlation coefficients.) Birkhead and Clarkson (1980) took sixteen samples in seventeen months; the sample sizes were in the range 30-58, and the correlation coefficients from 0.25-0.58; thirteen of the sixteen were significant. We have not heard the last of Birkhead and Clarkson, but the evidence, so far, all points to homogamy.

Gammarids have long precopulas (p. 115); larger females, it has been abundantly proved, lay more eggs (e.g. Heinze 1932; Cheng 1942; Berg 1948; Kinne 1952; Birkhead and Clarkson 1980). In natural populations, the males in pairs have a larger average size than unpaired males (Bjerknes 1974 for *G. lacustris*; Birkhead and Clarkson 1980 for *G. pulex*). This is presumably caused by male competition. Males have been seen taking over paired females in the laboratory (Heinze 1932; Clemens 1950; Hughes 1979). Birkhead and Clarkson (1980) surprisingly found that large males were no more effective than smaller males at taking over paired females, but Ward (in preparation) has confirmed the more usual result, that bigger males have an advantage, by more extensive observations. The gammarids, if this were all, would count as a strong case of predicted and observed homogamy.

But it is not all. These are at least two other explanations of homogamy in

Gammarus: the mechanics of mating (Crozier and Snyder 1923), and the distribution in space of different sized individuals (Birkhead and Clarkson 1980). Let us look at them in turn. Crozier and Snyder thought that homogamy in *Gammarus* was mechanically necessary because small males were just not capable of holding large females. Birkhead and Clarkson directly refuted this. They showed that males could pair with, and fertilize, females regardless of their relative sizes. They then present a more interesting non-adaptationist hypothesis of their own. They suggest that homogamy results from the distribution of *Gammarus* in space. Large individuals inhabit some places, small individuals others. If they then pair up at random, a large sample which included several sites would show homogamy. Males pair with females of a similar size because those females happen to be nearby, not because of sexual competition to fertilize as many eggs as possible. Birkhead and Clarkson gave some evidence for their theory. They took nine (necessarily small) samples from specific geographical sites within a total sample which showed homogamy. None of these nine subsamples gave significant correlation coefficients. Such was the evidence. But is it convincing?

I think not. Birkhead and Clarkson's theory is the most interesting challenge to the theory of homogamy which we are testing, so it is worth examining in some detail. Let us look at their test first, and then go on to their argument. First, just look at the sizes of their sub-samples. They are small: n from 9 to 15. It is hardly surprising that they could not reject the null hypothesis with these little samples. (Birkhead and Clarkson are in the epistemologically precarious position of trying to confirm a null hypothesis.) Look next at the actual correlation coefficients in their nine samples. Eight of them are positive (with r from +0.14 to +0.77). Although they are severally nonsignificant, if we combine them by Fisher's method, they turn out (as a whole) to provide evidence of homogamy which just reaches statistical significance ($\chi^2_{18} = 29.93$, $0.01 < p < 0.05$).

If their evidence does not stand up, what of their argument? It too, can be crumbled by criticism. It confuses proximate and ultimate causes. That homogamy is sexually selected adaptation is perfectly compatible with non-random spatial dispersion according to size: this non-random dispersion may be the mechanism (or some manifestation of the mechanism) by which homogamy is caused. Only if they can give another argument to explain the spatial dispersion can sexual selection be rejected. But they do not give such an argument. The most likely place to look for one would be in the relation of *Gammarus* to its substrate. An individual *Gammarus* may settle in niches that are the same size as itself. It would perhaps then be less likely to be swept downstream by the current. (Even if this is true it would suggest that the non-adaptive homogamy of *Gammarus* is exceptional, because most of the species discussed in this chapter do not live, in niches, with a stream swirling over head.) But this theory can be refuted by evidence which Birkhead and Clarkson themselves provide. Within one locality, they found, the males are larger than the females. Spatial dispersion is not controlled only by size: it is also controlled by sex. This is difficult for Birkhead and Clarkson to explain, but it is readily explained if the spatial distribution reflects sexual selection: the males move around until they find themselves among females that they are likely to be able to keep in the face of mate competition.

Gammarus, in all, fits the canons of evidence necessary to predict homogamy. Birkhead and Clarkson cannot convince us that it does not show homogamy as well. We shall score it in favour of the theory.

The following remark by Holmes (1903), p. 289 on another gammarid, *Hyalella azteca* (Hyalellidae), suggests that small males cannot mate with large females.

The males, as a rule are considerably larger than the females and usually get their partners into the desired position quite readily; but when a small male attempts to carry a large female he experiences much difficulty. I have observed a male *Hyalella* endeavouring to carry a female somewhat larger than himself . . . Owing to the larger size of his partner the male could not reach around her body to carry her away . . . After watching the further struggles of the male for over half an hour I became convinced, though he was not, that he had undertaken an impossible task, and discontinued my observations.

Hyalella may be homogamous because of the mechanical impossibility of matings between small males and large females. *Hyalella* has a precopula (p. 115), so if it is homogamous, sexual selection may be the reason.

Isopoda

Sphaeromidae

Shuster (1981, Fig. 4) measured the body lengths of 54 pairs of a freshwater isopod, *Thermosphaeroma thermophilum*. They were homogamous ($r = 0.69$, $p < 0.001$). He also records a lengthy precopula (p. 123), that larger females lay more eggs (p. 710), and that larger males are more successful at guarding females (p. 704). The species therefore satisfies all the conditions which predict homogamy, and it shows it too. Shuster (pp. 701-2) even gave some evidence of a male preference for larger females. Males, when given the choice, initiated precopulas more with larger than with smaller females.

Asellidae

Ridley and Thompson (1979) measured the body lengths of several samples of pairs of *Asellus aquaticus*. They took three large samples and found significant homogamy in all of them (1. $r = 0.37, n = 108, p < 0.001$; 2. $r = 0.27, n = 123, p < 0.01$; 3. $r = 0.57, n = 25, p < 0.01$). *Asellus aquaticus* has a precopula of several days (p. 124). Ridley and Thompson also proved that the paired males were larger, on average, than unpaired males, and that larger males could take over paired females from smaller males; they showed, and reviewed previous evidence (their Table 1) that larger females lay more eggs. The homogamy of *Asellus*, therefore, fits the theory.

Janiridae

Veuille (1980), working with laboratory populations, measured the lengths of pairs of four species of *Jaera*. Two of them, *J. albifrons* ($r = 0.09$, $n = 32$) and *J. istri* ($r = 0.09$, $n = 103$) were not homogamous; the other two, *J. italica* ($r = 0.32$, $n = 83$, $p < 0.01$) and *J. nordmanni* ($r = 0.59$, $n = 35$, $p < 0.01$) were. (There is a little discrepancy between the sample sizes in Veuille's table and his figures. The table gives them as 30, 103, 83, and 36 respectively, but the figures have 32? (it is difficult to count), 83, and 35 points on them. I checked the first with M. Veuille, and he kindly informed me, in a letter, that the correct sample size for *albifrons* was 32. None of the discrepancies affect the statistical significances.) We have two homogamous species, and two non-homogamous ones. What would the theory predict?

In Chapter 3 we saw (p. 124) that *J. albifrons* has a short precopula, but the other three have long ones. So we can already predict the non-homogamous *J. albifrons* correctly. In *J. albifrons*, at least, it has been confirmed that larger females lay more eggs (Jones and Naylor 1971; Jones 1974), and we can assume that they do in all four. But what of the third variable? In *italica* and *nordmanni* Veuille found that the paired males were larger on average than the unpaired males. These two should be homogamous: and they are. In *albifrons* and *istri*, however, the average sizes of the paired and unpaired males were the same. Homogamy in *albifrons* has already been ruled out once by its short mating, but now we can rule it out in *istri* as well. All four species fit the theory. *J. italica* and *J. nordmanni* satisfy all the conditions, and are homogamous. *J. albifrons* and *J. istri* fail two and one of the conditions, and are not.

Arachnida

Xiphosura

Pomerat (1933) measured seven aspects of the size of 100 pairs of *Limulus polyphemus*, which he collected between 1 June and 5 June 1932. He then indulged in a correlating frenzy, calculating all 49 possible direct and cross correlations for the sizes of the pairs. Most of the direct correlations were in the range 0.1 to 0.15 and were not statistically significant.

Limulus (as we have seen, p. 129) have a precopula of uncertain duration. I have found nothing on the relation of female size and fecundity, and nothing of use on males. Milne and Milne (1967, pp. 24-5) do mention that males may attempt to take females from other males: 'often a suitor who hurries to a particularly fragrant female finds that she already has a partner. He may try to displace the male who already has hold of her'. We do not know whether larger males are any more successful in the competition for females.

Fish

Cichlidae

First there is some Lorenzian evidence that cichlids pair up non-randomly (by size) in the aquarium. Any cichlid ethologists will say that monogamous cichlids will only pair in an aquarium if the male and female are of the correct relative sizes. Lorenz (1935, pp. 324-5) is an early example. In order for a pair to form, he tells us, the male and female have to display to each other (with *Imponiergehaben*, or 'demonstrative behaviour'). A smaller fish never dares to display to a larger one, but flees instead. Thus,

> In these fish, collapse of the demonstrative behaviour of the female in response to that of the male signifies the failure of pair formation: the male then responds unspecifically with pursuit, and possibly killing, of the female. For this reason, it is hardly ever possible to pair off cichlids of greatly differing sizes. In particular, it is virtually impossible to induce a smaller female to hold her ground before the demonstrative behaviour of a much larger male.

This statement is based on Lorenz's observations of *Aequidens pulcher* and *Hemichromis bimaculatus*; both species form pairs for a long time, at least long enough to take care of the eggs and young (Baerends and Baerends-van Roon 1950, chs. VI and VII; Breder and Rosen 1966, pp. 467-9). Baerends and Baerends-van Roon (1950, p. 139) themselves showed experimentally that males of *Sarotherodon mossambicus* (which they called *Tilapia natalensis*) would display aggressively to larger individuals, but would 'invite' smaller individuals to pair with them. If all the males in the population behaved like this, the mated pairs would all consist of a male with a female of the same size or smaller than the male. This alone would show up as a correlation of the mate sizes.

Barlow (1970) and Barlow and Green (1970) conducted more systematic experiments on the effect of size on pair formation. They used two species: *Etroplus maculatus*, which is monogamous, and *Sarotherodon melanotheron* (Barlow used the older generic name *Tilapia*), which has only a brief association of the sexes for mating. In *Etroplus maculatus* the males tend to attack individuals larger than themsleves, and females attack individuals smaller than themselves (Barlow 1970, p. 795). Males, therefore, mate most easily with a smaller female, and females with a larger male: 'the usual outcome is a relatively peaceable pair with the male about 20 per cent heavier than the female' (Barlow 1970, p. 779). In *Sarotherodon melanotheron*, by contrast, 'no obvious species-typical size-relationship could be detected by . . . direct observation', and it 'is so peaceable and so tolerant of the relative size of its mate' (Barlow and Green 1970, p. 86). Barlow and Green (their Fig. 3) measured the proportion of successful pairing in relation to the male/female weight ratio. Although they found a slight trend towards easier pair formation as the female was relatively heavier, there was no significant difference between the pairing success of the different weight-ratio groups. The results for the two species fit the theory: the species with lengthy pairing is fussy about the size of its mates, the species with a brief mating is not.

There is also one paper on a field population of a cichlid. The results are again as we have come to expect. The species, *Cichlasoma maculicauda*: the biologist, Perrone (1978). He measured the lengths of 117 pairs: they were homogamous ($r = 0.6$, $p < 0.01$, his Fig. 2 has single points on it for '4-6 pairs', for instance, so I had to make some assumptions to calculate the correlation coefficient). *Cichlasoma maculicauda* is monogamous. It pairs for at least long enough to set up a territory, mate, and rear the eggs and young. Larger females lay more eggs (Perrone 1978). We do not know whether the paired males are larger, on average, than unpaired males, but we do know that larger males are better at defending territories than smaller males (Perrone 1978). If (as is not unlikely) territorial intruders include male conspecifics then larger males would be better than smaller males at defending their females.

Another kind of anecdotal evidence deserves to be mentioned, but it is not conclusive so we will not go into it in detail. In monogamous species, the male of the pair is nearly always found to be the larger individual. However, males usually have a higher average size so it is not possible to decide whether mating is random or selective. Here are some field observations: Matthew (1961, pp. 6 and 8) measured six pairs of *Boulengerochromis microlepis*, and in five the male was larger. The six are not significantly correlated. This species is probably monogamous. The males mature at a larger size than females. Gosse (1963, p. 225) says that the male of pairs of the monogamous *Hemichromis fasciatus* is always larger. Again the average size of males is higher in the population as a whole. Lowe-McConnell (1969, pp. 267, 270, 273, 278, 284-6, and 294) mentions six species in which the male is generally the larger member of a pair. None of these observations demonstrate non-random pairing, but the recurring pattern suggests that non-random pairing may be common in monogamous cichlids.

Salmonidae

Hanson and Smith (1967) measured the sizes of pairs of sockeye salmon *Onchorhynchus nerka*. They published their measurements not as the number of pairs on particular size combinations, but as the number of 'observed' pairs. One observation was made per day, so if a pair stayed together for a week it would appear as seven 'observed pairs' in Hanson and Smith's test. Their results are inflated by an unknown amount: it is as if each frequency (of a size combination) has been multiplied by a number between one and about ten (we will come to the duration of pairing soon). However, their results are sufficient (I think) to confirm homogamy. Let us look at their numbers. They measured pairs in 1964 and 1965. In 1964 they divided the fish into 'large' and 'small', giving a 2 × 2 test of whether pairing was random ($\chi_1^2 = 12.42$, 'p' < 0.005, the non-randomness can be seen by inspection to be homogamous). In 1965 they divided the males into six 'age' (in fact assessed by size) classes, and the females into four ($\chi_{15}^2 = 107.25$, $p < 0.005$). The non-randomness can be shown to be due to homogamy by calculating a correlation 'coefficient' ($r = 0.16$, 'n' = 1167, 'p' < 0.01, the n, of course, is not the number of pairs; it is the number of pair-observations: but even if the real n were one tenth as much it would almost be significant; 150 pairs would be needed for $r = 0.16$.).

At mating the female sockeye salmon usually first sets up a territory and then is later joined on it by one (or more) males. The sexes stay together until they spawn. Spawning may take place in several stages, and after the last the male leaves to seek another female (Hanson and Smith 1967, p. 1966; Greeley 1932, p. 244, tells us that the male leaves about five minutes after spawning). And how long do the sexes spend together from forming a pair until spawning? The answer, according to Hanson and Smith, is variable. Sometimes it is long, sometimes short. 'A few males were observed each with the same female during daily visits totalling a week or more. These males appeared to be virtually monogamous during this period', while other 'males remained with females only during a short period of redd [nest] preparation and spawning (often less than 24 hr)' (p. 1964). Because we are not told the frequencies of the two types of spawning, we cannot calculate the average time a pair spend together. We must turn instead to the observations of Hartman, Merrell, and Painter (1964). They marked ten pairs and then recorded every day whether they were together. The average time spent by a male at a redd was 4.7 days. (This may be a slight underestimate because the male could have been at a redd for up to a day before being first marked.) The evidence therefore suggests that males spend long enough on each mating to predict homogamy, if the other two variables are satisfied.

But are they? Take male size first. While a male is paired to a female he often has to fight other males off. It pays to be large in these fights: 'when a paired male was permanently displaced from a redd by another male, it usually meant that a larger male had taken over at the expense of a smaller one' (Hanson and Smith 1967, p. 1964; Schroder 1981 proves that large size is an advantage to males in the related chum salmon *O. keta*). Larger female sockeye salmon lay more eggs (literature summarized by Foerster 1968, pp. 123-8, especially Fig. 21), but we need to know more to be certain that a male would fertilize more eggs by mating with a larger female. A female does not lay all her eggs at once: 'females often spawned more than once in a period of a few days and sometimes spawned with different mates on each occasion (Hanson and Smith 1967, p. 1964). As we said in the section on methods, in the absence of evidence we will ignore the possibility that larger females do not lay more eggs at any one spawning. The sockeye salmon we will (tentatively) count as a case of predicted and observed homogamy for size.

Cyprinodontidae

Males of the desert pupfish *Cyprinodon* defend territories, in groups, which are briefly visited by the females to spawn. Kodric-Brown (1977) found some evidence of homogamy for size. We will consider the evidence in some detail before concluding that it is inadequate. She studied mating at two sites (named Area I and Area II), in two years (1970 and 1972). (Areas I and II are two regions of the same lake (of which there is a map in Kodric-Brown 1978); Area II is an inferior area for breeding by pupfishes.) She divided the males and females into 'large' and 'small' (the actual frequency distribution of sizes is continuous and approximately normal: Kodric-Brown 1977, Fig. 1), to give four 2 × 2 tables of the number of matings of the four combinations of size and sex, place and year. She then combined the data for the two areas within each year and calculated two χ^2 values. Both were highly significant, and due to homogamy. We, however, are interested in mate choice at each site. Will Kodric-Brown's analysis serve for us as well? No, it will not: and here is why it will not. First we must notice (in the small print below Kodric-Brown 1977's Table 6) that the criterion for distinguishing small from large differed between the two areas, so we cannot combine the two. Second, there is, in Kodric-Brown's Table 6, an unknown amount of redundancy. The matings were observed over five days in 1970 and 16 days in 1972. Each female might mate more than once (Kodric-Brown 1977, p. 751), and each male mated many times. The total number of males, it appears, was only four to six in each area ('the number of spawnings for four to six territorial males in each of the two types of breeding habitat was recorded', p. 752). The number of independent observations was clearly too low to draw any conclusions about homogamy. Third, the overall homogamy is mainly due to the fact that large fish mate in Area I, and small fish in Area II. If we calculate a separate χ^2 for each of the four cases, we find that the two for Area I are not significant (1970 $\chi^2 = 2.48$; 1972 $\chi^2 = 0.86$) and the two for Area II have too low 'expecteds' to conclude anything. (If we ignore the low 'expected', the 1970 Area II $\chi^2 = 28.2$, 'p' < 0.001.)

The sexual association is very short: 'spawning was rapid, taking less than a minute' (Kodric-Brown 1978, p. 821). Large females probably spawn more eggs at each visit to a male, for 'the number of spawning acts per female with a single male is correlated with the size of the female, and larger females carry a greater number of ripe eggs than smaller ones (Soltz 1974 [a PhD thesis which I have not seen]' (Kodric-Brown 1977, p. 757). Larger males exclude the smaller males from the best sites for territories (Kodric-Brown 1977, 1978).

The theory clearly predicts that *Cyprinodon* should not be homogamous. The males should be selected to mate with any female that comes to them, and Kodric-Brown (1977, p. 762) thinks that they do: 'males do not select mates according to their size, but spawn with all receptive females'. The evidence, however, does not allow us to conclude whether *Cyprinodon* is or is not homogamous.

Anura

Pelobatidae

Woodward (1982) has measured amplexing pairs of three species of *Scaphiopus*. He found no evidence of homogamy in any of them. He measured three samples of *S. couchi* ($n = 28, 19, 13$: $r = 0.286, 0.143, 0.2$), one of *S. bombifrons* ($n = 10, r = 0.269$), and one of *S. multiplicatus* ($n = 9, r = 0.526$). Most of these sample sizes are too small to satisfy our criterion (n at least 20) for evidence for random mating. Taken together they might even suggest slight homogamy, but there would be little justification for combining measurements on three species. The group must be excluded from the test for another reason: we do not know the duration of their amplexus. Chapter 3 contains nothing on them. Woodward does not give the duration of amplexus, although he (and he is not the only such author) does describe an 'explosive' breeding season. He does also show that the mating males are, on average, larger than the non-mating males.

Leptodactylidae

Crump (1974, p. 38) incidentally took some measurements on mating pairs of several species of *Eleutherodactylus* in her general study of a tropical anuran community. She found no evidence of homogamy, but unfortunately all the sample sizes are well below 20. She calculated r for the three species of which she measured nine or more pairs: *E. altamazonicus* ($r = -0.068, n = 9$), *E. croceoinguinis* ($r = 0.093, n = 9$), and *E. vaiabilis* ($r = 0.092, n = 10$). They provide, at most, perhaps a suggestion of random mating. The duration of mating is not known, although the meagre facts for confamilials (p. 155) suggest that it is quick. We know nothing about the relation between size and fitness in males. Crump did measure the relation between size and fecundity in the females of these species. Interestingly, only one of them (*altamazonicus*) showed a significant positive correlation. Larger females of the other two were no more fecund than smaller females, even though the sample sizes were large (86 and 148). If forced to draw a conclusion, we would read into the facts a predicted and observed lack of homogamy. But the facts are not sufficient to use in the test.

Hylidae

We have measurements of six species: *Hyla garbei*, *H. versicolor*, *H. marmorata*, *H. crucifer*, *H. cinerea*, and *Triprion patasatus*. *Hyla garbei* was part of the study by Crump which we met in the previous section. She found no evidence of homogamy ($r = -0.014, n = 19$). Nor is there strong evidence for homogamy in any of the other four *Hyla*. For *H. versicolor*, small samples by Fellers

(1979) and Gatz (1982) provide no evidence of homogamy ($r = 0.44, 0.14, 0.76, n = 9$ (it seems), 9, 6; n.s.). Nor could Gerhardt (1982, p. 590) find evidence of homogamy in *H. cinerea* in a larger sample ($r = 0.18, n = 66$, n.s.), or Gatz (1982) in a small sample of *H. crucifer* ($r = 0.14, n = 8$, n.s.). The other two species were studied by Lee and Crump (1982). The four studies which we have just summarized, like most of the studies on anurans, measured only one aspect (the total body length) of the frogs; they then calculated a correlation coefficient. Lee and Crump did more. They measured 18 aspects of the frogs and performed a 'principal components' analysis on the results. Their method would be more sensitive to any homogamy than the simpler methods used by other authors. It is, in a sense, too good for our test, because it is not comparable with the results of other authors. Their small sample sizes (9 and 16 pairs) also argue against including their results in our final test. But we will mention it anyway, for completeness. *H. marmorata*, like its congeners, showed no homogamy. For the first principal component, $r = 0.54$ ($n = 9$, n.s.). The correlation for one of the three principal components in *Triprion patasatus* did just creep into statistical significance ($r = 0.55, n = 16, 0.01 < p < 0.05$).

The duration of amplexus, as we have seen (p. 155), is short in all the hylids which have been appropriately studied (which include only two, *H. cinerea* an *H. versicolor* of the five in this section). In a small sample, Crump (1974) found no evidence that larger female *H. garbei* lay more eggs ($r = -0.055$, $n = 21$). The relation between size and fitness in males has been measured in five of the six species. In three, *H. cinerea, H. crucifer*, and *T. patasatus*, the males in amplexus were no larger than the males out of it; in the fourth (*H. marmorata*) the males in amplexus were on average larger than those out of it; and the fifth (*H. versicolor*) the paired males were not larger in one small study (Fellers) but were in another (Gatz). How, then, shall we count the hylids in their test? We know so little, and what we do know shows so many differences between species, that we would do well only to use those species for which we know enough to predict, without fear of unknown variables, whether they should be homogamous or not. That rules out all but *H. cinerea*. It lacks homogamy. It has a short amplexus and larger males are at no advantage, so it ought to lack it. This prediction will not be altered whether larger females turn out to lay more eggs, or no more eggs, than smaller females. *Triprion patasatus* may be an exception, but we know little about it, and it only just showed homogamy in a small sample with an exceptionally sensitive technique. I have therefore not counted it against the theory.

Ranidae

Homogamy for size has been measured in two species of *Rana*, one with a long amplexus and the other with a short one. We will take the former one first. It is the American woodfrog *Rana sylvatica*. Four populations have been measured, three by Berven (1982) and one by Howard (1980). Individually, two were significant and two not. Berven's three are all quite high, and the non-significant one is probably only due to the low sample size ($r = 0.44, 0.22, 0.28, n = 99, 83, 38$); Howard's population had a low non-significant correlation ($r = 0.09, n = 35$): perhaps populations differ. But we will follow our method of combining different estimates for the same species and using them as a single entry in the test. The combined evidence favours homogamy ($\chi_8^2 = 32.54, p < 0.001$).

The amplexus is long (p. 158). Larger females lay more eggs (Berven 1982; and a PhD thesis, which I have not seen, by Collins 1975, cited by Howard 1980). The average size of males in amplexus was higher than the average for the whole male population in both Berven's and Howard's frogs. The homogamy, therefore, is predicted. Berven provides yet further evidence for the theory. In an experiment in the laboratory, he demonstrated that males (when given the choice between different sized females) preferentially joined larger females in amplexus.

Howard (1978) measured the length (from snout to ischium) of 55 pairs of *Rana catesbiana*. They were homogamous ($r = 0.34, p < 0.01$). Amplexus in this species, as we have seen (p. 158), lasts only about 50 minutes. Howard found a correlation between the size and number of matings of a male, so larger males have a mating advantage. This is partly due to the superior prowess of larger males in battle with smaller males (Emlen 1976; Howard 1978, p. 861). Larger females lay more eggs (Howard 1978, p. 861), as they do in other ranids (Kuramoto 1978).

The short amplexus must lead us to predict random mating. But we are mislead: mating is homogamous. Howard explains homogamy by a learned preference for larger males in females. While they are still young, females (according to Howard) have yet to learn how to distinguish big males from small: older females, which are bigger than young females, choose bigger males. Thus a graph of the sizes of paired males and females encloses a triangle of points as small females mate with all sizes of males, and large females mate only with large males. Howard's Fig. 9 is roughly of this shape. The correlation coefficient alone is in this case not adequate to describe the relation between mate sizes: we also need to know how the points are distributed about the graph.

Howard also provides a reason why female *Rana catesbiana* should choose large males: the larger males defend better territories. His Fig. 13 shows that eggs have a lower mortality on the territories of larger males. It is crucial for Howard's case that females lay their eggs on the territory of the males they mate with. Emlen (1976) had watched this species in exactly the same pond (Crane Pond, near Ann Arbor, Michigan) in 1965 and 1966. Emlen (p. 303) stated that 'in all three cases of amplexus, the pair left the chorus area moving into deeper water under the power of the female. Of six egg masses located, five were outside of active chorus areas. This suggests that male territory location and female oviposition site were not synonymous in this study.' Howard (1978, p. 864) came to exactly the opposite conclusion: 'female bullfrogs use a male's territory as an egg deposition site. Out of 73 observed copulations, only three instances occurred in which females did not deposit their eggs at the original site of amplexus. The exceptions include (1) a parasitic male . . . and (2) two cases in which I disturbed the amplexed pair by approaching too closely.' Howard's conclusion stands on much more comprehensive observations, and so must be preferred. It also delicately hints at an explanation of Emlen's observations.

Howard has shown that a certain kind of female choice would explain his results. He has provided a valid argument to show why females should so choose. But he has not provided any evidence that they do so. If he is correct, *Rana catesbiana* will violate the letter, but not the true spirit of our law: males should not discriminate among females and they do not. Female choice could produce homogamy but it has been excluded from the theory. It was excluded because it is difficult to predict, from indirect evidence, in many species. Howard's arguments may apply well to *Rana catesbiana* but they could not support a full comparative study. Nor were they intended to. In the end we must count this frog against the theory.

Bufonidae

We shall examine, in this section, twelve studies of five species of *Bufo*. We will take the five American species first, and then the European toad *Bufo bufo*.

First, *Bufo americanus*. Four samples have been measured. Willoughby and Pomerat (1932) found slight homogamy in a large sample, which was made up of samples from more than one population ($r \simeq 0.3, n = 86, p < 0.01$); Licht (1976) found homogamy (he did not give the correlation in his original paper, but privately communicated it to Wilbur, Rubenstein, and Fairchild 1978, who published it: $r = 0.78, p < 0.01, n = 42$, Licht tells me, not 84 as Wilbur *et al.* state); Wilbur *et al.* found no homogamy in a small sample ($r = -0.098, n = 7$); and Kruse (1981) found none in a large sample ($r = 0.18, n = 63$, n.s.). How are we to use these contradictory reports? If they were all from a single study we could combine them by Fisher's method, as we have done on other occasions. If they were different species we could include them all according to their deserts. But they are separate studies of different populations of the same species. Should we combine them or include them separately? Let us return to first principles. The reason for combining studies is that they are thought to be separate estimates of one quantity. The reason for taking them separately is that they are thought to be estimates of different quantities. The question is whether we think that different populations of the same species should all be the same, or whether they could differ. They may, of course, differ: all populations of *Bufo americanus* do not have to have the same correlation. They may differ . . . but do we expect them to? All the other variables are (as we shall see) the same in all populations. The best course, therefore, is to take the level of species as an arbitrary frontier: we will combine all studies of the same species into a single estimate unless the theory predicts differences between populations. Without an arbitrary frontier there is a danger that the study will be biased by lots of contradictory little studies of single species.

In fact the probability of Licht's result is so small that when combined with the other three samples by Fisher's method, it swamps them, and the combined probability is still strongly significant. *Bufo americanus* is homogamous.

Wilbur *et al.* (1978) also measured the pairs of three larger samples of *B. quercicus*, which are severally and jointly insignificant (respectively: $r = 0.11$, -0.23, 0.32: $n = 20, 25, 13$: n.s.). (They also measured tiny samples of four pairs each of two other species of *Bufo*, *B. terrestris* × *americanus* and *B. terrestris*. Neither showed any significant correlation.)

Third, *Bufo woodhousei woodhousei*. Three samples measured by Woodward (1982) yield no evidence of homogamy (respectively: $r = 0.07, 0.13, 0.33$: $n = 27, 80, 17$: combined $\chi_6^2 = 6.8, p > 0.1$). Fourth, Fowler's toad *Bufo woodhousei fowleri*. Fairchild (1981) states, in a footnote, that the correlation between the sizes of mates was not significant. He does not state what it was, however, and the sample size can only be inferred as about 40 or 50. Furthermore he measured the correlation coefficient at those times when the sex ratio was most female biased: it was not a random sample from the breeding population. Fairchild's footnote, therefore, is not much use to us. The fifth and final American species is *Bufo typhlonius*; it was the subject of Wells (1979). He measured 64 pairs and found no evidence of homogamy ($r = 0.13, n = 64$, n.s.). The average size of the males in amplexus was larger than that of those out of it, but we cannot use the species because the duration of amplexus is uncertain. It may be long. The breeding season is explosive, and Wells noticed pairs already formed making their way to the pond, but he does not give us anything precise enough to use.

The remaining species for us to consider is *Bufo bufo bufo*. Two samples have been measured. Davies and Halliday's (1977) was homogamous ($r = 0.57, n = 16$, $p < 0.05$); but Gittins, Parker and Slater's (1980) is not so straightforward. Gittins *et al.* (1980) measured pairs both as the toads arrived at the lake, and at the spawning site. The pairs arriving were homogamous ($r = 0.29, n = 78$, $p < 0.05$); those at the pond showed a similar correlation, but it was statistically insignificant because of the smaller sample size ($r = 0.24, n = 41$, $p > 0.05$). These two correlation coefficients are not significantly different ($t = 0.325$, n.s.). So far, all the measurements tell a consistent story. Now we have a surprise. Gittins *et al.* found that of those 41 pairs at the spawning site, the 11 that happened to be spawning when collected had a quite different correlation ($r = -0.42, n = 11$, n.s.), which is significantly different from that ($r = 0.34$) of the 30 non-spawning pairs that they were among ($t = -2.03$, $0.02 < p < 0.05$). The explanation of this difference is not known. I cannot think of any explanation, and with such small sample sizes, and probabilities only just creeping below 5 per cent, I would attribute it to the play of chance. It remains an anomaly. The 16 pairs measured by Davies and Halliday (1977, p. 56) were 'actually engaged in spawning'. Furthermore, for their toads, Davies and Halliday found that 'before the start of spawning there was no significant correlation between the lengths of males and females in amplexus (on one day, $r = 0.094$, $n = 32$; on another day, $r = 0.299$, $n = 26$)'. Thus Davies and Halliday and Gittins et al. directly contradict each other. If we ignore the strange difference between the spawning and non-spawning pairs in the study of Gittins et al., we may reasonably conclude that Bufo bufo is homogamous. The difference between the two studies mainly concerns how the correaltion builds up on route to the pond.

In summary, *Bufo bufo* and *Bufo americanus* are homogamous; *B. quercicus, B. w. woodhousei*, and (probably) *B. w. fowleri* are not. What do the other three variables allow us to predict? We will work through the three, variously asigning species to the test at each stage. First, the fecundity of females. In the *Bufo bufo* studied by Davies and Halliday (1977) larger females laid more eggs. The same is true of Kruse's *B. americanus*, and of other *Bufo* (Clarke 1974). But in the populations of *Bufo quercicus* studied by Wilbur et al. (1978, pp. 265-6) 'the regression of clutch size on female snout-ischium length does not account for a significant amount of variation ($F_{1,14} = 0.83 \ldots$)'. This is exceptional, but it is enough to explain the absence of homogamy in this species. One for the theory.

During the period of amplexus the males often fight over females. Larger males can take females from smaller males (Davies and Halliday 1977, 1979; Gittins et al. 1980: all for *B. bufo*). The relative sizes of males in and out of amplexus differ between species. In *Bufo americanus* the paired males were no larger than the unpaired males in any of the three studies. However, in the *B. quercicus* studied by Wilbur et al. the paired males were larger. So were they in Woodward's *B. w. woodhousei*, Fairchild's *B. w. fowleri*, and both studies of *B. bufo*. We will take these facts as they stand, and conclude that in some species larger males have an advantage, but in others they do not. Another possibility is that the differences between species are not real. This is suggested by the finding of Davies and Halliday and of Gittins et al. that *Bufo bufo* initially pair up at random, and then larger males gradually take over the females. The average size of the paired males gradually increases relative to the population. Perhaps those studies which found no difference between the sizes of paired and unpaired males sampled the toads earlier than the others. But this is only a suggestion. It will not affect our interpretation. So we can now judge *B. americanus*. It should not be homogamous, but it is. One against the theory.

The duration of amplexus has been discussed in Chapter 3 (pp. 155-7). Species with long breeding seasons tend to have short amplexuses, species with short seasons have long amplexuses. *Bufo bufo* has a long amplexus, so its homogamy fits the theory. The duration of the amplexus of *B. quercicus* is not known (but that does not matter here). In *B. americanus* it is quite long (but that cannot save the theory). In the remaining two species it is not known: we will therefore exclude them from the test. If we trusted the rule established in the last chapter, we might reason from the long season of *B. w. fowleri* that its amplexus is short, and from the short season of *B. w. woodhousei* that its amplexus is long. They both lack homogamy, so one would fit the theory, and the other not.

Birds

Anatidae

Ankney (1977) measured the culmen lengths (which are correlated with body weight) of 48 pairs of the lesser snow goose *Chen caerulescens*. There was non-random mating for size; the male was larger than the female in a significantly greater number of pairs than would be expected by chance ($p = 0.004$, by a binomial test). This non-randomness should show up as a correlation, but in fact the correlation coefficient was not statistically significant ($r = 0.21$, $n = 48$, n.s.). If the result of the binomial test is trusted, then the most likely explanation of the insignificance of the correlation is that the sample size was too small. The result, however, is ambiguous, so we will not use it in the test.

The lesser snow goose, like other geese, is monogamous. Heavier (at the time of arrival at the breeding area) females lay more eggs (Ankney and MacInnes 1978).

Laridae

That in gulls the larger member of a pair is the male is almost universally accepted among gull field-workers. This fact is used to sex the members of the pairs. Goethe (1937, p. 9) was the first to measure some independently sexed gulls. He measured the bill, wing, and leg length of seven shot pairs of herring gulls *Larus argentatus*: in all seven the male was larger than the female (for bill length, $r = 0.21$, $n = 7$, n.s.). Tinbergen (1953, p. 104) tells us that Goethe's observation (that 'in every pair the male is larger than the female') had been 'fully confirmed by us'. The distributions of male and female bill lengths overlap, but the male mean is higher than the female mean (Goethe 1937, pp. 7-8); so it is not clear whether or not mating is random for size. Harris (1964) measured the bill length (and five other characters) of ten pairs of great black-backed gulls *Larus marinus*: in all ten the male was larger. Again the distributions of bill lengths overlap, the male mean being higher. The correlation coefficient for bill lengths was quite high but not statistically significant ($r = 0.5$, $n = 10$, n.s.). Again it is difficult to decide whether mating is random. Harris and Jones (1969, p. 133; a paper on herring gulls and lesser black-backed gulls *L. fuscus*) 'were able to examine both individuals from 15 pairs of gulls and each male had a larger bill than his mate. Field observations of other pairs confirmed this in every case.' I suspect, in view of the widespread belief that the male of the pair is always larger, that the pairing of gulls is non-random for size. The insignificance of the correlation coefficients (as in geese) is probably only due to the small sample sizes. Homogamy, however, cannot be proved with the facts we have; the sample sizes are too small to prove the absence of homogamy; we will therefore leave gulls out of the test.

Gulls are monogamous, but I have not found any facts on the relation between size and fitness in males or females.

Falconidae

In the American kestrel (*Falco sparverius*), as in other birds of prey, the female is larger than the male. Willoughby and Cade (1964, pp. 87-90) thought, in reverse Lorenzian fashion, that pair formation might not be possible unless the female was larger than, and dominant over, the male. They tried to test this by experimentally setting up pairs of artificial size combinations. They put together pairs in which the male was larger, others in which the female was larger, and others in which the male and female were of the same weight. They found no difference in the breeding success of the three kinds of birds in captivity. These limited experiments suggest that American kestrels are indiscriminate about the sizes of their mates. But the birds were forced to breed together (or not at all); it remains as a possibility that they might choose a particular sized bird as a mate if given the choice to begin with.

Geospizidae

Boag and Grant (1978) measured several aspects of the size of 61 pairs of the finch *Geospiza fortis*. It is one of 'Darwin's finches', on the Galápagos.) They found low correlations for most of the measurements, and all in all they suggest weak homogamy. The species is monogamous; but I have found nothing on the relation between size and fitness in males or females.

Mammals

Hominidae

There are hundreds of studies of homogamy for physical (and mental) characteristics in humans. Fortunately the results, for physical characteristics, have been tabulated by Roberts (1977), so we do not have to go through them all again here. Size has been measured in these studies in almost all imaginable ways: 'xiphoepigastric length', chest girth, middle finger length, wrist circumference, etc. In the 49 studies of homogamy for gross stature, correlation coefficients vary from -0.26 to $+0.63$; 28 were statistically significant (the one case of heterogamy not among them). What should we conclude? In some studies there is homogamy, in others there is not.

Similarly, we do not know whether humans are adapted for monogamy or polygamy (Symons 1979). In our test we could play humans as jokers: homogamy may or may not be predicted, and it may or may not be observed. A case could be made for any of the four possibilities. I have left them out of the test.

LEGISLATION

In Table 4.1 I have summarized the variables for all the species for which a correlation coefficient can be calculated. We have first to decide which species are independent tests of the theory, and then we can carry out the several tests which, as we saw in the section on methods, can be applied.

How are we to recognize the independent trials in Table 4.1? Outgroup comparison should not be applied in exactly the same fashion as in Chapter 3. The evidence is rather different. In Chapter 3 the evidence was drawn, relatively thickly, from a few higher taxa. In this chapter we have summarized only a few sparsely drawn species from many higher taxa. In Chapter 3 we generally knew the states of neighbouring taxa, and so could use them as outgroups. In this chapter we generally do not know the states of neighbouring taxa. We may have a species drawn from the Nematocera (Diptera), and the nearest outgroup is the Coleoptera: different species of Coleoptera are in different states; some are homogamous, some not. The theory, also, is of a different kind from Chapter 3. Whereas that chapter's theory had only two variables, each with two states, this one has four variables. And we have only summarized the evidence for the species in which one of them (homogamy) is known: we have not summarized all the evidence on the other three variables. But there is an independent evolutionary trial of the theory when any of the variables changes. Two of the variables (relations between male and female sizes and fitness) are fairly conservative, as we have seen. But the third (mating duration) is not. It has often changed during evolutionary time. Strictly, we should have summarized all the evidence on mating duration in all species: then we would know for sure when it has changed from long to short. I have preferred to keep the study down to a manageable size. I have not reviewed all the facts on mating duration. When we come to go through the evidence, I will simply assert for some of the groups that outgroups could be found with a different character state. I do not think these assertions are controversial.

We have three tests. The first two are 2×2 tables, of homogamy predicted/not versus homogamy observed/not. The first includes only trials for which all four variables are known so we can be certain what the theory predicts. The second is more inclusive, admitting all trials for which the theory makes a reasonable prediction. The third is a $2 \times n$ test: it compares (non-parametrically) the correlation coefficients of the species which the theory predicts to be homogamous with those of the species which it predicts to mate at random.

We will take them in order. Let us turn to Table 4.1 and, working down from the top, see which species can be admitted to the first test. The first is the lovebug *Plecia*. It lacks homogamy, although the theory predicts that it should have it. (This can be symbolized −+: not observed, but predicted.) The next is *Drosophila melanogaster* (+−), the third *Trapezia* (++), the fourth *Gammarus*: now we come to a cladistic problem. The long precopula of *Gammarus* is probably a derived state within the Amphipoda (p. 118), so we can count it as an independent trial. But all the *Gammarus* species can only be counted once. *Asellus* (++) continues what may be the ancestral peracaridan condition, and so too do *Thermosphaeroma* and two species of *Jaera* (++): how many times shall we count these? The *Jaera* should probably not be counted independently of *Asellus*. There may be *Asellus* species without precopulas (p. 169), and *J. albifrons* may be the ancestral species of the group (in which case the long precopulas would be re-derived). But the conservative procedure is to score only one trial from the Asellota. *Thermosphaeroma* I will score as an independent trial; this is a dubious decision, but is justified by the flabelliferan species which probably lack precopulas, and the unlikelihood that all the species between *Thermosphaeroma* and *Asellus* have an advantage to large size in males. There can be no doubt that the two other *Jaera* species are independent, predicted losses of homogamy (−−). The next species on the list are the two fish, the cichlid and the salmon (both ++). They are independent trials: long pairing is relatively rare in fish (Breder and Rosen 1966) and the two are phylogenetically distant; there are sure to be outgroups between the two with short pairings. The remaining species are two *Rana* (++ and +−) and one *Bufo* (++). Long amplexus is derived in both *Rana* and *Bufo* (pp. 159 and 156), so both ++ count. The nearest +− is *Drosophila*, so this too is independent. The result (allowing the isopods to be uncertain) is Table 4.2. The trend does not reach

Table 4.2

		Homogamy	
		Observed	Not observed
Homogamy	Predicted	7-8	1
	Not predicted	2	2

statistical significance (Fisher exact test, $p = 0.22$). But perhaps the numbers will be high enough in the second test to confirm the theory.

We start again at the top of Table 4.1. This time we can use the columns at the far right: they were prepared with this test in mind. We only have to decide which species count as independent trials. The two molluscs do (one is a hermaphrodite, the other dioecious, and they are not closely related: there must be plenty of species with short pairings between them). *Pyrrhocoris* is a hemipteran, far from the diptera *Plecia* so we will have them as independent trials too. *Drosophila melanogaster* has a different state (+−) so is another trial. *Scatophaga* and *D. subobscura* are more difficult. Short pairing may be common in the group to which they belong: it may be the ancestral state. Let us count only one trial from them. The next difficulty comes from the crustaceans. Long pairing and so predicted homogamy may be the ancestral state in the Eucarida-Peracarida. How many trials here? Let us eliminate the easy ones first. *Paralithodes* has its own state (−+); long pairing is probably derived in *Uca* relative to the ancestral brachyuran (p. 104), although the prediction here is not certain; the isopods we dealt with for the previous test. What of *Trapezia, Paguritta*, and *Alpheus*? Clearly the three *Alpheus* and two *Trapezia* can be counted no more than once each, but can they be counted that much? I think so. *Alpheus* is in the Natantia, far removed from the Paguridae (Anomura, Reptantia) and the Xanthidae (Brachyura, Repantia). There are many species with short pairings scattered among those taxa, and long pairing may be derived in *Trapezia* within the Xanthidae (p. 100). Let us count them as three cases. The two fish we have discussed already, so we come next to the frogs and toads. *Hyla cinerea* and *Bufo quercicus* (both −−) are independent: *Bufo quercicus* is predicted by the fecundity of its females, *Hyla* by its short pairing and lack of advantage to size in males. *Bufo bufo* and *Rana sylvatica* we have dealt with before. There remain the two species with +−: they are independent as well, for in one larger males have an advantage while in the other they do not. *Geospiza* may certainly be scored, and whether *Chen* should be admitted is less certain. If *Uca* and *Chen* are uncertain, the result is Table 4.3. The trend,

Table 4.3

| | | Homogamy | |
		Observed	Not observed
Homogamy	Predicted	14–16	3
	Not predicted	3	6

which is in the direction predicted by the theory, is statistically significant whether we count that 14 or 16 (Fisher exact test, $p = 0.018$ or 0.012). The cladistic technique has acted to remove cases which support the theory, but none which go against it: the number of *species* in the four categories, using the criteria for inclusion used in Table 4.3, are: 24, 3, 3, 7. Once again, the insistence on independent trials has made it more difficult to confirm the trend relative to the naïve technique of counting species.

The third test is carried out in the Appendix to this chapter. It uses the absolute values of the correlation coefficients rather than dividing them into the statistically significant and not significant. It tests whether there is any difference between the correlations of the species which the theory (by the same criteria as were used for Table 4.3) predicts to be homogamous, and those which it predicts to mate at random. There is. Those which the theory predicts to be homogamous have significantly higher correlations than those which it predicts not to be homogamous ($p < 0.005$). This test uses the correlation coefficient rather than its significance, so there is no need to exclude the correlations based on small samples; more species can therefore be used, and the statistical significance is even higher than in previous tests.

So much for the formal tests. They lend strong support for the theory, but further support can be found. For we have reviewed some kinds of evidence which cannot be used in these formal tests: Lorenzian evidence. The theory was also supported by all the experiments (except for a tiny experiment on falcons) in which the success of pair formation was studied for different size ratios of male to female. The monogamous cichlid was fastidious about its mates: the polygamous species was not. Lobsters, and the monogamous shrimp *Stenopus*, were also fussy about the size of their mates: they too fit the theory.

Although the theory is supported by the evidence, there are several exceptions. Table 4.3 contains six exceptions out of 26 (or 28) trials. We will finish by looking at these exceptions in some more detail. Do they suggest any other factors which our theory overlooks? Or are they too feeble to suggest anything other than imminent self-destruction as soon as the next set of observations come in from the field? Let us turn to those six. What were they again? Two flies, *Plecia* and *Drosophila*, a bug (*Pyrrhocoris*), a crab (*Paralithodes*), and two anurans (*Rana* and *Bufo*). We will proceed in two stages, asking first how strong exceptions they are, and then what other factors they suggest.

The important variable underlying the theory is the incidence of male choice. Males should choose when they invest sufficiently in each mating. Thus there are two possible kinds of exception to the theory. The first, weaker, kind would disobey the theory but not its underlying variable. This could happen if some extra variable over-rode the three in the theory. A stronger kind of exception would be a species which either showed male choice when males invested little in mating, or did not show male choice when males invested a lot. Are our six, in this sense, strong or weak exceptions? They are probably all weak, or even more feeble. We know very little about *Pyrrhocoris*: the prediction of homogamy was based on a single casual word describing the duration of mating. Further observations might well bring it into line. They might also tame *Plecia*: I suspect that a larger sample would reveal significant homogamy in the lovebug. The observations on *Drosophila* and *Paralithodes* are suspect as well. The method used for *Drosophila* is questionable, and *Paralithodes* may have been upset by commercial fishing. That leaves the two anurans. *Bufo americanus* is puzzling, although Kruse's failure to find homogamy makes one wonder whether Licht's result may not have been the play of chance. *Rana catesbiana* appears to be a real exception, but only a weak one. It has homogamy but only a short mating, but the homogamy is probably caused by female, not male, choice: the underlying theory is not contradicted.

The theory may apply to some taxonomic groups more than others. It seems to work best with crustaceans (the group from which my own study started). It tends to get frogs wrong as often as not, and does not cover the insects satisfactorily. What else, we are led to ask, do we need to know to predict homogamy in these groups? There are three main possibilities. The first, female choice, we have already met. It probably causes (or contributes to cause) homogamy in some groups, of which frogs are the most obvious candidates. I explained in the hypothesis section why I left female choice out of the theory, and will say no more about it here.

A second additional factor is the degree of male competition. In a few of the insects which we reviewed (particularly *Plecia, Scatophaga*, and *Popillia*) we found evidence of exceptionally intense male competition for females. The degree of male competition could interact with mating duration in a more sophisticated mode of homogamy. If competition is so intense that a male may well not mate at all, he will be less easily selected to discriminate among females. The difficulty with this factor, like with female choice, is that it is not easy to predict from the kind of facts which are available in the literature. It cannot be used in a comparative study.

The final factor is mechanical constraint. It is an idea which powerfully influenced earlier biologists. It dominates the literature before the mid-nineteen seventies as much as the theory of sexual selection dominates the papers of the last five or so years. They invoked it; but they never tested it. For tests we have to wait for students of sexual selection, and they have found uniformly negative results. Whenever artificial pairs of various size combinations have been set up, they have been found all to be completely fertile (Ridley and Thompson 1979 for *Asellus*; Birkhead and Clarkson 1980 for *Gammarus*). The direct evidence therefore goes against the theory of mechanical constraint. But we can mention also indirect evidence, and a comparative test. There is some suggestive indirect evidence. Michelsen's (1964) observations on *Rhagium* particularly hint at mechanically imposed non-random mating. For the comparative test we return to that hypothesis of Pomerat which we met in the Introduction, and which seemed to fit the facts so well in the nineteen thirties. Pomerat's theory was that in some kinds of species there will be stronger mechanical constraints on the sizes of mates than in others. Species with internal fertilization and hard exoskeletons should be homogamous; species with external fertilization, or soft exoskeletons (or both) should not. Insects should be homogamous, frogs not, crustaceans (we may reason) will be homogamous if they have internal fertilization, but not if the spermatophore is fixed to the outside (as in primitive decapods), or fertilization is external (as in *Gammarus*). We have the makings here of a final 2 × 2 table, for trials which Pomerat would predict to give homogamy, and trials which he would predict to give random mating, against the facts. Ah, 'trials'! We can entangle ourselves in cladism one last time. If the theory predicts homogamy for whole taxonomic groups (such as insects, all of which have hard exoskeletons and internal fertilization), it can have few independent trials. Let us look first at the insects. If *Leptinotarsa* and *Popillia* fitted the theory in 1932, the insects which have been measured since have provided little further support. *Drosophila subobscura*, both species of *Magicicada*, *Pyrrhocoris*, *Plecia*, *Scatophaga*, *Anisoplia*, *Coptocephala*, and *Tetraopes* all disobey it: only *Chauliognathus* and *Drosophila subobscura* fit it. It is not possible to apply outgroup comparison to these species. They are too sparsely distributed. We can, in this case, get a better idea of how the evidence deals with Pomerat's theory by putting the numbers of species in a 2 × 2 table (although, of course, we will not calculate a spurious 'probability'. The result is shown in Table 4.4. (I have left the two birds out of the Table.) By removing

Table 4.4 Numbers of species fitting and not fitting Pomerat's theory

		Homogamy	
		Observed	Not observed
Homogamy	Predicted	13	13
	Not predicted	15	5

the more grossly non-independent species (congeneric *Gammarus* and *Alpheus*, one each of the homogamous and the non-homogamous *Jaera*, and one *Trapezia*, one *Rana* and one *Bufo*) we do not improve the trend much (11, 12, 8, 4 are the resulting figures). The evidence clearly provides no support for Pomerat. A fanatical adherent of the methods of Chapter 1 would argue that it is at present untestable. A more flexible comparative biologist would look at the figures of Table 4.4 and below and conclude that the evidence goes against Pomerat.

A modern biologist will not be surprised by the fate of the theory of mechanical constraints. It is rejected by the comparative evidence; it has been directly refuted by experiment: but worse than that it is theoretically untenable. If a large male really were prevented from mating with small females because his genitals were too big, or a small male could not inseminate a large female because his apparatus were too small, then natural selection would rapidly adjust their sizes. The allometric relation is not inevitable. Even if there were an allometric relation of genital and body sizes (which has not been shown), it could be changed if it were causing a relatively reduced fitness for males with too big or too small genitalia.

The simple sexual selection theory which we have tested in this chapter appears to contain a part of the truth about homogamy; but it is probably not the only factor. As we search for a more general theory of homogamy we will have to take other factors into account. But first we will have to find out how they can be included in a comparative hypothesis which can be tested with the kind of facts that are likely to be available for many species.

APPENDIX TO CHAPTER 4, NON-PARAMETRIC COMPARISON OF CORRELATION COEFFICIENTS BETWEEN SPECIES PREDICTED AND NOT PREDICTED TO BE HOMOGAMOUS

At the beginning of the last section it is explained which species count as independent trials of the hypothesis. The following list is obtained by reading down the right-hand column ('Homogamy pred.') of Table 4.1: for each species with a + or − the correlation coefficient is taken from the third column (r). (If there is more than one estimate of r, I have taken the median.)

Predicted +			Predicted −		
Species	r	Rank	Species	r	Rank
Paralithodes	−0.05	1	Drosophila	0.006	2
Pyrrhocoris	0.03	3	Bufo quercicus	0.05	4
Onchorhynchus	0.16	10.5	Scatophaga	0.08	5
Chen	0.21	12	Jaera albifrons	0.09	7
Rana sylvatica	0.28	13	Jaera istri	0.09	7
Plecia	0.3	14.5	Hyla	0.09	7
Geospiza	0.3	14.5	Tetraopes	0.15	9
Chauliognathus	0.31	16	Drosophila	0.16	10.5
Larus	0.35	18	Rana	0.34	17
Bufo bufo	0.4	20	Bufo americanus	0.39	19
Asellus/Jaera	0.43	21			
Chomodoris	0.49	22			
Gammarus	0.6	23.5			
Cichlasoma	0.6	23.5			
Paguritta	0.63	25			
Uca	0.65	26			
Trapezia	0.69	27.5			
Thermosphaeroma	0.69	27.5			
Alpheus	0.73	29			

It is not perfectly obvious how to combine *Drosophila subobscura* and *Scatophaga*, which are both predicted −, but should probably not be scored independently. We will calculate the statistic twice, once including them both, and once excluding *Drosophila* (if one is to be removed, it is more conservative to exclude the lower. The ranks written down are for the first test; they should be altered for the second. The first test gives $U_{19,10} = 157.5$, $p < 0.05$, the second $U_{19,9} = 139.5$, $p < 0.005$. The facts fit the theory.

References

Aiken, D. E. and Waddy, S. L. (1980). Reproductive biology. In *The biology and management of lobsters* (2 Vols) (ed. J. S. Cobb and B. F. Phillips) Vol. I, pp. 215-76. Academic Press, New York.

Alcock, A. (1905). *Catalogue of the Indian Decapod Crustacea in the collection of the Indian Museum. Part II. Anomura. Fasciculus I. Pagurides.* Indian Museum, Calcutta.

Alcock, J., Barrows, E. M., Gordh, G., Hubbard, L. J., Kirkendall, L., Pyle, D. W., Porder, T. L., and Zalcom, F. G. (1978). The ecology and evolution of male reproductive behaviour in the bees and wasps. *Zoological Journal of the Linnean Society* **64**, 293-326.

Alpatov, W. W. (1925). Über die homogame und pangame Paarung im Tierreiche. *Zoologischer Anzeiger* **62**, 329-331.

Altmann, S. A. (1974). Baboons, space, time and energy. *American Zoologist* **14**, 221-48.

Amano, H. and Chant, D. A. (1978). Mating behaviour and reproductive mechanisms of two species of predacious mites, *Phytoseiulus persimilis* Athias-Henriot and *Amblyseius andersoni* (Chant) (Acarina: Phytoseiidae). *Acarologia* **XX**, 196-213.

— (1978a). Some factors affecting reproduction and sex ratios in two species of predacious mites, *Phytoseiulus persimilis* Athias-Henriot and *Amblyseius andersoni* (Chant) (Acarina: Phytoseiidae). *Canadian Journal of Zoology* **56**, 1593-607.

Ameyaw-Akumpfi, C. and Hazlett, B. A. (1975). Sex recognition in the crayfish *Procambarus clarkii*. *Science* **190**, 1225-6.

Anderton, T. (1909). The lobster (*Homarus vulgaris*). *Report of the Marine Department of New Zealand* 1908-9, 17-23.

Andrews, E. A. (1895). Conjugation in an American crayfish. *American Naturalist* **XXIX**, 867-73.

— (1904). Breeding habits of crayfish. *American Naturalist* **XXXVIII**, 165-206.

Ankney, C. D. (1977). Mate size and mate selection in lesser snow geese. *Evolutionary Theory* **3**, 143-7.

— and MacInnes, C. D. (1978). Nutrient reserves and reproductive performance of female lesser snow geese. *Auk* **95**, 459-71.

Antheunisse, L. J., van den Hoven, N. P., and Jefferies, D. J. (1968). The breeding characters of *Palaemonetes varians* (Leach) (Decapoda, Palaemonidae). *Crustaceana* **14**, 259-70.

Arakawa, K. Y. (1964). On mating behavior of giant Japanese crab, *Macrocheira kaempferi* de Haan. *Researches on Crustacea* (Carcinological Society of Japan, Tokyo) **1**, 41-6.

Arcangeli, A. (1948). Appunti sulla riproduzione degli Isopodi terrestri (Crostacei). *Bollettino dell'istituto e museo di zoologia della Università di Torino* **1**, 5-13.

Arthur, D. R. (1962) *Ticks and disease.* Pergamon, Oxford.

Atema, J., Jacobson, S., Karnofsky, E., Oleszko-Skuts, S., and Stein, L. (1979). Pair formation in the lobster, *Homarus americanus*: behavioral development, pheromones and mating. *Marine Behaviour and Physiology* **6**, 277-96.

Austad, S. N. (1983). The evolution of sperm priority patterns in spiders. In *Sperm competition and the evolution of animal mating systems* (ed. R. L. Smith). Academic Press, New York.

Ayer, A. J. (1972). *Probability and evidence.* Macmillan, London.

Bacesco, M. (1940). Les Mysidacés des eaux roumaines. *Annales Scientifiques de l'Université de Jassy* **26**, 453–803.

Baerends, G. P. and Baerends-von Roon, J. M. (1950). An introduction to the study of the ethology of cichlid fishes. *Behaviour* Suppl. 1.

Baird, W. (1850). *The natural history of the British Entomostraca.* Ray Society, London.

Baker, R. R. and Parker, G. A. (1979). The evolution of bird coloration. *Philosophical Transactions of the Royal Society of London* **B287**, 63–130.

Balinsky, B. I. (1969). The reproductive ecology of amphibians of the Transvaal highveld. *Zoologica Africana* **4**, 37–93.

— and Balinsky, J. B. (1954). On the breeding habits of the South African bullfrog, Pyxicephalus adspersus. *South African Journal of Science* **51**, 55–8.

Balss, H. (1957). Decapoda. VIII. Systematik. *Bronn's Klassen und Ordnungen der Tierreichs* **5**, I, 7 (12), 1505–672.

Banta, A. M. (1914). Sex recognition and the mating behavior of the wood frog, *Rana sylvatica. Biological Bulletin* **XXVI**, 171–83.

Barlow, G. W. (1970). A test of appeasement and arousal hypotheses of courtship behavior in a cichlid fish, *Etroplus maculatus. Zeitschrift für Tierpsychologie* **27**, 779–806.

— and Green, R. F. (1970). The problems of appeasement and of sexual roles in the courtship of the blackchin mouthbreeder, *Tilapia melanotheron* (Pisces: Cichlidae). *Behaviour* **XXXVI**, 84–115.

Barrows, E. M. and Gordh, G. (1978). Sexual behaviour in the Japanese beetle, *Popillia japonica. Behavioural Biology* **23**, 341–54.

Bateson, W. and Brindley, H. H. (1892). On some cases of variation in secondary sexual characters, statistically examined. *Proceedings of the Zoological Society of London* 585–94.

Bauer, R. T. (1976). Mating behaviour and spermatophore transfer in the shrimp *Heptacarpus pictus* (Stimpson) (Decapoda: Caridea: Hippolytidae) *Journal of Natural History* **10**, 415–40.

— (1979). Sex attraction and recognition in the caridean shrimp *Heptacarpus paludicola* Holmes (Decapoda: Hippolytidae) *Marine Behaviour and Physiology* **6**, 157–74.

Baumann, H. (1961). Der Lebensablauf von *Hysibius (H.) convergens* Urbanowicz (Tardigrada). *Zoologischer Anzeiger* **167**, 362–81.

Beavers, J. B. and Hampton, R. B. (1971). Growth, development and mating behavior of the citrus mite *Tetranychus* (Acarina: Tetranychidae). *Annals of the Entomological Society of America* **64**, 1804–6.

Beer, C. G. (1959). Notes on the behaviour of two estuarine crab species. *Transactions of the Royal Society of New Zealand* **86**, 197–203.

Bell, T. (1853). *A history of the British stalk-eyed Crustacea.* Van Voorst, London.

Berg, K. (1938). Studies on the bottom animals of Esrom Lake. *Kongelige Danske Videnskabernes Selskabs Skrifter* **9** (8), 1–255.

— (1948). Biological studies on the River Susaa. *Folia Limnologica Scandanavica* **4**, 1–318.

Berland, L. (1927). Contributions à l'étude des Arachnides. *Archives de Zoologie Experimentale et générale* **66**, *Notes et Revue* 7-29.
Berry, P. F. (1970). Mating behaviour, oviposition and fertilization in the spiny lobster Palinurus Homarus (Linaeus). *Investigational Report, Oceanographic Research Institute, South African Association for Marine Research* No. 24.
— and Hartnoll, R. G. (1970). Mating in captivity of the spider crab *Pleistacantha moseleyi* (Miers) (Decapoda, Majidae). *Crustaceana* **19**, 214-15.
— and Heydorn, A. E. F. (1970). A comparison of the spermatophoric masses and mechanisms of fertilization in southern African spiny lobsters (Palinridae). *Investigational Report, Oceanographic Research Institute, South African Association for Marine Research* No. 25.
Berven, K. A. (1981). Mate choice in the woodfrog, *Rana sylvatica. Evolution* **35**, 707-22.
Bethe, A. (1897). Das Centralnervensystem von Carcinus Maenas. (3 Mittheilung.) Ein anatomischphysiologischer Versuch. II Theil. *Archiv für Mikroskopische Anatomie* **51**, 382-452.
Betz, K.-H. (1974). Phänologie, Reproduktion und Wachstum der valviferen Assel *Idotea chelipes* (PALLAS, 1766) in der Schlei. *Kieler Meeresforshungen* **XXX**, 65-79.
Bhimachar, B. S. (1965). Life history and behaviour of Indian prawns. *Fishery Technology* **2**, 1-11.
Binford, R. (1913). The germ-cells and the process of fertilization in the crab, Menippe mercenaria. *Journal of Morphology* **24**, 147-'204'.
Bird, N. T. (1968). Effects of mating on subsequent development of a parasitic copepod. *Journal of Parasitology* **54**, 1194-6.
Birkhead, T. R. and Clarkson, K. (1980). Mate selection and guarding in *Gammarus pulex. Zeitschrift für Tierpsychologie* **52**, 365-80.
Bishop, M. J. (1982). Criteria for the determination of the direction of character state changes. *Zoological Journal of the Linnean Society* **74**, 197-206.
Bjerknes, Y. (1974). Life cycle and reproduction of *Gammarus lacustris* G. O. Sars (Amphipoda) in a lake at Hardangervidda, western Norway. *Norwegian Journal of Zoology* **22**, 39-43.
Blades, P. I. (1977). Mating behavior of *Centropages typicus* (Copepoda: Calanoida). *Marine Biology* **40**, 57-64.
— and Youngbluth, M. J. (1979). Mating behavior of *Labidocera aestiva* (Copepoda: Calanoida). *Marine Biology* **51**, 339-55.
Blair, W. F. (1961). Calling and spawning seasons in a mixed population of anurans. *Ecology* **42**, 99-110.
Blanke, R. (1974). Rolle der Beute beim Kopulationsverhalten von *Meta segmentata* (cl.) (Araneae, Araneidae). *Forma et Functio* **7**, 83-94.
Blegvad, H. (1922). On the biology of some Danish gammarids and mysids (*Gammarus locusta, Mysis flexuosa, M. neglecta, M. inermis*). *Report of the Danish Biological Station to the Board of Agriculture* **XXVIII**, 1-103.
Bles, E. J. (1905). The life-history of *Xenopus laevis*, Daud. *Transactions of the Royal Society of Edinburgh* **XLI**, 789-821.
Bliss, D. E. (1968). Transition from water to land in decapod crustaceans. *American Zoologist* **8**, 355-92.
Block, F. (1935). Contribution à l'étude des gametes, et de la fécondation chez les Crustacés décapodes. *Travaux de la Station Zoologique de Wimereux* **XII**, 185-277.

Boag, P. T. and Grant, P. R. (1978). Heritability of external morphology in Darwin's finches. *Nature, Lond.* **274**, 793-4.
Bock, K.-D. (1967). Experimente zue Ökologie von *Orchestia platensis* Kröyer. *Zeitschrift für Morphologie und Ökologie der Tiere* **58**, 405-28.
Bock, W. J. (1959). Preadaptation and multiple evolutionary pathways. *Evolution* **13**, 194-211.
— (1967). The use of adaptive characters in avian classification. *Proceedings of the XIV International Ornithologial Congress* 61-74.
— (1976). Adaptation and the comparative method. In *Major patterns in vertebrate evolution* (ed. M. K Hecht, P. C. Goody, and B. M. Hecht) pp. 57-82. Plenum Press, New York.
Böhme, W. (1981-). *Handbuch der Reptilien und Amphibien Europas.* Akademische Verlagsgesellschaft, Wiesbaden.
Bonnet, P. (1929). Les Araignées exotiques en Europe. II. Elevage à Toulouse de la grande Araignée fileuse de Madagascar et considérations sur l'Aragnéiculture. *Bulletin de la Société Zoologique de France* **54**, 501-23.
— (1935). Theridion tepidariorum C. L. Koch. Araignée cosmopolite. Répartition—Cycle vital—Moeurs. *Bulletin de la Société d'Histoire Naturelle de Toulouse* **68**, 335-86.
Bonnier, J. (1900). Contribution à l'étude des Epicarides. Les Bopyridae. *Travaux de la Station Zoologique de Wimereux* VIII. (Also published separately by Imprimerie L. Danel, Lille.)
Boolootian, R. A., Giese, A. C., Farmanfarmaian, A. and Tucker, J. (1959). Reproductive cycles of five west coast crabs. *Physiological Zoology* **32**, 213-20.
Borgia, G. (1981). Mate selection in the fly *Scatophaga stercoraria*: female choice in a male-controlled system. *Animal Behaviour* **29**, 71-80.
Borowsky, B. (1978). The relationship between tube-sharing and the time of the female's molt in *Microdeutopus gryllotalpa* and *Ampithoe valida* (Amphipoda). *American Zoologist* **18**, 621. [Abstract.]
— (1980). The pattern of tube-sharing in *Microdeutopus gryllotalpa* (Crustacea: Amphipoda). *Animal Behaviour* **28**, 790-7.
Borradaile, L. A. (1903). Marine crustaceans. Parts IV-VII. In *The fauna and geography of the Maldive and Laccadive Archipelagoes* (ed. J. Stanley Gardner). Cambridge University Press.
— (1907). On the classification of the Decapoda. *Annals and Magazine of Natural History*, 7th series **19**, 457-86.
Bott, R. (1940). Begattung und Eiblage von *Eupagurus prideauxi* Leach. *Zeitschrift für Morphologie und Ökologie der Tiere* **36**, 651-67.
Böttger, K. (1962). Zur Biologie und Ethologie der einheimischen Wassermilben *Arrhenurus* (*Megalurasarus*) *globator* (Müll.), 1776, *Piona nodata* (Müll.) 1776, and *Eylais infundibulifera meridionalis* (THON), 1899 (Hydrachnellae, Acari). *Zoologische Jahrbücher. Abteilung für Systematik, Ökologie und Geographie der Tiere* **89**, 501-84.
Boudreaux, H. B. (1963). Biological aspects of some phytophagous mites. *Annual Review of Entomology* **8**, 137-54.
Boulenger, G. A. (1896-7). *The Tailless Batrachians of Europe*, 2 Vols. Ray Society, London.
Bourdon, R. (1962). Observations préliminaires sur la ponte des Xanthidae. *Bulletin. Société Lorraine des Sciences* **2**, 3-28.
Bousfield, E. L. (1973). *Shallow-water Gammaridean Amphipoda of New England.* Cornell University Press, Ithaca.

Bousfield, E. L. (1978). A revised classification and phylogeny of amphipod crustaceans. *Transactions of the Royal Society of Canada*, 4th series **XVI**, 343-90.
Bowman, T. E. (1960). Description and notes on the biology of *Lironeca puhi*, n. sp. (Isopoda: Cymothoidae), parasite of the Hawaiian moray eel, *Gymnothorax eurostus* (Abbott). *Crustaceana* **1**, 84-91.
Bragg, A. N. (1936). Notes on the breeding habits, eggs, and embryos of *Bufo cognatus* with a description of the tadpole. *Copeia* 14-20.
— (1945). The spadefoot toads in Oklahoma with a summary of our knowledge of the group. II. *American Naturalist* **79**, 52-72.
Brandes, G. (1897). Zur Begattung der Dekapoden. *Biologisches Centralblatt* **17**, 346-50.
Breder, C. M and Rosen, D. E. (1966). *Modes of reproduction in fishes*. Natural History Press, New York.
Brian, M. V. and Brian, A. D. (1955). On the two forms macrogyna and microgyna of the ant *Myrmica rubra* L. *Evolution* **9**, 280-90.
Bristowe, W. S. (1929). The mating habits of spiders, with special reference to the problems surrounding sex dimorphism. *Proceedings of the Zoological Society of London* 309-58.
— (1958). *The world of spiders*. Collins, London.
Brocchi, P.-L.-A. (1875). Recherches sur les organes génitaux males des crustacés décapodes. *Annales des Sciences Naturelles* 6me sér. **II**, no. 2, 1-131.
Broekhuysen, G. J. (1936). On development, growth and distribution of Carcinides maenas (L.). *Archives Néerlandaises de Zoologie* **2**, 257-400.
— (1937). Some notes on sex recognition in Carcinides maenas (L.). *Archives Néerlandaises de Zoologie* **3**, 156-64.
— (1941). The life-history of *Cyclograpsus punctatus*, M. Edw.: breeding and growth. *Transactions of the Royal Society of South Africa* **28**, 331-66.
— (1955). The breeding and growth of *Hymenosoma orbiculare* Desm. (Crustacea, Brachyura). *Annals of the South African Museum* **41**, 313-43.
Brooks, W. F. and Herrick, F. H. (1891). The embryology and metamorphosis of the Macroura. *Memoirs of the National Academy of Sciences* **5**, 319-576.
Brusca, R. C. (1978a). Studies on the cymothoid fish symbionts of the eastern Pacific (Isopoda, Cymothoidae). I. Biology of *Nerocila californica*. *Crustaceana* **34**, 141-54.
— (1978b). Studies on the cymothoid fish symbionts of the eastern Pacific (Crustacea: Cymothoidae). II. Systematics and biology of *Lironeca vulgaris* Stimpson 1857. *Occasional Papers of the Allan Hancock Foundation*, n.s. No. 2.
— (1981). A monograph on the Isopoda Cymothoidae (Crustacea) of the eastern Pacific. *Zoological Journal of the Linnean Society* **73**, 117-99.
Buchli, H. (1968). Notes sur la mygale terricole *Cteniza moggridgei* (Pick. Cambr. 1874). *Revue d'Ecologie et de Biologie du Sol* **5**, 1-40.
Bückle Ramírez, L. F. (1965). Untersuchungen über die Biologie von *Heterotanais oerstedi* Kröyer (Crustacea, Tanaidacea). *Zeitschrift für Morphologie und Ökologie der Tiere* **55**, 714-82.
Bullar, J. F. (1876). The generative organs of the parasitic Isopoda. *Journal of Anatomy and Physiology* **XI**, 118-23.
Bunnell, P. (1973). Vocalizations in the territorial behavior of the frog *Dendrobates pumilio*. *Copeia* 277-84.
Burkenroad, M. D. (1939). Further observations on Penaeidae of the Northern

Gulf of Mexico. *Bulletin of the Bingham Oceanographic Collection (Peabody Museum of Natural History, Yale Unviersity)* **VI** (6), 1-62.

Burkenroad, M. D. (1947). Reproductive activities of decapod Crustacea. *American Naturalist* **LXXXI**, 392-8.

Butler, T. H (1960). Maturity and breeding of the Pacific edible crab, *Cancer magister* Dana. *Journal of the Fisheries Research Board of Canada* **17**, 641-6.

Butlin, B. K., Read, I. L. and Day, T. H. (1982). The effect of a chromosomal inversion on adult size and male mating success in the seaweed fly, *Coelopa frigida. Heredity* **49**, 51-62.

Cain, A. J. (1964). The perfection of animals. In *Viewpoints in biology* (ed. J. D. Carthy and C. L. Duddington.) Vol. 3, pp. 36-63. Butterworth, London.

— (1982). On homology and convergence. In *Problems of Phylogenetic Reconstruction* (ed. K. A. Joysey and A. E. Friday) pp. 1-19. Academic Press, London.

Cameron, W. P. L. (1925). The fern mite (*Tarsonemus tepidariorum*, Warburton). *Annals of Applied Biology* **XII**, 93-112.

Camin, J. H. (1953). Observations on the life history and sensory behaviour of the snake mite, *Ophionyssus natricis* (Gervais) (Acarina: Macronyssidae). *Special Publications of the Chicago Academy of Science* No. 10.

Campbell, F. M. (1883). On the pairing of Tegenaria guyoni, Guer., with a description of certain organs in the abdominal sexual region of the male. *Journal of the Linnean Society of London* **17**, 162-74.

Campbell-Parmentier, F. (1963). Vitellogenèse, maturation des ovocytes, accouplement et ponte en relation avec l'intermue chez *Orchestia gammarella* Pallas (Crustacé amphipode *Talitridae*). *Bulletin de la Société Zoologique de France* **LXXXVIII**, 474-88.

Carefoot, T. H. (1973). Studies on the growth, reproduction, and life cycle of the supralittoral isopod *Ligia pallasii. Marine Biology* **18**, 302-11.

Carlisle, D. B. (1957). On the hormonal inhibition of moulting in decapod Crustacea. II. The terminal anecdysis in crabs. *Journal of the Marine Biological Association* **36**, 291-307.

Castro, P. (1971). The natantian shrimps (Crustacea, Decapoda) associated with invertebrates in Hawaii. *Pacific Science* **25**, 395-403.

— (1978). Movements between coral colonies in *Trapezia ferruginea* (Crustacea, Brachyura), an obligate symbiont of scleractinian corals. *Marine Biology* **46**, 237-45.

Cavolini, Ph. (1787). *Memorie sulla generazione dei pesci e dei granchi.* Naples.

Champion, G. C. ['on behalf of Mr J. Edwards'] (1907). Forms of Osphya and concurrent species. *Proceedings of the Entomological Society of London* xxiv-xxv.

Chance, M. R. A. (1959). What makes monkeys sociable? *New Scientist* **5**, 520-2.

Chantran, M. S. (1870). Observations sur l'histoire naturelle des Ecrevisses. *Comptes rendues hebdomadaires des séances de l'Académie des Sciences* **LXXI**, 43-5.

— (1872). Sur la fécondation des Ecrevisses. *Comptes rendus hebdomadaires des séances de l'Academie des Sciences* **LXXIV**, 201-2.

Charniaux-Cotton, H. (1957). Croissance, régéneration et déterminisme endocrinien des characlères sexuels d'*Orchestia gammarella* (Pallas) Crustacé Amphipode. *Annales des Sciences Naturelles* **19**, 411-560.

Cheesman, L. E. (1923). Notes on the pairing of the land-crab, *Cardisoma armatum. Proceedings of the Zoological Society of London* 173.

Cheng, C. (1942). On the fecundity of some gammarids. *Journal of the Marine Biological Association* **25**, 467-75.

Cheung, T. S. (1966). An observed act of copulation in the shore crab, *Carcinus maenas* (L.). *Crustaceana* **11**, 107-8.

— (1968). Trans-molt retention of sperm in the female stone crab, *Menippe mercenaria*, *Crustaceana* **15**, 117-20.

Chhapgar, B. F. (1956). On the breeding habits and larval stages of some crabs of Bombay. *Records of the Indian Museum* **54**, 33-52.

Chidester, F. E. (1911). The mating habits of four species of the Brachyura. *Biological Bulletin* **21**, 235-48.

Chittleborough, R. G. (1976). Breeding of *Panulirus longipes cygnus* George under natural and controlled conditions. *Australian Journal of Marine and Freshwater Research* **27**, 499-516.

Churchill, E. P. (1917-18). Life history of the blue crab. *Bulletin of the United States Bureau of Fisheries* **36**, 91-128.

Clark, D. J. (1969). Notes on the biology of *Atypus affinis* Eichwald (Araneae-Atypidae). *Bulletin of the British Arachnological Society* **1**, 36-9.

Clarke, R. D. (1974). Postmetamorphic growth rates in a natural population of Fowler's toad, *Bufo woodhousei fowleri*. *Canadian Journal of Zoology* **52**, 489-98.

Claus, C. (1863). *Die Frei Lebenden Copepoden*. Wilhelm Engelmann, Leipzig.

— (1889). *Copepodenstudien. I. Peltidien*. Alfre Hölder, Vienna.

Cleaver, F. C. (1949). Preliminary results of the coastal crab (Cancer Magister) investigation. *Biological Report. Washington State Department of Fisheries* **49A**, 47-82.

Clemens, H. P. (1950). Life cycle and ecology of Gammarus fasciatus Say. *Contribution. The Fransz Theodore Stone Institute of Hydrobiology, Ohio State University* No. 12.

Clutter, R. I. (1969). The microdistribution and social behavior of some pelagic mysid shrimps. *Journal of Experimental Marine Biology and Ecology* **3**, 125-55.

— and Theilacker, G. H. (1971). Ecological efficiency of a pelagic mysid shrimp; estimates from growth, energy budget, and mortality studies. *Fishery Bulletin* **69**, 93-115.

Clutton-Brock, T. H. and Harvey, P. H. (1977). Primate ecology and social organization. *Journal of Zoology* **183**, 1-39.

— — (1977a). Species differences in feeding and ranging behaviour in primates. In *Primate ecology* (ed. T. H. Clutton-Brock) pp. 557-84. Academic Press, London.

— — (1978a). Mammals, resources and reproductive strategies. *Nature, Lond.* **273**, 191-5.

— — (1978b). Introduction. In *Reading in sociobiology* (ed. T. H. Clutton-Brock and P. H. Harvey) p. 314. W. H. Freeman, San Francisco.

— — (1979). Comparison and adaptation. *Proceedings of the Royal Society of London* **B205**, 547-65. [Also printed in Maynard Smith and Holliday (1979), *q.v.*]

— — (1980). Primates, brains and ecology. *Journal of Zoology* **190**, 309-23.

— — and Rudder, B. (1977). Sexual dimorphism, socioeconomic sex ratio and body weight in primates. *Nature, Lond.* **269**, 797-9.

Clyne, D. (1971). The mating of a huntsman. *Victorian Naturalist* **88**, 244-8.

Coker, R. E. (1923). Breeding habits of Limnoria at Beaufort, N.C. *Journal of the Elisha Mitchell Scientific Society* **39**, 95-100.

Collin, B. (1909). La conjugation d'Anoplophyra branchiarum (Stein) (A. circulans Balbiani). *Archives de Zoologie Expérimenale et Générale* (5me sér.) **I**, 345-88.

Compton, G. L. and Krantz, G. W. (1978). Mating behaviour and related morphological specialization in the uropodine mite, *Caminella peraphora*. *Science* **200**, 1300-1.

Cone, W. E., McDonough, L. M., Maitlen, J. C. and Burdajewicz, S. (1971a). Pheromone studies of the twospotted spider mite. 1. Evidence of a sex pheromone. *Journal of Economic Entomology* **64**, 355-8.

— Predki, S. and Klostermeyer, E. C. (1971b). Pheromone studies of the twospotted spider mite. 2. Behavioral response of males to quiescent deutonymphs. *Journal of Economic Entomology* **64**, 379-82.

Cooper, W. K. (1937). Reproductive behavior and haploid parthenogenesis in the grass mite, *Pediculopsis graminum* (Reut.) (Acarina, Tarsonemidae) *Proceedings of the National Academy of Science* **23**, 41-4.

Corkett, C. J. and McLaren, I. A. (1978). The biology of *Pseudocalanus*. *Advances in Marine Biology* **15**, 1-231.

Costa, M. (1966). Notes on macrochelids associated with manure and coprid beetles in Israel. II. Three new species of the *Macrocheles pisentii* complex, with notes on their biology. *Acarologia* **9**, 304-29.

Coste, M. (1948). Faits pour servir à l'histoire de la fécondation chez les Crustacés [report of M. Gerbe]. *Comptes rendus hebdomadaires des séances de l'Académie des Sciences* **XLVI**, 432-3.

Coutière, H. (1899). les Alphéidae, morphologie externe et interne, formes larvaire, bionomie. *Annales des Sciences Naturelles, Zoologie*, 8me ser. **9**, 1-599.

Cowles, R. P. (1913). The habits of some tropical Crustacea. *Philippine Journal of Science* **8**, 119-25.

Crane, J. (1947). Eastern Pacific expeditions of the New York Zoological Society. XXXVIII. Intertidal brachygnathous crabs from the west coast of tropical America, with special reference to ecology. *Zoologica* **32**, 69-95.

— (1949). Comparative biology of salticid spiders at Rancho Grande, Venezuela. Part IV. An analysis of display. *Zoologica* **34**, 159-214.

— (1975). *Fiddler crabs of the world*. Princeton University Press.

Crawford, G. I. (1937). A review of the amphipod genus *Corophium*, with notes on the British species. *Journal of the Marine Biological Association* **21**, 589-630.

Crisci, J. V. and Stuessy, T. F. (1980). Determining primitive character states for phylogenetic reconstruction. *Systematic Botany* **5**, 112-35.

Croker, R. A. (1971). A new species of *Melita* (Amphipoda: Gammaridae) from the Marshall Islands, Micronesia. *Pacific Science* **25**, 100-8.

Crook, J. H. (1964). The evolution of social organization and visual communication in the weaver birds (Ploceinae). *Behaviour* Suppl. 10.

— (1965). The adaptive significance of avian social organization. *Symposium of the Zoological Society of London* **14**, 181-218.

— and Gartlan, (1966). Evolution of primate societies. *Nature, Lond.* **210**, 1200-3.

Crowson, R. A. (1965). Observations on the constitution and subfamilies of the family Melandryidae. *Eos* **41**, 507-13.

Crozier, W. J. and Snyder, L. H. (1923). Selective coupling of gammarids. *Biological Bulletin* **45**, 97-104.

Crump, M. L. (1972). Territoriality and mating behaviour in *Dendrobates granuliferens* (Anura: Dendrobatidae). *Herpetologica* 28, 195-8.
— (1974). Reproductive strategies in a tropical anuran community. *Miscellaneous Publication. University of Kansas. Museum of Natural History* No. 61.
Cullen, J. M. (1972). Some principles of animal communication. In *Non-verbal Communication* (ed. R. A. Hinde) pp. 101-25. Cambridge University Press.
Curio, E. (1973). Towards a methodology of teleonomy. *Experientia* 29, 1045-58.
Daguerre de Hureaux, N. (1966). Etude du cycle biologique de *Sphaeroma serratum* au Maroc. *Bulletin de la Société des Sciences de Maroc* 46, 19-52.,
Dalens, H. (1977). Comportement constructeur chez l'Isopode *Nesiotoniscus corsicus* Racovitza. *Bulletin de la Société d'Histoire Naturelle de Toulouse* 113, 181-2.
Darlington, C. D. and Mather, K. (1949). *Elements of genetics*. George Allen & Unwin, London.
Darwin, C. R. (1868). *The variation of animals and plants under domestication*, 2 Vols. John Murray, London.
— (1871). *The descent of man, and selection in relation to sex*. John Murray, London.
Darwin, F. (ed.) (1887). *The Life and Letters of Charles Darwin*, 3 Vols. John Murray, London.
Darwin, F. and Seward, A. C. (eds.) (1903). *More letters of Charles Darwin*, 2 Vols. John Murray, London.
Daum, J. (1954). Zur Biologie einer Isopodenart unterirdischer Gewässer: *Caecosphaeroma (Virei) burgundum Dollfus*. *Annales Universitatis Saraviensis* III, 104-59.
David, R. (1936). Recherches sur la biologie et l'intersexualité de *Talitrus saltator* Mont. *Bulletin biologiques de la France et de la Belgique* 70, 332-57.
Davies, N. B. and Halliday, T. R. (1977). Optimal mate selection in the toad *Bufo bufo*. *Nature, Lond.* 269, 56-8.
— — (1979). Competitive mate searching in male common toads, *Bufo bufo*. *Animal Behaviour* 27, 1253-67.
Dawkins, R. (1976). *The selfish gene*. Oxford University Press. Oxford.
— (1982). *The extended phenotype*. W. H. Freeman, Oxford.
— and Carlisle, T. R. (1976). Parental investment, mate desertion and a fallacy. *Nature, Lond.* 262, 131-3.
De Beer, G. (1960). Darwin's notebooks on transmutation of species. Part I. *Bulletin of the British Museum (Natural History). Historical Series* 2.
De Geer, C. (1752-1778). *Mémoire pour Servir à l'Histoire des Insectes*, 7 Vols. Pierre Hesselberg, Stockholm.
De Jong, R. (1980). Some tools for evolutionary and phylogenetic studies. *Zeitschrift für Zoologische Systematik und Evolutionsforschung* 18, 1-23.
De Lafresnaye, F. (1848). Observations sur l'accouplement du crabe commun de nos cotes du Calvados, le *Cancer maenas* de Linné (aujourd'hui *Carcinus maenas* des auteurs). *Revue Zoologique, par la Société Cuvierienne* 279-82.
De L'Isle, A. (1876). Mémoire sue les moeurs et l'accouchement de l'*Alyetes obstetricians*. *Annales des Sciences Naturelles, Zoologie*, 6me sér. 3, no. 7, 1-51.
Della Valle, A. (1889). Deposizione, fecondazione, e segmentazione delle uova del Gammarus pulex. *Atti della Società dei Naturalisti e Matematici. Modena*, ser. 3 VIII, 107-20.
— (1893). *Gammarini del Golfo di Napoli*, 2 Vols. *Fauna und Flora des Golfes von Neapel* No. 20.

Desportes, C. and Andrieux, L. H. (1944). Sur la biologie de *Lepidurus apus* (L. 1761). *Bulletin de la Société Zoologique de France* **LXIX**, 61-8.

Devine, C. E. (1966). Ecology of *Callianassa filholi* Milne-Edwards 1878 (Crustacea. Thalassinidea). *Transactions of the Royal Society of New Zealand. Zoology* **8**, 93-110.

Dexter, D. M. (1971). Life history of the sandy beach *Neohaustorius schmidtzi* (Crustacea: Haustoriidae). *Marine Biology* **8**, 232-7.

Dingle, H. and Caldwell, R. L. (1972). Reproductive and maternal behavior of the mantis shrimp *Gonodactylus bredini* Manning, (Crustacea: Stomatopoda). *Biological Bulletin* **142**, 417-26.

Dobzhansky, T. (1956). What is an adaptive trait? *American Naturalist* **XC**, 337-47.

— (1970). *Genetics of the evolutionary process.* Columbia University Press, New York.

Dole, J. W. and Durant, P. (1974). Movements and seasonal activity patterns of *Atelopus oxyrhynchus* (Anura: Atelopodidae) in a Venezuelan cloud forest. *Copeia* 230-5.

Downes, J. A. (1966). Observations on the mating behaviour of the crab hole mosquito *Deinocerites cancer* (Diptera: Culicidae). *Canadian Entomologist* **98**, 1169-77.

— (1978). Feeding and mating in the insectivorous Ceratopogonidae (Diptera). *Memoirs of the Entomological Society of Canada* No. 104.

Downing, W. (1936). The life-history of *Psoroptes communis* var. *ovis* with particular reference to latent or suppressed scab. *Journal of Comparative Pathology and Therapeutics* **49**, 163-80 and 183-209.

Ducruet, J. (1973). Comportement sexuel spécifique et interspécifique chez les Gammares du group *pulex* (Crustacés Amphipodes). *Comptes rendus hebdomadaires des séances de l'Académie des Sciences* **276D**, 1037-9.

Dudich, E. (1925). Über die artliche Zugehörigkeit des *Asellus* von Ungarn, Polen, Dalmatien und Italien. *Zoologischer Anzeiger* **63**, 1-7.

Duellman, W. E. (1975). On the classification of frogs. *Occasional papers of the Museum of Natural History, University of Kansas,* No. 42.

— and Maness, S. J. (1980). The reproductive behavior of some hylid marsupial frogs. *Journal of Herpetology* **14**, 213-2.

Dufour, L. (1844). Anatomie générale des insectes. *Annales des Sciences Naturelles* 3me sér. **1**, 244-64.

Dunham, P. J. (1978). Sex pheromones in Crustacea. *Biological Reviews* **53**, 555-83.

— and Skinner-Jones, D. (1978). Intermolt mating in the lobster (*Homarus americanus*). *Marine Physiology and Behaviour* **5**, 209-14.

Durant, P. and Dole, J. W. (1974). Food of *Atelopus oxyrhynchus* (Anura: Atelopodidae) in a Venezuelan cloud forest. *Herpetologica* **30**, 183-7.

Du Réau de la Gaigonnière, L. (1908). Note sur l'apparition fréquente de *Lepidurus productus* (Leach) aux environs d'Angers. *Bulletin de la Société des Sciences Naturelles de l'Ouest de la France,* 2me sér. **VIII**, 187-91.

Duteutre, M. (1929). La promenade nuptiale chez les Crabes. *Bulletin de la Société d'Océanographie de France* **9**, 841-3.

— (1930). Mensurations de Carcinus moenas en promenade pré-nuptiale. *Compte rendu. Association française de l'avancement des sciences* **54**, 249-50.

Dybas, H. S. and Lloyd, M. (1972). Isolation by habitat in two synchronized species of periodical cicadas (Homoptera: Cicadidae: *Magicicada*). *Ecology* **43**, 444-59.

Eanes, W. F., Gaffney, P. M., Koehn, R. K., and Simon, C. M. (1977). A study of sexual selection in a natural population of the milkweed beetle *Tetraopes tetraophthalmus*. In *Measuring selection in natural populations* (ed. F. B. Christiansen and T. M. Fenchel) pp. 49–64. Springer-Verlag, Berlin.
Ebeling, W. (1959). Citrus pests in the United States. In *Subtropical fruit pests* (ed. W. Ebeling) pp. 144–7. University of California Press, Berkeley.
Edwards, D. C. (1968). Reproduction in *Olivella biplicata*. *Veliger* 10, 297–304.
— (1969). Zonation by size as an adaptation for intertidal life in *Olivella biplicata*. *American Zoologist* 9, 399–417.
Edwards, E. (1966). Mating behaviour in the European edible crab (*Cancer pagurus* L.). *Crustaceana* 10, 23–30.
Efford, I. E. (1967). Neoteny in the sand crabs of the genus *Emerita* (Anomura, Hippidae). *Crustaceana* 13, 81–93.
Eibl-Eibesfeldt, I. (1950). Ein Beitrag zur Paarungsbiologie der Erdkröte (*Bufo bufo* L.). *Behaviour* II, 217–36.
— (1956). Vergleichende Verhaltensstudien an Anuren. 2. Zur Paarungsbiologie der Gattungen Bufo, Hyla, Rana und Pelobates. *Zoologischer Anzeiger*. Suppl. 19, 315–23.
El-Badry, E. A. and Elbenhawy, E. M. (1968). Studies on the mating behaviour of the predaceous mite *Amblyseius gossipi* [Acarina, Phytoseiidae]. *Entomophaga* 13, 159–62.
— and Zaher, M. A. (1961). Life-history of the predator mite, *Typhlodromus (Amblyseius) cucumeris* Oudemans [Acarina: Phytoseiidae]. *Bulletin de la Societe Entomologique d'Egypte* XLV, 427–34.
Eldredge, N. and Cracraft, J. (1980). *Phylogenetic patterns and the evolutionary process*. Columbia University Press, New York.
Ellis, R. J. (1961). A life history study of *Asellus intermedius* Forbes. *Transactions of the American Microscopical Society* 80, 80–102.
— (1971). Notes on the biology of the isopod *Aellus tomalensis* Harford in an intermittent pond. *Transactions of the American Microscopical Society* 90, 51-61.
Eltringham, S. K. and Hockley, A. R. (1961). Migration and reproduction of the wood-boring isopod, *Limnoria*, in Southampton Water. *Limnology and Oceaography* 6, 467–82.
Embody, G. C. (1911). A preliminary study of the distribution, food and reproductive capacity of some fresh-water amphipods. *Internationale Revue der gesamten Hydrobiologie und Hydrographie*, 4, Biologische Supplement, III Serie, 1–32.
Emerton, J. H. (1878). *The structure and habits of spiders*. S. E. Cassino, Salem, Mass.
— (1890). New England spiders of the families Drassidae, Agalenidae and Dysderidae. *Transactions of the Connecticut Academy of Arts and Sciences* VIII (1888–1892) 166–206.
Emlen, S. T. (1976). Lek organization and mating strategies in the bullfrog. *Behavioral Ecology and Sociobiology* 1, 283–313.
— and Oring, L. W. (1977). Ecology, sexual selection, and the evolution of mating systems. *Science* 197, 215–23.
Engelmann, F. (1970). *The physiology of insect reproduction*. Pergamon, Oxford.
Enock, F. (1885). The life-history of *Atypus piceus*, Sulz. *Transactions of the Entomological Society of London* 389–420.
Enriques, P. (1908). Die Conjugation und sexuelle Differenzung der Infusorien.

Zweite Abhandlung: Wiederconjugante und Hemisexe bei *Chilodon*. *Archiv für Protistenkunde* **XII**, 213-76.

Evans, G. O., Sheals, J. G., and MacFarlane, D. (1961). *The terrestrial Acari of the British Isles*, Vol. 1. British Museum, London.

Ewing, A. (1961). Body size and courtship behaviour in *Drosophila melanogaster*. *Animal Behaviour* **IX**, 93-9.

— (1964). The influence of wing area on the courtship behaviour of *Drosophila melanogaster*. *Animal Behaviour* **XII**, 316-20.

Ewing, H. E. (1914). The common red spider or spider mite. *Bulletin. Oregon Agricultural Experimental Station* No. 121.

— (1918). The life history and behavior of the house spider. *Proceedings of the Iowa Academy of Sciences* **XXV**, 177-204.

Fage, L. (1951). Cumacés. *Faune de France* **54**, 1-136.

Fahrenbach, W. H. (1962). The biology of a harpacticoid copepod. *Cellule* **62**, 301-76.

Fain-Maurel, M. A. (1966). Contribution à l'histologie et à la caryologie de quelques isopodes. Spermatogenèse et infrastructure du spermatozoide des oniscides et des cymothoides. *Annales des Sciences Naturelles, Zoologie*, sér. 12, **8**, 1-188.

Fairchild, L. (1981). Mate selection and behavioral thermoregulation in Fowler's toads. *Science* **212**, 950-1.

Farmer, A. S. D. (1974). Reproduction in *Nephrops norvegicus* (Decapoda: Nephropidae). *Journal of Zoology* **174**, 161-83.

Feldman-Musham, B. and Borut, S. (1971). Copulation in ixodid ticks. *Journal of Parasitology* **57**, 630-4.

Fellers, G. M. (1979). Mate selection in the gray treefrog, *Hyla versicolor*. *Copeia* 286-90.

Fielder, D. R. and Eales, A. J. (1972). Observations on courtship, mating and sexual maturity in *Portunus pelagicus* (L., 1766) (Crustacea, Portunidae). *Journal of Natural History* **6**, 273-7.

Fish, J. D. and Mills, A. (1979). The reproductive biology of *Corophium volutator* and *C. arenarium* (Crustacea: Amphipoda). *Journal of the Marine Biological Association* **59**, 355-68.

Fishelson, L. (1966). Observations on the littoral fauna of Israel, V. On the habitat and behaviour of *Alpheus frontalis* H. Milne-Edwards (Decapoda, Alpheidae). *Crustaceana* **11**, 98-104.

Fisher, R. A. (1930). *The genetical theory of natural selection*. Clarendon Press, Oxford.

Fitzgerald, A. (1927). *The essays and hymns of Synesius of Cyrene*, 2 Vols. Oxford University Press, London.

Fleming, W. E. (1972). Biology of the Japanese beetle. *Technical Bulletin*, United States Department of Agriculture, no. 1449, 129 pp.

Foerster, R. E. (1968). *The sockeye salmon*. Fisheries Research Board of Canada, Ottawa. (Bulletin No. 62.)

Forsman, B. (1938). Untersuchungen über die Cumaceen des Skageraks. *Zoologiska Bidrag från Uppsala* **XVIII**, 1-162.

— (1944). Beobachtungen über *Jaera albifrons* Leach an der schwediscen Westküste. *Arkiv för Zoologi, Stockholm* 35A, No. 11.

— (1951). Studies on Gammarus duebeni Lillj. with notes on some rock pool organisms in Sweden. *Zoologiska Bidrag från Uppsala* **XXIX** (1949-1952), 215-37.

Forsman, B. (1956). Notes on the invertebrate fauna of the Baltic. *Arkif för Zoologi*, Ser. 2 **9**, 389–419.
Forster, R. R. and Forster, L. M. (1973). *New Zealand spiders*. Collins, Auckland.
Foxon, G. E. H. (1936). Notes on the natural history of certain sand-dwelling Cumacea. *Annals and Magazine of Natural History*, 10 Ser. **XVII**, 377–93.
Fraser, J. II. (1936). The occurrence, ecology and life history of *Tigriopus fulvus* (Fischer). *Journal of the Marine Biological Association* **XX** (1935–6), 523–36.
Fricke, H.-W. (1967). Garnelen als Kommensalen der tropischen Seeanemone *Discosoma*. *Natur und Museum* **97**, 53–8.
Friedrich, H. (1883). Die Geschlechtsverhaltnisse der Onisciden. *Zeitschrift für Naturwissenschaften* **LVI**, 447–74.
Fryer, G. (1960). The spermatophores of *Dolops ranarum* (Crustacea, Branchiura): their structure, formation, and transfer. *Quarterly Journal of Microscopical Science* **101**, 407–32.
Gadd, C. H. (1946). Observations on the yellow tea-mite *Hemitarsonemus latus* (Banks) Ewing. *Bulletin of Entomological Research* **37**, 157–62.
Garth, J. S. (1964). The Crustacea Decapoda (Brachyura and Anomura) of Eniwetok Atoll, Marshall Islands, with special reference to the obligate commensals of branching corals. *Micronesica* **1**, 137–44.
Garton, J. S. and Brandon, R. A. (1975). Reproductive ecology of the green treefrog, *Hyla cinerea*, in southern Illinois (Anura: Hylidae). *Herpetologica* **31**, 150–61.
Gatz, A. J. (1982). Size selective mating in *Hyla versicolor* and *Hyla crucifer*. *Journal of Herpetology* **15**, 114–6.
Gauld, D. T. (1957). Copulation in calanoid copepods. *Nature, Lond.* **180**, 510.
Gaylor, D. (1921). A study of the life history and productivity of *Hyalella knickerbockeri* Bate. *Proceedings of the Indiana Academy of Science* 239–50.
Geisselmann, B., Flindt, R. and Hemmer, H. (1971). Studien zur Biologie, Ökologie und Merkmalsvariabilität der beiden Braunfroscharten, *Rana temporaria* L. und *Rana dalmatina* Bonaparte. *Zoologische Jahrbücher. Abteilung für Systematik, Ökologie und Geographie der Tiere* **98**, 521–68.
Gerhardt, H. C. (1973). Reproductive interactions between *Hyla crucifer* and *Pseudacris arnata* (Anura: Hylidae). *American Midland Naturalist* **89**, 81–8.
— (1982). Sound pattern recognition in some North American treefrogs (Anura: Hylidae): implications for mate choice. *American Zoologist* **22**, 581–95.
Gerhardt, U. (1924). Weitere Studien über die Biologie der Spinnen. *Archiv für Naturgeschichte* **90a**, 85–192.
Gerlach, A. C. (1857). *Kratze und Rande*. Berlin.
Ghiselin, M. T. (1969). *The triumph of the Darwinian method*. University of California Press, Berkeley.
— (1974). *The economy of nature and the evolution of sex*. University of California Press, Berkeley.
Giese, A. C. and Pearse, J. S. (eds.) (1975–). *Reproduction in marine invertebrates*, 5 vols. to date. Academic Press, New York.
Gilat, E. (1962). The benthonic Amphipoda of the Mediterranean coast of Israel. II. Ecology and life history. *Bulletin of the Research Council of Israel, Section B, Zoology* **11B**, 71–92.

Ginet, R. (1967). Écologie, éthologie et biologie de *Niphargus* (Amphipodes Gammarides hypogés). *Annales de Spéléologie* **XV**, 127-376.

Gittins, S. P., Parker, A. G. and Slater, F. M. (1980). Mate assortment in the common toad (*Bufo bufo*). *Journal of Natural History* **14**, 663-8.

Gittleman, J. (1981). The phylogeny of parental care in fishes. *Animal Behaviour* **29**, 936-41.

Glatz, L. (1972). Der Spinnapparat haplogyner Spinnen (*Arachnida, Araneae*). *Zeitschrift für Morphologie der Tiere* **72**, 1-25.

— (1973). Der Spinnapparat der Orthognatha (Arachnida, Araneae). *Zeitschrift für Morphologie der Tiere* **75**, 1-50.

Gnewuch, W. T. and Croker, R. A. (1973). Macroinfauna of northern New England marine sand. I. the biology of *Mancocuma stellifera* Zimmer, 1943 (Crustacea: Cumacea). *Canadian Journal of Zoology* **51**, 1011-20.

Goethe, F. (1937). Beobachtungen und Untersuchungen zur Biologie der Silbermöwe (*Larus a. argentatus* Pontopp.) auf der Vogelinsel Memmertsand. *Journal für Ornithologie* **85**, 1-119.

Goodhard, C. B. (1939). Notes on the bionomics of the tube-dwelling amphipod, *Leptocheirus pilosus* Zaddach. *Journal of the Marine Biological Association* **XXIII** (1938-9), 311-25.

Goodman, N. (1955). *Fact, fiction and forecast.* Harvard University Press, Cambridge, Mass.

Gosse, J. P. (1963). Le milieu aquatique et l'écologie des poissons dans la région de Yangambi. *Annales du Musée Royal de l'Afrique Centrale, Tervuren Belgique*, série in-8° No. 116.

Gould, S. J. (1970). Dollo on Dollo's Law: irreversibility and the status of evolutionary laws. *Journal of the History of Biology* **3**, 189-212.

— and Lewontin, R. C. (1979). The spandrels of San Marco and the Panglossian paradigm: a critique of the adaptationist programme. *Proceedings of the Royal Society of London* **B205**, 581-98. [Also in Maynard Smith and Holliday (1979), *q.v.*]

Grafen, A. and Ridley, M. (1983). A model of mate guarding. *Journal of Theoretical Biology* **103** in press.

Grandjean, M. F. (1938). *Octodectes cynotis* (Hering) et les prétendues trachées des Acaridiae. *Bulletin de la Société Zoologique de France* **62**, 280-90.

Gravier, Ch. and Matthias, P. (1930). Sur la reproduction d'un Crustacé Phyllopode du groupe des Conchostracés (*Cyzicus cycladoides*, Joly). *Comptes rendus hebdomadaires de séances de l'Académie des Sciences* **151**, 183-5.

— — (1932). Sur la mode d'accouplement d'un Crustacé Phyllopode, *Cyzicus cycladoides* (Joly). *Archivio Zoologico Italiano* **XVI**, 1127-33.

Gray, G. W. and Powell, G. C. (1966). Sex ratios and distribution of spawning king crabs in Alitak Bay, Kodiak Island, Alaska (Decapoda, Anomura, Lithodidae). *Crustaceana* **10**, 303-9.

Greeff, R. (1866). Untersuchungen über den Bau und die Naturgeschichte der Bärthierchen. *Archiv für Mikroskopische Anatomie* **II**, 102-31.

Greeley, J. R. (1932). The spawning habits of brook, brown and rainbow trout and the problem of egg predators. *Transactions of the American Fish Society* **62**, 239-48.

Green, J. M. (1970). Observations on the behavior and larval development of *Acanthomysis sculpta* (Tatersall), (Mysidacea). *Canadian Journal of Zoology* **48**, 289-92.

Greenspan, B. N. (1980). Male size and reproductive success in the communal

courtship system of the fiddler crab *Uca rapax*. *Animal Behaviour* **28**, 387-92.
Gross, M. and Shine, R. (1981). Parental care and mode of fertilization in ectothermic vertebrates. *Evolution* **35**, 775-93.
Guest, W. C. (1979). Laboratory life history of the palaemonid shrimp *Macrobrachium amazonicum* (Heller) (Decapoda, Palaemonidae). *Crustaceana* **37**, 141-52.
Guinot, D. (1978). Principes d'une classification evolutive des Crustaces Decapodes Brachyoures. *Bulletin biologique de la France et de la Belgique* **CXII**, 211-92.
—— (1979). Données nouvelles sur la morphologie, la physiologie et la taxonomie des Crustacés Décapodes Brachyoures. *Memoires du Muséum National d'Histoire Naturelle*, ser. A. 112.
Haeger, J. S. and Phinizee, J. (1959). The biology of the crabhole mosquito Deinocerites cancer Theobald. *Report. Florida Antimosquito Association* **30**, 34-7.
—— and Provost, M. W. (1965). Colonization and biology of *Opifex fuscus*. *Transactions of the Royal Society of New Zealand, Zoology* **6**, 21-31.
Haempel, O. (1908). Über die Fortpflanzung und künstliche Zucht des gemeinen Klohkrebses (G. pulex und fluviatilis). *Allgeneine Fisch-Zeitung* 33.
Hailman, J. (1977). *Optical signals*. Indiana University Press, Bloomington.
—— (1981). Comparative studies. In *The Oxford companion to animal behaviour* (ed. D. J. McFarland) pp. 92-7. Oxford University Press.
Hanson, A. J. and Smith, H. D. (1967). Mate selection in a population of sockeye salmon (*Onchorhynchus nerka*) of mixed age-groups. *Journal of the Fisheries Research Board of Canada* **24**, 1955-77.
Haq, S. M. (1972). Breeding of *Eupertina acutifrons*, a harpacticoid copepod, with special reference to dimorphic males. *Marine Biology* **15**, 221-35.
Harris, M. P. (1964). Measurements and weights of great black-backed gulls. *British Birds* **57**, 71-5.
—— and Jones, P. H. (1969). Sexual differences in measurement of herring and lesser black-backed gulls. *British Birds* **62**, 129-33.
Hart, R. C. and McLaren, I. (1978). Temperature acclimation and other influences on embryonic duration on the copepod *Pseudocalanus* sp. *Marine Biology* **45**, 23-30.
Hartman, W. L., Merrell, T. R. and Painter, R. (1964). Mass spawning behavior of sockeye salmon in Brooks River, Alaska. *Copeia* 362-8.
Hartnoll, R. G. (1965). Notes on the marine grapsid crabs of Jamaica. *Proceedings of the Linnean Society* **17C**, 113-47.
—— (1968). Morphology of the genital ducts in female crabs. *Journal of the Linnean Society, Zoology* **47** (1967-68), 279-300.
—— (1968a). The female reproductive organs of *Lucifer* (Decapoda, Servestidae). *Crustaceana* **15**, 263-71.
—— (1968b). Reproduction in the burrowing crab, *Corystes cassivelaunus* (Pennant, 1777) (Decapoda, Brachyura). *Crustaceana* **15**, 165-70.
—— (1969). Mating in the Brachyura. *Crustaceana* **16**, 161-81.
—— and Smith, S. M. (1977). Pair formation and the reproductive cycle in *Gammarus duebeni*. *Journal of Natural History* **12**, 501-11.
Harvey, P. H. and Arnold, S. J. (1982). Female mate choice and runaway sexual selection. *Nature, Lond.* **297**, 533-4.

Harvey, P. H., Kavanagh, M., and Clutton-Brock, T. H. (1978). Sexual dimorphism in primate teeth. *Journal of Zoology* **186**, 475-85.
— and Mace, G. M. (1982). Comparisons between taxa and adaptive trends: problems of methodology. In *Current problems in sociobiology* (ed. King's College Sociobiology Group) pp. 343-61. Cambridge University Press.
— and Martin, R. D. (1983). The comparative method in evolutionary biology. In *Evolution, adaptation and behavioral science* (ed. I. B. DeVore and J. Tooby). Aldine, Chicago.
Harvey, W. H. (1860). Darwin on the origin of species. *Gardener's Chronicle and Agricultural Gazette* 18 February 1860, 145-6.
— (1861). [Review of the *Origin*.] *Dublin Hospital Gazette* 15 May 1861, 150.
Havinga, B. (1930). Der Granat (*Crangon vulgaris* Fabr.) in den holländischen Gewässern. *Journal du Conseil* **V**, 57-87.
Hay, W. P. (1905). The life history of the blue crab (*Callinectes sapidus*). *Report of the Bureau of Fisheries* 395-414.
Hazlett, B. A. (1966). Social behavior of the Paguridae and Diogenidae of Curaçao. *Studies on the Fauna of Curaçao and other Caribbean Islands* **XXIII**, No. 88.
— (1968). The sexual behavior of some European hermit crabs (Anomura: Paguridae). *Pubblicazioni della Stazione Zoologica di Napoli* **36**, 238-52.
— (1970). Tactile stimuli in the social behavior of *Pagurus bernhardus* (Decapoda, Paguridae). *Behaviour* **36**, 20-48.
— (1972). Shell fighting and sexual behavior in the hermit crab genera *Paguristes* and *Calcinus*, with comments on *Pagurus*. *Bulletin of Marine Science* **22**, 806-23.
— (1975). Ethological analyses of reproductive behavior in marine Crustacea. *Pubblicazioni della Stazione Zoologica di Napoli* **39**, Suppl. 677-95.
Heinze, K. (1932). Fortpflanzung und Brutpflege bei Gammarus pulex L. und Carinogammarus Roeselii Gerv. *Zoologische Jahrbücher. Abteilung für allgemeine Zoologie und Physiologie der Tiere* **51**, 397-440.
Heinzmann, U. (1970). Untersuchungen zur Bio-Akustik und Ökologie der Geburtshelferkröte, *Alyetes o. obstreticians* (Laur.) *Oecologia* **5**, 19-55.
Heldt, H. (1931). Observations sur la ponte, la fécondation et les premiers stades du développement de l'oeuf chez *Penaeus caramote* Risso. *Comptes rendus hebdomadaires de séances de l'Académie des Sciences* **193**, 1039-41.
Helle, W. (1967). Fertilization in the two-spotted mite (*Tetranychus urticae*: Acari). *Entomologica Experimentitia et Applicata* **10**, 103-10.
— and Overmeer, W. P. J. (1973). Variability in tetranychid mites. *Annual Review of Entomology* **18**, 97-120.
Heller, C. (1864). Horae dalmatinae. Bericht über eine Reise nach der Ostküste des adriatischen Meeres. *Verhandlungen der kaiserlich-königlichen zoologisch-botanischen Gesellschaft in Wien* **XIV**, 17-64.
Henderson, J. T. (1924). The gribble: a study of the distribution factors and life-history of *Limnoria lignorum* at St Andrews, N.B. *Contributions to Canadian Biology*, n.s. **2**, 309-25.
Hennecke, J. (1911). Beiträge zur Kenntnis der Biologie und Anatomie der Tardigraden (Macrobiotus macronyx Duj.) *Zeitschrift für Wissenschaftliche Zoologie* **97**, 721-32.
Hennig, W. (1966). *Phylogenetic systematics.* University of Illinois Press, Urbana.
Henning, H. G. (1975). Kampf-, Fortpflanzungs- und Häutungsverhalten-Wachstum

und Geschlechtsreife von *Cardisoma gaunhumi* Latreille (Crustacea, Brachyura). *Forma et Functio* **8**, 463-510.

Henry, J.-P. (1964). Contribution à l'étude de la biologie d'*Asellus cavaticus* Leydig. *International Journal of Speleology* **1**, 279-86.

Herreid, C. F. and Kinney, S. (1967). Temperature and development of the wood frog, *Rana sylvatica*, in Alaska. *Ecology* **48**, 579-90.

Herrick, F. H. (1895). The American lobster: a study of its habits and development. *Bureau of the US Fisheries Commission* **XV**.

Hess, W. N. (1941). Factors influencing moulting in the crustacean, *Crangon armillatus*. *Biological Bulletin* **81**, 215-20.

Hetrick, L. A. (1970). Biology of the 'love-bug', *Plecia nearctica* (Diptera: Bibionidae). *Florida Entomologist* **53**, 23-6.

Heusser, H. (1961). Die Bedeutung der ässeren Situation im Verhalten einiger Amphibienarten. *Revue Suisse de Zoologie* **68**, 1-39.

— (1963). Die Ovulation des Erdkrötenweibchens im Rahmen der Verhaltensorganisation von *Bufo bufo* L. *Revue Suisse de Zoologie* **70**, 741-58.

Heydorn, A. E. F. (1969). Notes on the biology of *Panulirus homarus* and on length/weight relationships of *Jasus lalandii*. *Investigational Report, Division of Sea Fisheries, South Africa* No. 69.

Hiatt, R. W. (1948). The biology of the lined shore crab, *Pachygrapsus crassipes* Randall. *Pacific Science* **2**, 135-213.

Hill, L. L. and Coker, R. E. (1930). Observations on mating habits of Cyclops. *Journal of the Elisha Mitchell Scientific Society* **45**, 206-20.

Hinde, R. A. and Tinbergen, N. (1958). The comparative study of species-specific behavior. In *Behavior and evolution* (ed. A. Roe and G. G. Simpson) pp. 251-68. Yale University Press, New Haven.

Hinsch, G. W. (1968). Reproductive behavior in the spider crab, *Libinia emarginata* (L.). *Biological Bulletin* **135**, 273-8.

Hipeau-Jacquotte, R. (1973*a*). Manifestation d'un comportement territorial chez les crevettes Pontoniinae associées aux Mollusques Pinnidae a Tuléar (Madagascar). *Journal of Experimental Marine Biology and Ecology* **13**, 63-71.

— (1973*b*). Etude des crevettes Pontoniinae (Palaemonidae) associées aux Mollusques Pinnidae a Tuléar (Madagascar). 5. L'infestation dans les conditions naturelles. *Téthys* **5**, 383-402.

— (1973*c*). Etude des crevettes Pontoniinae (Palaemonidae) associées aux Mollusques Pinnidae a Tuléar (Madagascar). 6. Comportement sexuel. *Téthys* **5**, 403-8.

— (1974). Etude des crevettes Pontoniinae (Palaemonidae) associées aux Mollusques Pinnidae a Tuléar (Madagascar). *Archives de Zoologie Expérimentale et Générale* **115**, 359-86.

Hiraiwa, Y. K. (1935-6). Studies on a bopyrid, *Epipenaeon japonica* Thielemann. III. Development and life-cycle, with special reference to the sex differentiation in the bopyrid. *Journal of Science of the Hiroshima University*, Series B, Div. 1 (Zoology) **IV**, 101-41.

Hoestlandt, H. (1948). Recherches sur la biologie de l'*Eriocheir sinensis* H. Milne-Edwards (Crustacé Brachyoure). *Annales de l'Institut Océanographique. Monaco* **XXIV**.

Hoffman, D. L. (1973). Observed acts of copulation in the protandric shrimp, *Pandalus platyceros* Brandt (Decapoda, Pandalidae). *Crustaceana* **24**, 242-4.

Höglund, H. (1943). On the biology and larval development of *Leander squilla*

(L.) forma *typica* de Man. *Svenska Hydrografisk-Biologiska Kommissionens Skrifter*, n.s., II No. 6.

Holdich, D. M. (1968). Reproduction, growth and bionomics of *Dynamene bidentata* (Crustacea: Isopoda). *Journal of Zoology* **156**, 137–53.

Holmes, S. J. (1901). Observations on the habits and natural history of Amphithoe longimana Smith. *Biological Bulletin* **II**, 165–93.

— (1903). Sex recognition among amphipods. *Biological Bulletin* **V**, 288–92.

— (1909). Sex recognition in Cyclops. *Biological Bulletin* **XVI**, 313–15.

Homsher, P. J. and Sonenshine, D. E. (1976). The effect of presence of females on spermatogenesis and early mate seeking behaviour in two species of *Dermacentor* ticks (Acari: Ixodidae). *Acarologia* **18**, 226–33.

Hoogland, J. H. and Sherman, P. W. (1976). Advantages and disadvantages of bank swallow (*Riparia riparia*) coloniality. *Ecological Monographs* **46**, 33–58.

Horn, H. S. (1968). The adaptive significance of colonial nesting in the Brewer's blackbird (*Euphagus cyanocephalus*). *Ecology* **49**, 682–94.

Hotovy, R. (1937). Zur Kopulation von Triops cancriformis Bosc. *Zoologischer Anzeiger* **120**, 29–32.

Howard, R. D. (1978). The evolution of mating strategies in bullfrogs, *Rana catesbiana*. *Evolution* **32**, 850–871.

— (1980). Mating behaviour and mating success in woodfrogs, *Rana sylvatica*. *Animal Behaviour* **28**, 705–16.

Hudinaga, M. (1942). Reproduction, development and rearing of *Penaeus Japonicus* Bate. *Japanese Journal of Zoology* **X**, 305–93.

Hughes, D. A. (1973). On mating and the 'copulation burrows' of crabs of the genus *Ocypode* (Decapoda, Brachyura). *Crustaceana* **24**, 72–6.

Hughes, J. T. and Matthiessen, C. (1962). Observations on the biology of the American lobster, *Homarus americanus*. *Limnology and Oceanography* **7**, 414–21.

Hughes, R. V. (1979). Precopula in *Gammarus pulex*. Unpublished PhD thesis, University of Liverpool.

Hull, D. L. (1978). A matter of individuality. *Philosophy of Science* **45**, 335–60.

Husson, R. and Daum, J. (1953). Sur la biologie de *Caecosphaeroma burgundum*. *Comptes rendus hebdomadaires des séances de l'Académie des Sciences* **236**, 2345–7.

Huxley, J. S. (1932). *Problems of relative growth*. Methuen, London.

Hyman, L. S. (1955). *The invertebrates*, Vol. 5 Echinoderms. McGraw-Hill, New York.

Hynes, H. B. N. (1955). The reproductive cycle of some British freshwater Gammaridae. *Journal of Animal Ecology* **24**, 352–87.

Idyll, C. P. (1971). The crab that shakes hands. *National Geographic Magazine* **139**, 254–71.

Ikeda, S. (1897). Notes on the breeding habit and development of Racophorus schlegelii, Günther. *Annotationes Zoologicae Japanenses* **I**, 113–22.

Inger, R. F. (1967). The development of a phylogeny for frogs. *Evolution* **21**, 369–84.

Ingle, R. W. and Thomas, W. (1974). Mating and spawning of the crayfish *Austropotamobius pallipes* (Crustacea: Astacidae). *Journal of Zoology* **173**, 525–38.

Itô, T. (1970). The biology of the harpacticoid copepod, *Tigriopus japonicus* Mori. *Journal of the Faculty of Science, Hokkaido University (Zoology)* **17**, 474–500.

Jackson, H. H. T. (1912). A contribution to the natural history of the amphipod, *Hyalella knickerbockeri* (Bate). *Bulletin. Wisconsin Natural History Society* **10**, 49-60.

Jackson, R. R. (1977). Web sharing by males and females of dictynid spiders. *Bulletin of the British Arachnological Society* **4**, 109-12.

— (1978). The mating strategy of *Phidippus johnsoni* (Aranae, Salticidae). I. Pursuit time and persistence. *Behavioral Ecology and Sociobiology* **4**, 123-32.

— (1982). The biology of *Portia fimbriata*, a web-building jumping spider (Araneae, Salticidae) from Queensland: intraspecific interactions. *Journal of Zoology* **196**, 295-305.

Jacobs, J. (1961). Laboratory cultivation of the marine copepod *Pseudodiamptomus coronatus* Williams. *Limnology and Oceanography* **6**, 443-6.

Jacquotte, R. (1963). Habitat électif des *Pontoniinae* commensales des *Pinnidae* de Tuléar (Madagascar). *Recuil des Travaux. Station Marine d'Endoume* **29**, 59-62.

Jameson, D. L. (1955). Evolutionary trends in the courtship and mating behavior of Salientia. *Systematic Zoology* **4**, 105-19.

Jennings, H. S. (1911). Computing correlation in cases where symmetrical tables are commonly used. *American Naturalist* **XLV**, 123-8.

— (1911a). Assortative mating, variability and inheritance of size, in the conjugation of Paramecium. *Journal of Experimental Zoology* **11**, 1-134.

— and Lashley, K. S. (1913). Biparental inheritance of size in Paramecium. *Journal of Experimental Zoology* **15**, 193-9.

Jensen, J. P. (1956). Biological observations on the isopod *Sphaeroma hookeri* Leach. *Videnskabelige Meddeleser fra Dansk Naturhistorisk Forening i Kjobenhavn* **117**, 305-39.

Johnson, D. S. and Liang, M. (1966). On the biology of the watchman prawn, *Anchistus custos* (Crustacea; Decapoda; Palaemonidae), an Indo-West Pacific commensal of the bivalve *Pinna*. *Journal of Zoology* **150**, 433-55.

Johnson, L. A. S. (1970). Rainbow's end: the quest for an optimal taxonomy. *Systematic Zoology* **19**, 203-39.

Johnson, L. K. (1982). Sexual selection in a brentid beetle. *Evolution* **36**, 251-62.

Johnson, M. W. and Olson, J. B. (1948). The life history and biology of a marine copepod, Tisbe furcata (Baird). *Biological Bulletin* **95**, 320-32.

Johnson, V. R. (1969). Behaviour associated with pair formation in the banded shrimp *Stenopus hispidus* (Olivier). *Pacific Science* **23**, 40-50.

Johnson, W. S. (1976). Biology and population dynamics of the intertidal isopod *Cirolana harfordi*. *Marine Biology* **36**, 343-50.

Jones, D. A. (1970). Population densities and breeding in *Eurydice pulchra* and *Eurydice affinis* in Britain. *Journal of the Marine Biological Association* **50**, 635-55.

Jones, D. F. (1928). *Selective fertilization*. University of Chicago Press.

Jones, M. B. (1974). Breeding biology and seasonal population changes of *Jaera nordmanni nordica* Lemercier [Isopoda, Asellota]. *Journal of the Marine Biological Association* **54**, 727-36.

— and Naylor, E. (1971). Breeding and bionomics of the British members of the *Jaera albifrons* group of species (Isopoda: Asellota). *Journal of Zoology* **165**, 183-99.

Jordan, K. (1896). On mechanical selection and other problems. *Novitates Zoologicae* **III**, 426-525. [Esp. pp. 518-22.]

Jurine, B. (1820). *Histoire des monocles, qui se trouvent aux environs de Genève*. J. J. Paschoud, Geneva.
Kaestner, A. (1968). *Invertebrate zoology*, 3 Vols. McGraw-Hill, New York.
Kamalaveni, S. (1949). On the ovaries, copulation and egg-formation in the hermit-crab, *Clibanarius olivaceous* Henderson (Crustacea, Decapoda). *Journal of the Zoological Society of India* 1, 120-8.
Kamiguchi, Y. (1972). Mating behavior in the freshwater prawn, *Palaemon paucidens*. A study of the sex pheromone and its effects on males. *Journal of the Faculty of Science, Hokkaido University*, Series VI, *Zoology* 18, 347-55.
Kanneworff, E. (1965). Life cycle, food, and growth of the amphipod *Ampelisca macrocephala* Liljeborg from the Øresund. *Ophelia* 2, 305-18.
Kaston, B. J. (1948). Spiders of Connecticut. *Bulletin. Connecticut Geological and Natural History Survey* No. 70.
Katona, S. K. (1975). Copulation in the copepod *Eurytemora affinis* (Poppe, 1880). *Crustaceana* 28, 89-95.
Kinn, D. N. and Witcosky, J. J. (1977). The life cycle and behaviour of *Macrocheles boudreauxi* Krantz. *Zeitschrift für Angewandte Entomologie* 84, 136-44.
Kinne, O. (1952). Zum Lebenszyclus von *Gammarus duebeni* Lillj. nebst einigen Bemerkungen zur Biologie von *Gammarus zaddachi* Sexton subsp. *zaddachi* Spooner, *Veröffentlichungen des Instituts für Meeresforschung in Bremerhaven* 1, 187-203.
— (1955). *Neomysis vulgaris* Thompson, eine autökologisch-biologische Studie. *Biologisches Zentralblatt* 74, 160-202.
— (1959). Ecological data on the amphipod Gammarus duebeni. A monograph. *Veröffentlighungen des Instituts für Meeresforschung in Bremerhaven* 6, 177-202.
Kirk, H. B. (1923). Notes on the mating-habits and early life-history of the culicid *Opifex fuscus* Hutton. *Transactions and Proceedings of the New Zealand Institute* 54, 400-6.
Kjennerud, J. (1950). Ecological observations on *Idothea neglecta* G. O. Sars. *Universitet I Bergen Arbok 1950 Naturvitenskapelig rekke* 7.
Klaasen, F. (1975). Ökologische und ethologische Untersuchungen zue Fortpflanzungsbiologie von Gecarinus lateralis (Decapoda, Brachuyra). *Forma et Functio* 8, 101-74.
Knoepffler, L.-P. (1962). Contribution a l'étude de genre *Discoglossus* (Amphibiens, Anoures). *Vie et Milieu* 13, 1-94.
Knowlton, N. (1980). Sexual selection and dimorphism in two demes of a symbiotic, pair-bonding snapping shrimp. *Evolution* 34, 161-73.
Knudsen, J. W. (1960). Reproduction, life history, and larval ecology of the Californian Xanthidae, the pebble crabs. *Pacific Science* 14, 3-17.
— (1964). Observations on the reproductive cycles and ecology of the common Brachyura and crablike Anomura of Puget Sound, Washington. *Pacific Science* 18, 3-33.
— (1967). *Trapezia* and *Tetralia* (Decapoda, Brachyura, Xanthidae) as obligate ectoparasites of pocilloporid and acroporid corals. *Pacific Science* 21, 57-7.
Kodric-Brown, A. (1977). Reproductive success and the evolution of breeding territories in pupfish (*Cyprinodon*). *Evolution* 31, 750-66.
— (1978). Establishment and defence of breeding territories in a pupfish (Cyprinodontidae: *Cyprinodon*). *Animal Behaviour* 26, 818-34.
Kostalos, M. S. (1979). Life history and ecology of *Gammarus minus* Say (Amphipoda, Gammaridae). *Crustaceana* 37, 113-22.

Kottler, M. J. (1980). Darwin, Wallace, and the original of sexual dimorphism. *Proceedings of the American Philosophical Society* 124, 203-6.
Kramer, P. (1967). Beobachtungen zur Biologie und zum Verhalten der Klippenkrabbe *Grapsus grapsus* L. (Brachuyra Grapsidae) auf Galàpagos und am ekuadorianischen Festland. *Zeitschrift für Tierpsychologie* 24, 385-401.
Krauss, O. (1968). Isolationsmechanismen und Genitalstrukturen bei wirbellosen Tieren. *Zoologischer Anzeiger* 181, 22-38.
Krebs, J. R. and Davies, N. B. (1978). Introduction: ecology, natural selection and social behaviour. In *Behavioural ecology* (ed. J. R. Krebs and N. B. Davies) pp. 1-18. Blackwell, Oxford.
— — (1981). *An introduction to behavioural ecology*. Blackwell, Oxford.
Krishnan, L. and John P. A. (1974). Observations on the breeding biology of *Melita zeylanica* Stebbing, a brackish water amphipod. *Hydrobiologia* 44, 413-30.
Kroger, R. L. and Guthrie, J. F. (1972). Incidence of the parasitic isopod, *Olencira praegustator*, in juvenile Atlantic menhaden. *Copeia* 370-4.
Kruse, K. C. (1981). Mating success, fertilization potential, and male body size in the American toad (*Bufo americanus*). *Herpetologica* 37, 228-33.
Kühne, H. and Becker, G. (1964). Der Holz-Flohkrebs *Chelura terebrans* Philippi *(Amphipoda, Cheluridae)*. *Zeitschrift für Angewandte Zoologie* Beihefte 1, 1-141.
Kuramoto, M. (1978). Correlations of quantitiative parameters of fecundity in amphibians. *Evolution* 32, 287-96.
Labat, R. (1954). Observations sur l'accouplement et la ponte de Paramysis nouveli. *Bulletin de la Société d'Histoire Naturelle de Toulouse* 89, 406-9.
Lack, D. (1947). *Darwin's finches*. Cambridge University Press.
— (1968). *Ecological adaptations for breeding in birds*. Methuen, London.
Laing, J. E. (1969). Life history and life table of *Metaseiulus occidentalis* Athias-Henriot. *Annals of the Entomological Society of America* 62, 978-82.
Lanciani, C. A. (1973). Mating behavior of water mites of the genus *Eylais*. *Acarologia* 14, 631-7.
Lang, K. (1948). *Monographie der Harpacticiden*, 2 Vols. Hakkan Ohlssons Boktryckeri, Lund.
Langenbeck, C. (1898). Formation of the germ layers in the amphipod Microdeutopus Gryllotalpa, Costa. *Journal of Morphology* 14, 301-26.
Lasker, R. Wells, J. B. J. and McIntyre, A. D. (1970). Growth, reproduction, respiration and carbon-utilization of the sand-dwelling harpacticoid copepod, *Asellopsis intermedia*. *Journal of the Marine Biological Association* 50, 147-60.
Le Goffe, M. (1939). Observations et expériences sur la biologie du *Lepidurus apus* Leach. Crustacé Phyllopode Notostracé: éthologie, croissance, régénération. *Bulletin de la Société scientifique de Bretagne* 16, 35-50.
Le Roux, M.-L. (1933). Recherches sur la sexualité des gammariens. *Bulletin Biologique de France et de Belgique* Suppl. XVI.
Lee, C. M. (1972). Structure and function of the spermatophore and its coupling device in the Centropagidae (Copepoda: Calanoida). *Bulletins of Marine Ecology* VIII, 1-20.
Lee, J. C. and Crump, M. L. (1981). Morphological correlates of male mating success in *Triprion petasatus* and *Hyla marmorata* (Anura: Hylidae). *Oecologia* 50, 153-7.
Lee, M. S. and Davis, D. W. (1968). Life history and behavior of the predatory

mite *Typhlodromus occidentalis* in Utah. *Annals of the Entomological Society of America* **61**, 251-55.

Legrand, J.-J. (1958). Comportement sexuel et modalités de la fécondation chez l'Oniscoide *Porcellio dilatatus* Brandt. *Comptes rendus hebdomadaires des séances de l'Académie des Sciences* **246**, 3120-2. And: Induction de la maturité ovrienne et de la mue parturielle par la fécondation chez l'Oniscoide, *Porcellio dilatatus. Ibid.* **247** 754-7.

Lehr, R. and Smith. F. F. (1957). The reproductive capacity of three strains of the two-spotted mite complex. *Journal of Economic Entomology* **50**, 634-6.

Leichmann, G. (1890). Über die Eiblage und Befruchtung bei *Asellus aquaticus. Zoologischer Anzeiger* **13**, 715-16.

— (1891). Beiträge zur Naturgeschichte der Isopoden. *Bibliotheca Zoologica* **3** (10), 1-44.

Leuken, W. (1968). Mehrmaliges Kopulieren von *Armadillidium*-Weibchen (Isopoda) während einer Parturialhäutung. *Crustaceana* **14**, 113-18.

Levi, H. W. (1982). Araneae. In *McGraw Hill synopsis and classification of living organisms*, pp. 77-95. McGraw Hill, New York.

Levin, D. A. and Kerster, H. W. (1973). Assortative pollination for stature in *Lythrum salicaria. Evolution* **27**, 144-52.

Lewbel, G. S. (1978). Sexual dimorphism and intraspecific aggression, and their relationship to sex ratios in *Caprella gorgonia* Laubitz & Lewbel (Crustacea: Amphipoda: Caprellidae). *Journal of Experimental Marine Biology and Ecology* **33**, 133-51.

Lewontin, R. C. (1974). *The genetic basis of evolutionary change.* Columbia University Press, New York.

— and White, M. J. D. (1960). Interaction between inversion polymorphisms of two chromosome pairs in the grasshopper, *Moraba scurra. Evolution* **14**, 116-29.

Leydig, F. (1851). Ueber Artemia salina und Branchipus stagnalis. *Zeitschrift für Wissenschaftliche Zoologie* **III**, 280-307.

Liang, Y. S. and Wang, C.-S. (1978). A new tree frog, *Rhacophorus taipeianus*, (Anura: Rhacophoridae) from Taiwan (Formosa). *Quarterly Journal of the Taiwan Museum* **XXXI**, 185-202.

Licht, L. E. (1969). Comparative breeding behavior of the red-legged frog (*Rana aurora aurora*) and the western spotted toad (*Rana pretiosa pretiosa*) in southwestern British Columbia. *Canadian Journal of Zoology* **47**, 1287-99.

— (1976). Sexual selection in toads (*Bufo americanus*). *Canadian Journal of Zoology* **54**, 1277-84.

Limerick, S. (1980). Courtship behavior and oviposition of the poison arrow frog *Dendrobates pumilio. Herpetologica* **36**, 69-71.

Lincoln, R. J. (1979). *British marine Amphipoda: Gammaridea.* British Museum (Natural History), London.

Linsenmair, K. E. and Linsenmair, C. (1971). Paarbildung und Paarzusammenhalt bei der monogamen Wüstenassel *Hemilepistus reaumuri* (Crustacea, Isopoda, Oniscidea). *Zeitschrift für Tierpsychologie* **29**, 134-55.

Liu, C.-C. (1930). Secondary sexual characters and sexual behavior in Peking toads and frogs. *Peking Natural History Bulletin* **5**, 49-52.

— (1931). Sexual behavior in the Siberian toad, *Bufo raddei* and the pond frog, *Rana nigromaculata. Peking Natural History Bulletin* **6**, 43-60.

— (1950). Amphibians of western China. *Fieldiana. Zoology Memoirs* No. 2.

Lloyd, A. J. and Yonge, C. M. (1947). The biology of *Crangon vulgaris* L.

in the Bristol Channel and Severn Estuary. *Journal of the Marine Biological Association* **XXVI**, 626-61.

Locket, G. H. (1926). Observations on the mating habits of some web-spinning spiders. *Proceedings of the Zoological Society of London* 1125-46.

— and Millidge, A. F. (1951, 1953). *British spiders*, 2 Vols. Ray Society, London.

Lockington, W. N. (1878). Remarks upon the Thalassinidea and Astacidea of the Pacific Coast of North America, with a description of a new species. *Annals and Magazine of Natural History*, 5th series **II**, 299-304.

Longhurst, A. R. (1955). The reproduction and cytology of the Notostraca (Crustacea, Phyllopoda). *Proceedings of the Zoological Society of London* **125**, 671-80.

Lorenz, K. (1935). Der Kumpan in der Umwelt des Vogels. *Journal für Ornithologie*, **LXXXIII** 137-213 and 289-413.

Lowe-McConnell, R. H. (1969). The cichlid fishes of Guyana, South America, with notes on their ecology and breeding behaviour. *Zoological Journal of the Linnean Society* **48**, 255-302.

Lynch, J. D. (1973). The transition from primitive to advanced frogs. In *Evolutionary biology of the Anurans* (ed. J. L. Vial) pp. 133-82. University of Missouri Press, Columbia.

McCauley, D. E. (1979). Geographic variation in body size and its relation to the mating structure of *Tetraopes* populations. *Heredity* **42**, 143-8.

— and Wade, M. J. (1978). Female choice and mating structure of a natural population of the soldier beetle, *Chauliognathus pennsylvanicus*. *Evolution* **32**, 771-5.

McCook, H. C. (1889-93). *American spiders and their spinning work*, 3 Vols. Published by the author, Philadelphia.

McCrone, J. D. and Levi, H. W. (1964). North American widow spiders of the *Latrodectus curacaviensis* group (Araneae: Theridiidae). *Psyche* **71**, 12-27.

MacDougall, M. S. (1925). Cytological observations on gymnostomatous ciliates, with a description of the maturation phenomena in diploid and tetraploid forms of Chilodon uncinatus. *Quarterly Journal of Microscopical Science*, n.s. **69**, 361-84.

Mace, G. M., Harvey, P. H. and Clutton-Brock, T. H. (1981). Brain size and ecology in small mammals. *Journal of Zoology* **193**, 333-54.

MacGinitie, G. E. (1930). The natural history of the mud shrimp *Upogebia pugettensis* (Dana). *Annals and Magazine of Natural History*, 10th series **VI**, 36-44.

— (1935). Ecological aspects of a Californian marine estuary. *American Midland Naturalist* **16**, 630-765.

— (1937). Notes on the natural history of several marine Crustacea. *American Midland Naturalist* **18**, 1031-7.

— (1938). Movements and mating habits of the sand crab, Emerita analoga. *American Midland Naturalist* **19**, 471-81.

Mackay, I. (1951). Observations on the amphipod *Eucrangonyx gracilis* S. I. Smith. *Journal of the Association of School Natural History Societies* No. 4, 14-19. (Also in *Annual Report of the Oundle School Natural History Society* (1951). pp. 5-10.)

McKeown, K. C. (1936). *Spider wonders of Australia*. Angus & Robertson, Sydney.

Mackie, J. (1976). *The cement of the universe*. Oxford University Press.

MacKinnon, D. L. and Hawes, R. S. J. (1961). *An introduction to the study of Protozoa*. Clarendon Press, London.
McLaren, I. and Corkett, C. J. (1978). Unusual genetic variation in body size, development times, oil storage, and survivorship in the marine copepod *Pseudocalanus*. *Biological Bulletin* **155**, 347-59.
McMullen, J. C. (1969). Effects of delayed mating on the reproduction of king crab, *Paralithodes camtschatica*. *Journal of the Fisheries Research Board of Canada* **36**, 2737-40.
McNab, B. I. (1963). Bioenergetics and the determination of home range size. *American Naturalist* **97**, 133-40.
Maercks, H. H. (1930). Sexualbiologische Studien an Asellus aquaticus L. *Zoologische Jahrbücher. Abteilung für Allgemeine Zoologie und Physiologie der Tiere* **48**, 399-508.
Magnus, D. B. E. (1967). Zur Ökologie sedimentbewohnender *Alpheus*-Garnelen (Decapoda, Natantia) des Roten Meeres. *Helgolander Wissenschaftliche Meeresuntersuchungen* **15**, 506-22.
Manning, J. T. (1975). Male discrimination and investment in *Asellus aquaticus* (L.) and *A. meridianus* Racovitsza (Crustacea: Isopoda). *Behaviour* **55**, 1-14.
— (1980). Sex ratio and optimal time investment strategies in *Asellus aquaticus* (L.) and *A. meridianus* Racovitza. *Behaviour* **74**, 264-73.
Manton, S. M. (1977). *The Arthropoda*. Clarendon Press, Oxford.
Marcus, E. (1929). Tardigrada. *Bronn's Klassen und Ordnungen des Tierreichs* **5**, iv, 3.
Markus, H. C. (1930). Studies on the morphology and life history of the isopod, Mancasellus macrouris. *Transactions of the American Microscopical Society* **49**, 220-37.
Martin, R. F. (1972). Evidence from osteology. *Evolution in the genus* Bufo (ed. W. F. Blair). University of Texas Press.
Martof, B. S. and Thompson, E. F. (1958). Reproductive behavior of the chorus frog, *Pseudacris nigrita*. *Behaviour* **XIII**, 243-57.
Marukawa, H. (1933). Biological and fishery research on Japanese king-crabs *Paralithodes camtschatica (Tilesius)*. *Journal of the Imperial Fisheries Experimental Station, Tokyo* No. 4 (paper no. 37). (Pp. 1-122 text in Japanese; pp. 123-52 English summary).
Maslin, T. (1952). Morphological criteria of phyletic relationships. *Systematic Zoology* **1**, 49-70.
Mason, J. C. (1970). Copulatory behavior of the crayfish, *Pacifastacus trowbridgii* (Stimpson). *Canadian Journal of Zoology* **48**, 969-76.
Mason, L. G. (1964). Stabilizing selection for mating fitness in natural populations of *Tetraopes*. *Evolution* **18**, 492-7.
— (1969). Mating selection in the Californian oak moth (Lepidoptera, Dioptidae). *Evolution* **23**, 55-8.
— (1972). Natural insect populations and assortative mating. *American Midland Naturalist* **88**, 150-7.
— (1980). Sexual selection and the evolution of pair-bonding in soldier beetles. *Evolution* **34**, 174-80.
Mathias, P. (1937). Biologie des Crustacés Phyllopodes. *Actualités Scientifiques et Industrielles* No. 447.
— and Bouat, M. (1934). Observations sur l'accouplement chez le *Branchipus stagnalis* L. (Crustacé Phyllopode). *Bulletin de la Société Zoologique de France* **59**, 326-33.

Mathisen, O. A. (1953). Some investigations of the relect crustaceans in Norway with special reference to Pontoporeia affinis Lindstrøm and Pallasea quadrispinosa G. O. Sars. *Nytt Magasin for Zoologi* **1**, 49–86.

Matthes, H. (1961). *Boulengerochromis microlepis*, a Lake Tanganika fish of economical importance. *Bulletin of Aquatic Biology* **3** (No. 24), 1–15.

Matthews, D. C. (1956). The probable method of fertilization in terrestrial hermit crabs based on a comparative study of spermatophores. *Pacific Science* **10**, 303–9.

— (1959). Observations on ova fixation in the hermit crab *Eupagurus prideauxii*. *Pubblicazione della Stazione Zoologica di Napoli* **31** (1959–60), 248–63.

Mayer, P. (1877). Zur Entwicklungsgeschichte der Dekapoden. *Jenaische Zeitschrift für Naturwissenschaft* **11**, 188–269.

— (1879). Über den Hermaphroditismus bei einigen Isopoden. *Mittheilungen aus der Zoologische Station zu Neapel*, 165–79.

Maynard, Smith, J. (1977). Parental investment: a prospective analysis. *Animal Behaviour* **26**, 1–9.

— (1982). *Evolution and the theory of games*. Cambridge University Press.

— and Holliday, R. (1979). Preface In *The evolution of adaptation by natural selection* (ed. J. Maynard Smith and R. Holliday) pp. v–vii. The Royal Society, London.

Mayr, E. (1947). Ecological factors in speciation. *Evolution* **1**, 263–88.

— (1954). Change of genetic environment and evolution. In *Evolution as a Process* (ed. J. S. Huxley, A. C. Hardy, and E. B. Ford) pp. 157–80. George Allen & Unwin, London.

— (1963). *Animal species and evolution*. Harvard University Press, Cambridge, Mass.

Mead, F. (1964). Sur l'existence d'une chevauchée nuptiale de longue durée chez l'isopode terrestre *Helleria brevicornis* Ebner. *Comptes rendus hebdomadaires des séances de l'Académie des Sciences* **258**, 5268–70.

— (1965). Observations sur l'accouplement chez l'Isopode terrestre *Helleria brevicornis* Ebner. *Comptes rendus hebdomadaires des séances de l'Académie des Sciences* **261**, 1752–5.

— (1967). Observations sur l'accouplement et la chevauchée nuptiale chez l'Isopode *Tylos latreillei* Audouin. *Comptes rendus hebdomadaires des séances de l'Académie des Sciences* **D264**, 2154–7.

— (1970). Observations sur le comportment sexuel de l'Isopode terrestre *Metoponorthus sexfasciatus* Budde-Lund. *Bulletin de la Société Zoologique de France* **95**, 55–60.

— (1977). La place de l'accouplement dans le cycle de reproduction des Isopodes terrestres (Oniscidea). *Crustaceana* **31**, 27–41.

Meeks, D. E. and Nagel, J. W. (1973). Reproduction and development of the wood frog, *Raba sylvatica*, in eastern Tennessee. *Herpetologica* **29**, 188–91.

Meisenheimer, J. (1921). *Geschlect und Geschlecther*. Gustav Fischer, Jena.

Menge, A. (1866). Preussische Spinnen. Erste Abteilung. *Schriften der Naturforschenden Gesellschaft in Danzig* **1**, 1–152.

Menon, M. Krisna (1934). A note on the males of *Emerita (Hippa) asiatica*. *Journal of the Bombay Natural History Society* **XXXVII**, 499–501.

Menzies, R. J. (1954). The comparative biology of reproduction in the wood-boring isopod crustacean *Limnoria*. *Bulletin of the Museum of Comparative Zoology* **112**, 361–88.

— Bowman, T. E. and Alverson, F. G. (1955). Studies of the biology of the fish parasite Livoneca convexa Richardson (Crustacea, Isopoda, Cymothoidae). *Wasmann Journal of Biology* **13**, 277-295.

Merrell, D. J. (1968). A comparison of the estimated size and the 'effective size' of breeding populations of the leopard frog, *Rana pipiens*. *Evolution* **22**, 274-83.

Michelsen, A. (1957). Undersøgelse af *Rhagium mordax* De G. og *bifasciatum* F.'s udbredelsesforhold på Nordfyn og deres parringsbiologie (*Cerambycidae*). *Entomologiske Meddelelser* **XVIII**, 77-83.

— (1958). *Rhagium*artenes udbredelse i Danmark og parringsbiologien hos *Rhagium mordax* De G. (Ceramb.). *Entomologiske Meddelelser* **XXVIII**, 338-51.

— (1964). Observations on the sexual behaviour of some longicorn beetles, subfamily Lepturinae (Coleoptera, Cerambycidae). *Behaviour* **22**, 152-66.

Miller, M. A. (1938). Comparative ecological studies on the terrestrial isopod Crustacea of the San Francisco Bay region. *University of California Publications in Zoology* **43**, 113-42.

Miller, N. (1909). The American toad (Bufo lentiginosus americanus, LeConte). *American Naturalist* **XLIII**, 641-68 and 730-45.

Mills, E. L. (1967). The biology of an ampeliscid amphipod crustacean sibling species pair. *Journal of the Fisheries Research Board of Canada* **24**, 305-55.

Milne, L. and Milne, M. (1967). *The crab that crawled out of the past*. G. Bell & Sons, London.

Molenock, J. (1975). Evolutionary aspects of communication in the courtship behavior of four species of anomuran crabs (*Petrolisthes*). *Behaviour* **53**, 1-30.

Monclus, M. and Prevosti, A. (1971). The relationship between mating speed and wing length in *Drosophila subobscura*. *Evolution* **25**, 214-17.

Monod, T. (1926). Les Gnathiidae. *Memoires de la Société des Sciences Naturelles et Physiques du Maroc* No. 13.

— (1930). Notes isopodologiques. III—Sur un *Cassidinopsis* peu connu des Iles Kerguelen. *Bulletin de la Société Zoologique de France* **55**, 437-46.

Montgomery, T. H. (1903). Studies on the habits of spiders, particularly those of the mating period. *Proceedings of the Academy of Natural Sciences of Philadelphia* **55**, 59-149.

— (1909). Further studies on the activities of araneads. II. *Proceedings of the Academy of Natural Sciences of Philadelphia* **LXI**, 548-69.

— (1910). The significance of the courtship and secondary sexual characters of araneads. *American Naturalist* **34**, 151-77.

Moreira, P. S. (1973). The biology of species of *Serolis* (Crustacea, Isopoda, Flabellifera): reproductive behavior of *Serolis polaris* Richardson, 1911. *Boletin Instituto Oceanografico, Universidade de Sao Paulo* **22**, 109-22.

Moutia, L. A. (1958). Contribution to the study of some phytophagous Acarina and their predators in Mauritius. *Bulletin of Entomological Research* **49**, 59-75.

Moznette, G. F. (1917). The cyclamen mite. *Journal of Agricultural Research* **X**, 373-90.

Murray, A. (1860). On Mr Darwin's theory of the origin of species. *Proceedings of the Royal Society of Edinburgh* **IV** (1857-62), 274-91.

Myers, A. A. (1971). Breeding and growth in laboratory-reared *Microdeutopus gryllotalpa* Costa (Amphipoda: Gammaridea). *Journal of Natural History* **5**, 271-7.

Nair, K. B. (1939). The reproduction, oogenesis and development of *Mesopodopsis orientalis* Tatt. *Proceedings of the Indian Academy of Science* **9B**, 175-223.

Nataraj, S. (1947). Preliminary observations on the bionomics, reproduction and embryonic stages of *Palaemon idae* Heller (Crustacea, Decapoda). *Records of the Indian Museum* **XLV**, 89-96.

Nayer, K. N. (1956). The life-history of a brackish water amphipod *Grandidierella bonnieri* Stebbing. *Proceedings of the Indian Academy of Science* **43B**, 178-89.

Naylor, E. (1955). The life cycle of the isopod *Idotea emarginata* (Fabricius). *Journal of Animal Ecology* **24**, 270-81.

— Quénisset, D. (1964). The habitat and life history of *Naesa bidentata* (Adams). *Crustaceana* **7**, 212-16.

Needler, A. B. (1931). Mating and oviposition in Pandalus danae. *Canadian Field-Naturalist* **XLV**, 107-8.

Nelson, K. and Hedgecock, D. (1977). Electrophoretic evidence of multiple paternity in the lobster *Homarus americanus* (Milne-Edwards). *American Naturalist* **III**, 361-5.

Nicholls, A. G. (1931). Studies on Ligia oceanica. I. A. Habitat and effect of change of environment on respiration. B. Observations on moulting and breeding. *Journal of the Marine Biological Association*, n.s. **XVII** (1930-1), 655-73.

Nickerson, R. B., Ossiander, F. J. and Powell, G. C. (1966). Change in size-class structure of populations of Kodiak Island commercial male king crabs due to fishing. *Journal of the Fisheries Research Board of Canada* **23**, 729-36.

Nielsen, E. (1932). *The biology of spiders*, 2 Vols. (Vol. I is English summary of Danish Vol. II.) Levin & Munksgaard, Copenhagen.

Noble, G. K. and Noble, R. C. (1923). The Anderson tree frog (*Hyla andersonii* Baird). Observations on its habits and life history. *Zoologica* **2**, 413-55.

Nolan, B. A. and Salmon, M. (1970). The behaviour and ecology of snapping shrimps (Crustacea: *Alpheus heterochaelis* and *Alpheus normanni*). *Forma et Functio* **2**, 289-335.

Nouvel, H. (1937). Observation de l'accouplement chez une espèce de Mysis, *Praunus flexusus*. *Comptes rendus hebdomadaires des séances de l'Académie des Sciences* **205**, 1184-6.

— (1940). Observations sur la sexualité d'un Mysidacé, *Heteromysis armoricana* n.sp. *Bulletin de l'Institut Océanographique* No. 789.

— and Nouvel, L. (1937). Recherches sur l'accouplement et la ponte chez les Crustacés Décapodes Natantia. *Bulletin de la Société Zoologique de France* **62**, 208-21.

— — (1939). Observations sur la biologie d'une Mysis: *Praunus flexuosus* (Müller, 1788). *Bulletin de l'Institut Océanographique* No. 761.

Nouvel, L. (1939). Observation de l'accouplement chez une éspèce de crevette, *Crangon crangon*. *Comptes rendus hebdomadaires des séances de l'Académie des Sciences* **209**, 639-41.

— (1940). Observations sur la biologie de *Lysmata seticaudata* Risso. *Comptes rendus hebdomadaires des séances de l'Académie des Sciences* **210**, 266-9.

Nussbaum, R. A. (1980). Phylogenetic implications of the amplectic behavior in sooglossid frogs. *Herpetologica* **36**, 1-5.

Nuttall, G. H. F. and Merriman, G. (1911). The process of copulation in *Ornithodorus moubata*. *Parasitology* **4**, 39-45.

Nye, P. A. (1977). Reproduction, growth and distribution of the grapsid crab *Helice crassa* (Dana, 1851) in the southern parts of New Zealand. *Crustaceana* **33**, 75-89.

Oehlert, B. (1958). Kampf und Paarbildung einiger Cichliden. *Zeitschrift für Tierpsychologie* **15**, 141-74.

Oka, H. (1930). Morphologie und Ökologie von *Clunion pacificus* Edwards (Diptera, Chironomidae). *Zoologische Jahrbücher. Abteilung für Systematik, Ökologie und Geographie der Tiere* **59**, 253-80.

Oldham, R. S. (1966). Spring movements in the American toad, *Bufo americanus*. *Canadian Journal of Zoology* **44**, 63-100.

Oliver, J. H. (1966). Notes on reproductive behavior in the Dermanyssidae (Acarina: Mesostigmata). *Journal of Medical Entomology* **3**, 29-35.

— (1974). Symposium on reproduction of arthropods of medical and vetinary importance. IV. Reproduction in ticks (Ixodoidea). *Journal of Medical Entomology* **11**, 26-34.

Oshima, S. (1938). Biological and fishery research in Japanese blue crab *Portunus trituberculatus*) (Miers). *Journal of the Imperial Fisheries Experimental Station* **9**, 208-12.

Ospovat, D. (1981). *The development of Darwin's theory*. Cambridge University Press.

Oudemans, A. C. (1926). Étude du genre *Notodres* Railliet 1893 et de l'espèce *Acarus bubulus* Qudms 1926. *Archives néerlandaises des sciences*, Ser. 3B, **IV**, 145-262.

Overmeer, W. P. J. (1972). Notes on mating behaviour and sex ratio control of *Tetranychus urticae* Koch (Acarina: Tetranychidee). *Entomologische Berichten, Amsterdam* **32**, 240-4.

Packard, A. S. (1878). A monograph of the phyllopod Crustacea of North America, with remarks on the order Phyllocarida. [12th?] *Annual Report of the United States Geological Survey [of Wyoming and Idaho], Section II, Zoology*, pp. 295-592. [Pp. 420-31 by C. F. Gissler.]

Packer, C. (1978). Reciprocal altruism in *Papio anubis*. *Nature, Lond.* **265**, 441-3.

— and Pusie, A. (1982). Cooperation and competition within coalitions of lions: kin selection or game theory. *Nature, Lond.* **296**, 740-2.

Palombi, A. (1939). Note biologiche sui peneidi. La fecondazione e la deposizione uova in *Eusicyonia carinata*. *Bollettino di Zoologia, Torino* **10**, 223-7.

Parker, G. A. (1970). Sperm competition and its evolutionary consequences in the insects. *Biological Reviews* **45**, 525-67.

— (1970a). The reproductive behaviour and the nature of sexual selection in *Scatophaga stercoraria* L. (Diptera: Scatophagidae). IV. Epigamic recognition and competition between males for the possession of females. *Behaviour* **XXXVII**, 113-39.

— (1970b). Sperm competition and its evolutionary effect on copula duration in the fly *Scatophaga stercoraria*. *Journal of Insect Physiology* **16**, 1301-28.

— (1972). Reproductive behaviour of *Sepsis cynipsea* (L.) (Diptera: Sepsidae). I. A preliminary analysis of the reproductive strategy and its associated behaviour patterns. *Behaviour* **XLI**, 172-206.

— (1974). Courtship persistence an female-guarding as male time investment strategies. *Behaviour* **XLVIII**, 157-84.

— (1978). Searching for mates. In *Behavioural ecology* (ed. J. R. Krebs and N. B. Davies) pp. 214-44. Blackwell, Oxford.

Parker, G. A. and Smith, J. L. (1975). Sperm competition and the evolution of the precopulatory passive phase behaviour in *Locusta migratoria migratorioides*. *Journal of Entomology* A49, 155-71.
Parsons, P. A. (1961). Fly size, emergence time and sternopleural chaeta number in *Drosophila*. *Heredity* 16, 455-73.
— (1965). Assortative mating for a metrical characteristic in *Drosophila*. *Heredity* 20, 161-7.
Partridge, L. and Farquhar, M. (1983). Lifetime mating success of male fruitflies (*Drosophila melanogaster*) is related to their size. *Animal Behaviour* 31, in press.
Patanè, L. (1959). Richerche sui fenomeni della sessualità negli isopodi. III. Ulteriori osservazioni sulla biologia sessuale di *Porcellio laevis* Latreille. *Bollettino delle Sedute. Accademia Gioenia di scienze naturali in Catania* 5, 113-32.
— (1962). Richerche sui fenomeni della sessualità negli isopodi. IV. Osservazioni sulla biologia sessuale di *Idotea baltica basteri* Aud. *Bollettino delle Sedute. Academia Gioenia di scienze naturali in Catania* 6, 361-72.
Patel, B. and Crisp, D. J. (1961). Relation between the brooding and moulting cycles in cirripedes. *Crustaceana* 2, 89-107.
Patton, W. K. (1974). Community structure among the animals inhabiting the coral Pocillopora damicornis at Heron Island, Australia. In *Symbiosis in the sea* (ed. W. B. Vernberg) pp. 219-43. University of South Carolina Press, Columbia, South Carolina.
— and Robertson, D. R. (1980). Pair formation in a coral inhabiting hermit crab. *Oecologia* 47, 267-9.
Pauly, F. (1956). Zur Biologie einiger Belbiden (Oribatei, Moosmilben) und zur Funktion ihrer pseudostigmatischen Organe. *Zoologische Jahrbücher. Abteilung für Systematik, Ökologie und Geographie der Tiere* 84, 275-328.
Pearl, R. (1905). A biometrical study of conjugation in Paramecium. *Proceedings of the Royal Society of London* B77, 377-83.
— (1907). A biometrical study of conjugation in Paramecium. *Biometrika* V, 213-97.
Pearse, A. S. (1912). Observations of the behavior of *Eubranchipus dadayi*. *Bulletin of the Wisconsin Natural History Society* 10, 109-17.
— (1914). On the habits of *Uca pugnax* (Smith) and *U. pugilator* (Bosc.). *Transactions of the Wisconsin Academy of Science, Arts and Letters* 17, 791-801.
Pearson, K. (1899). Data for the problem of evolution in man. III. *Proceedings of the Royal Society* 66, 23-32.
— and Lee, A. (1902). On the laws of inheritance in man. I. Inheritance of physical characteristics. *Biometrika* 2, 357-462.
Peckham, G. W. and Peckham, E. G. (1889). Observations on sexual selection in spiders of the family Attidae. *Occasional Papers of the Natural History Society of Wisconsin* 1, 1-60.
Perrone, M. (1978). Mate size and breeding success in a monogamous cichlid fish. *Environmental Biology of Fishes* 3, 193-201.
Peters, N., Panning, A., and Schnackenbeck, (1933). Die chinesische Wollhandkrabbe (*Eriocheir sinensis* H. Milne-Edwards) in Deutschland. *Zoologische Anzeiger* 104, Suppl. 1-180.
Petrunkevitch, A. (1911). Sense of sight, courtship and mating in *Dugesiella hentzi* (Girard), a theraphosid spider from Texas. *Zoologische*

Jahrbücher. Abteilung für Systematik, Ökologie und Geographie der Tiere **31**, 355–76.

Pike, R. B. (1947). Galathea. *Memoir. Liverpool Marine Biological Committee* No. 34.

— and Williamson, D. I. (1959). Observations on the distribution and breeding of British hermit crabs and the stone crab (Crustacea: Diogenidae, Paguridae, and Lithodidae). *Proceedings of the Zoological Society of London* **132**, 551–67.

Pillai, P. R. P. and Winston, P. W. (1969). Life history and biology of *Caloglyphus anomalus* Nesbitt (Acarina: Acaridae). *Acarologia* **11**, 295–303.

Platnick, N. I. (1979). Philosophy and the transformation of cladistics. *Systematic Zoology* **28**, 537–46.

— and Gertsch, W. J. (1976). The suborders of spiders: a cladistic analysis (Arachnida: Araneae). *American Museum Novitates* No. 2607.

Pomerat, C. M. (1932). Mating behaviour in the Japanese beetle, *Popillia japonica* Newm. *Journal of General Psychology* **7**, 16–33.

— (1933). Mating in *Limulus polyphemus*. *Biological Bulletin* **64**, 243–52.

Popper, K. (1957). *The poverty of historicism*. Routledge and Kegan Paul, London.

Porter, J. P. (1966). The habits, instincts, and mental powers of spiders, genera, *Argiope* and *Epeira*. *American Journal of Psychology* **17**, 306–57.

Potter, D. A. (1978). Functional sex ratio in the carmine spider mite. *Annals of the Entomological Society of America* **71**, 218–22.

— Wrensch, D. L. and Johnston, D. E. (1976). Guarding, aggressive behavior, and mating success in male twospotted spider mites *Annals of the Entomological Society of America* **69**, 707–11.

Powell, G. C. and Nickerson, R. B. (1965). Reproduction of king crabs *Paralithodes camtschatica* (Tilesius). *Journal of the Fisheries Research Board of Canada* **22**, 101–11.

— Shafford, B. and Jones, M. (1973). Reproductive biology of young adult king crabs *Paralithodes camtschatica* (Tilesius) at Kodiak Island, Alaska. *Proceeedings of the National Shellfisheries Association* **63**, 77–87.

Preston, E. M. (1973). A computer simulation of competition among five sympatric congeneric species of xanthid crabs. *Ecology* **54**, 469–83.

Prevost, B. (1803). Histoire d'un Insecte (ou d'un Crustacé). *Journal de Physique, de Chémie et d'Histoire Naturelle* **57**, 37–54 and 89–106.

— (1820). Mémoire sur le chirocéphale. In L. Jurine *Histoire des Monocles* (q.v.), pp. 201–44.

Provost, M. W. and Haeger, J. S. (1967). Mating and pupal attendance in *Deinocerites cancer* and comparisons with *Opifex fuscus* (Diptera: Culicidae). *Annals of the Entomological Society of America* **60**, 565–74.

Putnam, W. L. (1966). Insemination in *Balaustium* sp. (Erythraeidae). *Acarologia* **8**, 424–6.

Quayle, H. J. (1912). Red spiders and mites of citrus trees. *Bulletin, California University Experimental Station* **234**, 483–530.

— (1938). *Insects of citrus and other subtropical fruits*. Comstock, Ithaca.

Rabaud, E. (1911). *L'Adaptation et l'Évolution*. Chiron, Paris.

— (1925). Les phénomènes de convergence en biologie. *Bulletin biologique de France et Belgique* Suppl. No. 7.

Rabb, G. B. and Rabb, M. S. (1963). On the behavior and breeding biology of the African pupid frog *Hymenochirus boettgeri*. *Zeitschrift für Tierpsychologie* **20**, 215–41.

Rack, G. (1972). Pyemotiden an Gramineen in schwedischen landwirtschaftlichen Betrieben. Ein Beitrag zur Entwicklung von *Siteroptes graminum* (Reuter, 1900) (Acarina, Pymotidae). *Zoologischer Anzeiger* **188**, 157–74.

Racovitza, E.-G. (1919). Notes sur les isopodes. 1. *Asellus aquaticus auct.* est une erreur taxonomique. 2. *Asellus aquaticus* L. et *A. meridianus* n.sp. *Archives de Zoologie Expérimentale et Générale* **58**, *Notes et revue*, 31–43.

Radinovsky, S. (1965). The biology and ecology of granary mites of the Pacific northwest. IV. Various aspects of the reproductive behavior of *Leidinychus krameri* (Acarina: Uropodidae). *Annals of the Entomological Society of America* **58**, 267–72.

Raja Bai Naidu, K. G. (1954). A note on the courtship in the sand crab [*Philyra scabriuscula* (Fabricius)]. *Journal of the Bombay Natural History Society* **52**, 640–1.

Ramazzotti, G. (1972). Il phylum Tardigrada. *Memorie del'Istituto Italiano di Idrobiologia* No. 28.

Rao, R. M. (1965). Breeding behaviour in *Macrobrachium rosenbergii* (de Man). *Fishery Technology* **2**, 19–25.

Rapp, A. (1959). Zur Biologie und Ethologie der Käfermilbe *Parasitus coleoptratorum* L. 1758 (Ein Beitrag zum Phoresie-Problem). *Zoologischer Jahrbücher, Abteilung für Systematik, Ökologie und Geographie der Tiere* **86** (1958-9), 304–66.

Rathbun, M. J. (1895). The genus Callinectes. *Proceedings of the United States National Museum* **18**, 349–75.

Raup, D. M. (1977). Stochastic models in evolutionary palaeontology. In *Patterns of evolution* (ed. A. Hallam) pp. 59–78. Elsevier, Amsterdam.

Rehberg, H. (1880). Beiträge zur Kenntnis der freilebenden Süsswassercopepoden. *Abhandlungen herausgegeben von naturwissenschaftlichen Vereine zu Bremen* **VI**, 533.

Rensch, B. (1959). *Evolution above the species level.* Methuen, London.

Reverberi, G. and Pitotti, M. (1943). Il ciclo biologico e la determinazione fenotipica del sesso di *Ione thoracica* Montagu, Bopiride parassita di *Callianassa latucauda* Otto. *Pubblicazioni della Stazione Zoologica di Napoli* **XIX**, 112–84.

Rice, A. L. (1980). Crab zoeal morphology and its bearing on the classification of the Brachyura. *Transactions of the Zoological Society of London* **35**, 271–424.

Richman, D. B. and Cutler, B. (1978). A list of the jumping spiders (Aranae: Salticidae) of the United States and Canada. *Peckhamia* **1**, 82–110.

Richter, G. (1978). Einige Beobachtungen zur Lebensweise des Flohkrebs *Siphonoecetes della-vallei*. *Natur und Museum* **108**, 259–66.

Ridley, M. (1978). Paternal care. *Animal Behaviour* **26**, 904–32.

— and Thompson, D. J. (1979). Size and mating in *Asellus aquaticus* (Crustacea: Isopoda). *Zeitschrift für Tierpsychologie* **51**, 380–97.

Roberts, D. F. (1977). Assortative mating in man. *Bulletin of the Eugenics Society* Suppl. No. 2.

Robertson, F. W. (1957). Studies on quantitative inheritance XI. Genetic and environmental correlation between body size and egg production in *Drosophila melanogaster*. *Journal of Genetics* **55**, 428–43.

Robinson, M. H. and Robinson, B. (1973). Ecology and behavior of the giant woodspider *Nephila maculata* (Fabricius) in New Guinea. *Smithsonian Contributions to Zoology* No. 149.

Robinson, M. H. and Robinson, B. (1976). The ecology and behaviour of *Nephila maculata*: a supplement. *Smithsonian Contributions to Zoology* No. 218.
— — (1978). The evolution of courtship systems in tropical araneid spiders. *Symposium of the Zoological Society of London* **42**, 17-29.
— — (1980). Comparative studies of the courtship and mating behavior of tropical araneid spiders. *Pacific Insects Monographs* No. 36. (Bishop Museum Press, Hawaii, Honolulu.)
Rodrigues, S, S. de A. (1976). Sobre a reprodução, embriologia e desenvolvimento larval de *Callichirus major* Say, 1918 (Crustacea, Decapoda Thalassinidea). *Boletim Zoologia, Universidade de Sao Paulo* **1**, 85-103.
Roff, J. C. (1972). Aspects of the reproductive biology of the planktonic copepod *Limnocalanus macrurus* Sars, 1863. *Crustaceana* **22**, 155-60.
Rosen, M. and Lemon, R. E. (1974). The vocal behavior of spring peepers, *Hyla crucifer*. *Copeia* 940-50.
Rosenstadt, B. (1888). Beiträge zur Kenntnis der Organisation von *Asellus aquaticus* und verwandter Isopoden. *Biologisches Centralblatt* **8**, 452-62.
Röttger, R., Astheimer, H., Splindler, M., and Steinborn, J. (1972). Ökologie von *Asterocheres lilljeborgi*, eines auf *Henricia sanguinolenta* parasitisch lebenden Copepoden. *Marine Biology* **13**, 259-66.
Roth, L. M. and Willis, E. R. (1955). Intra-uterine nutrition of the 'beetle-roach' *Diploptera discoides* (Serv.) during embryogenesis, with notes on its biology in the laboratory (Blattaria: Diplopteridae). *Psyche* **62**, 55-68.
Rubenstein, D. (1981). Competition and reproductive success in the spider *Meta segmentata*. Typescript of IEC talk.
Ruello, N. V., Moffitt, P. F. and Phillips, (1973). Reproductive behaviour in captive freshwater shrimp *Macrobrachium australiense* Holthuis. *Australian Journal of Marine and Freshwater Research* **24**, 197-202.
Ryan, E. P. (1967). Structure and function of the reproductive system of the crab *Portunus sanguinolentus* (Herbst) (Brachyura: Portunidae). 1. The male system. And 2. The female system. *Symposium series. Marine Biological Association of India* **2**, 506-21; 522-44.
Ryan, M. J. (1980). Reproductive behavior of the bullfrog (*Rana catesbiana*). *Copeia* 108-14.
Salemaa, H. (1979). Ecology of *Idotea* spp. (Isopoda) in the northern Baltic. *Ophelia* **18**, 133-50.
Salfi, M. (1939). Ricerche etologiche ed ecologiche sugli Anfipodi tubicoli del Canale delle Saline di Cagliari. *Archivio Zoologico Italiano* **XXVII**, 31-62.
Salthe, S. N. and Mecham, J. S. (1974). Reproductive and courtship patterns. In *The physiology of the amphibia*, 3 Vols. (ed. B. Lofts) Vol. 2, pp. 209-521. Academic Press, New York.
Salvat, B. (1967). La macrofaune carcinologiques endogée des sediments meubles intertidaux (Tanaidaces, Isopodes et Amphipodes). Ethologie, bionomie et cycle biologique. *Memoirs des Muséum National d'Histoire Naturelle, Paris A* 45.
Sameoto, D. D. (1968). Comparative ecology, life histories, and behaviour of intertidal sand-burrowing amphipods (Crustacea: Haustoriidae) at Cape Cod. *Journal of the Fisheries Research Board of Canada* **26**, 361-88.
Samter, M. and Weltner, W. (1904). Biologische Eigentümlichkeiten der Mysis relicta, Pallasiella quadrispinosa und Pontoporeia affinis, erklärt aus ihrer eiszeitlichen Entstehung. *Zoologischer Anzeiger* **XXVII**, 676-94.
Samuel, D. E. (1971). Field methods for determining the sex of barn swallows (*Hirondo rustica*). *Ohio Journal of Science* **71**, 125-8.

Sars, G. O. (1900). *An account of the Crustacea of Norway*. 3. *Cumacea*. Christiana.
— (1903). *An account of the Crustacea of Norway*. 5. *Copepoda harpacticoida*. The Bergen Museum, Bergen.
Savage, T. (1971). Mating of the stone crab, *Menippe mercenaria* (Say) (Decapoda, Brachyura). *Crustaceana* **20**, 315-16.
Savory, T. H. (1928). *The biology of spiders*. Sidgwick & Jackson, London.
Say, T. (1817). An account of the Crustacea of the United States, *Journal of the Academy of Natural Sciences of Philadelphia* **1**, 57-460.
Schaub, D. L. and Larsen, J. H. (1978). The reproductive ecology of the Pacific treefrog (*Hyla regilla*). *Herpetologica* **34**, 409-16.
Schein, H. (1975). Aspects of the aggressive and sexual behaviour of *Alpheus heterochelis* Say. *Marine Behaviour and Physiology* **3**, 83-96.
Scheiring, J. F. (1977). Stabilizing selection for size as related to mating fitness in *Tetraopes*. *Evolution* **31**, 447-9.
Schlosser, (1756). In M. Gautier *Observations périodiques sur la physique* Paris.
Schmeil, O. (1892). Deutschlands freilebende Süsswasser-Copepoden. Teil I: Cyclopidae. *Bibliotheca Zoologica* **4** (11), 1-192.
— (1893). Deutschlands freilebende Süsswasser-Copepoden. II Teil: Harpacticidae. *Bibliotheca Zoologica* **5** (15), 1-103.
Schneider, H. (1973). Die Paarungsrufe einheimischer Ranidae (Anura: Amphibia). *Bonn Zoologischer Beitrage* **24**, 51-61.
Schneider, P. (1971). Lebensweise und socialen Verhalten der Wüstenassel *Hemilepistis aphganicus* Borutzky 1958. *Zeitschrift für Tierpsychologie* **29**, 121-33.
Schöbl, J. (1880). Ueber die Fortpflanzung isopoder Crustaceen. *Archiv für Mikroskopische Anatomie* **17**, 125-40.
Schöne, H. (1968). Agonistic and sexual display in aquatic and semi-terrestrial brachyuran crabs. *American Zoologist* **8**, 641-54.
Schoeman, A. S. D. (1977). The biology of *Pardosa crassipalpis* Purcell (Aranae: Lycosidae). *Journal of the Entomological Society of South Africa* **40**, 225-36.
Schoener, T. W. (1968). Sizes of feeding territories among birds. *Ecology* **49**, 123-41.
Schroder, S. M. (1981). The role of sexual selection in determining overall mating patterns and mate choice in chum salmon. Unpublished PhD thesis, University of Washington, Seattle.
Schultz, G. A. (1979). Aspects of the evolution and origin of the deep-sea isopod crustaceans. *Sarsia* **64**, 77-83.
Scudamore, H. H. (1948). Factors influencing molting and the sexual cycles in the crayfish. *Biological Bulletin* **95**, 229-37.
Segerstråle, S. G. (1937). Studien uber die Bodentierwelt in südfinnlandischen küstengewässern. IV. Zur Morphologie und Biologie des Amphipoden *Pontoporeia affinis*, nebst einer Revision der *Pontoporeia* Systematik. *Commentationes Biologicae, Societas Scientarum Fennica* **VII**, No. 1.
— (1947-8). New observations on the distribution and morphology of the amphipod, *Gammarus zaddachi* Sexton, with notes on related species. *Journal of the Marine Biological Association* **XXVII**, 219-44.
— (1950). The amphipods of the coasts of Finland—some facts and problems. *Commentationes Biologicae, Societas Scientarum Fennica* **X**, No. 14.
Seibt, U. and Wickler, W. (1979). The biological significance of the pair-

bond in the shrimp *Hymenocera picta. Zeitschrift für Tierpsychologie* **50**, 166-79.

Seidel, F. (1924). Der Geschlechtsorgane in der embryonalen Entwicklung von Pyrrhocoris apterus L. *Zeitschrift für Morphologie und Okologie der Tiere* **1**, 429-506.

Sexton, E. W. (1924). The moulting and growth-stages of Gammarus, with descriptions of the normals and intersexes of *G. chevreuxi. Journal of the Marine Biological Association* **13**, 340-401.

— (1928). On the rearing and breeding of Gammarus in laboratory conditions. *Journal of the Marine Biological Association* **15**, 33-55.

— (1935). Fertilization of successive broods of *Gammarus chevreuxi. Nature, Lond.* **CXXXVI**, 477.

— and Reid, D. M. (1951). The life-history of the multiform species *Jassa falcata* (Montagu) (Crustacea Amphipoda) with a review of the bibliography of the species. *Journal of the Linnean Society. Zoology* **42**, 29-91.

Sexton, O. (1958). Observations on the life history of a Venezuelan frog, *Atelopus cruciger. Acta Biologica Venezuelica* **2**, 235-42.

Seyfarth, R. M. (1978). Social relationships among adult male and female baboons. I. Behaviour during consortship. *Behaviour* **LXIV**, 204-26.

Sheader, M. (1977). Breeding biology of *Idotea pelagica* (Isopoda: Valvifera) with notes on the occurrence and biology of its parasite *Clypeoniscus hanseni* (Isopoda: Epicaridea). *Journal of the Marine Biological Association* **57**, 659-74.

— (1977*a*). Breeding and marsupial development in laboratory-maintained *Parathemisto gaudichaudi* (Amphipoda). *Journal of the Marine Biological Association* **57**, 943-54.

— and Chia, F.-S. (1970). Development, fecundity and brooding behaviour of the amphipod, *Marinogammarus obtusatus. Journal of the Marine Biological Association* **50**, 1079-99.

Sheard, K. (1949). The marine crayfishes (spiny lobsters), family Palinuridae, of Western Australia. *Bulletin. Commonwealth Scientific and Industrial Research Organization, Australia* No. 247.

Shimoizumi, M. (1952). The breeding habits of *Metoponorthus pruinosus* Brandt. *Journal of the Gakugei College, Tokushima University (Natural Science)* **II**, 31-5. [In Japanese, English summary p. 31.]

Shuster, S. M. (1981). Sexual selection in the Socorro isopod, *Thermosphaeroma thermophilum* (Cole) *(Crustacea: Peracarida). Animal Behaviour* **29**, 698-707.

Siddig, M. A. and Elbadry, E. A. (1971). Biology of the spider mite *Eutetranychus sudanicus. Annals of the Entomological Society of America* **64**, 806-9.

Siewing, R. (1963). Studies in Malacostracan phylogeny: results and problems. In *Phylogeny and evolution of Crustacea* (ed. H. B. Whittington and W. D. I. Rolfe), pp. 85-104. Museum of Comparative Zoology, Cambridge, Mass.

Silberbauer, B. I. (1971). The biology of the South African rock lobster *Jasus lalandii* (H. Milne-Edwards). 2. The reproductive organs, mating and fertilization. *Investigational Report, Division of Sea Fisheries, South Africa* No. 93.

Silverstone, P. A. (1976). A revision of the poison-arrow frogs of the genus *Phyllobates* Bibron *in* Sagra (Family Dendrobatidae). *Science Bulletin. Natural History Museum of Los Angeles County* No. 27.

Simpson, G. G. (1944). *The tempo and mode of evolution.* Columbia University Press, New York.

— (1947). *The meaning of evolution.* Yale University Press, New Haven.

Simpson, G. G. (1950). Evolutionary determinism. *Scientific Monthly* **LXXI**, 262–7. (Reprinted in G. G. Simpson, *This view of life*, New York, 1964.)
— (1953). *The major features of evolution.* Columbia University Press, New York.
— (1963). The historical factor in science. In *The fabric of geology* (ed. C. W. Albritton) Addison-Wesley, Reading, Mass. (Revised version in G. G. Simpson, *This view of life*, New York, 1964. Page references are to the revision.)
— G. G. (1978). [Book review of Hecht *et al.* q.v. under Bock (1976).] *Paleobiology* **4**, 217–21.
— (1979). *Splendid isolation.* Yale University Press, New Haven.
Smallwood, M. E. (1905). The salt-marsh amphipod: *Orchestia palustris*. *Cold Spring Harbor Monographs* No. 3.
Smith, E. W. (1953). The life history of the crawfish *Orconectes (Faxonella) cypeatus* (Hay). *Tulane Studies in Zoology* **1**, 79–96.
Smith, M. (1969). *The British amphibians and reptiles.* Collins, London.
Sneath, P. H. and Sokal, R. (1973). *Numerical taxonomy.* W. H. Freeman, San Francisco.
Snetsinger, R. (1955). Observations on two species of *Phidippus* (jumping spiders). *Entomological News* **LXVI**, 9–15.
Snow, C. D. and Neilsen, J. R. (1966). Premating and mating behavior of the Dungeness crab (*Cancer magister* Dana). *Journal of the Fisheries Research Board of Canada* **23**, 1319–23.
Snyder, L. H. and Crozier, W. J. (1922). Selective pairing in gammarids. *Proceedings of the Society for Experimental Biology and Medicine* **19**, 327–9.
Sømme, O. M. (1941). A study of the life history of the gribble *Limnoria lignorum* (Rathke) in Norway. *Nytt Magasin for Naturvidenskapene* **71**, 145–205.
Sörensen, W. (1880). Sur la rapprochement des sexes chez quelques Araignées. *Entomologisk Tidskrift* **1**, 171–4.
Sokal, R. R. and Rohlf, F. J. (1969). *Biometry.* W. H. Freeman, San Francisco.
Sokol, M. M. (1977). A subordinal classification of frogs. *Journal of Zoology* **182**, 505–8.
Solignac, M. (1972). Le comportement sexuel des *Jaera (albifrons) posthirsuta* (Isopodes Asellotes). *Comptes rendues hebdomadaires de séances de l'Académie des Sciences* **274D**, 1570–2.
— (1972a). Comparaison des comportements sexuels specifiques dans la superespèce *Jaera albifrons* (Isopodes Asellotes). *Comptes rendus hebdomadaires des séances de l'Académie des Sciences* **274D**, 2236–9.
Spalding, J. F. (1942). The nature and function of the spermatophore and sperm plug in *Carcinus maenas*. *Quarterly Journal of Microscopical Science* **83**, 399–422.
Spett, G. (1929). Zur Frage der Homogamie und Pangamie bei Tieren. Untersuchungen an einigen Coleopteren. *Biologisches Zentralblatt* **49**, 385–92.
Spickett, S. G. (1961). Studies on *Demodex folliculorum* Simon (1842). I. Life history. *Parasitology* **51**, 181–92.
Spieth, H. T. (1952). Mating behavior within the genus *Drosophila* (Diptera). *Bulletin of the American Museum of Natural History* **99**, 395–474.
Squires, H. J. (1966). Reproduction in *Sphyrion lumpi*, a copepod parasitic on redfish (*Sebastes* spp.). *Journal of the Fisheries Research Board of Canada* **23**, 521–6.
Starrett, P. (1967). Observations on the life history of frogs of the family Atelopodidae. *Herpetologica* **23**, 195–204.

Stauffer, R. C. (1975). *Charles Darwin's natural selection.* Cambridge University Press.

Staveley, E. F. (1866). *British spiders.* London.

Steel, E. A. (1961). Some observations on the life history of *Asellus aquaticus* (L.) and *Asellus meridianus* Racovitza. *Proceedings of the Zoological Society of London* **137**, 71–87.

Stephenson, K. (1942). The Amphipoda of N. Norway and Spitzbergen with adjacent waters. *Tromsø Museum Skrifter* **3**, 363–526.

Stevens, P. F. (1980). Evolutionary polarity of character states. *Annual Review of Ecology and Systematics* **11**, 333–58.

Storm, R. M. (1960). Notes on the biology of the red-legged frog (*Rana aurora aurora*). *Herpetologica* **16**, 251–9.

Strong, D. R. (1972). Life history variation among populations of an amphipod (*Hyalella azteca*). *Ecology* **53**, 1103–11.

— (1973). Amphipod amplexus, the significance of ecotypic variation. *Ecology* **54**, 1383–8.

Styron, C. E. and Burbanck, W. D. (1967). Ecology of an aquatic isopod, Lirceus fontinalis Raf., emphasizing radiation effects. *American Midland Naturalist* **78**, 389–415.

Subramoniam, T. (1977). Aspects of sexual biology of the anomuran crab *Emerita asiatica*. *Marine Biology* **43**, 369–77.

— (1979). Heterosexual raping in the mole crab, *Emerita asiatica*. *International Journal of Invertebrate Reproduction* **1**, 197–9.

Suppe, F. (ed.) (1978). *The structure of scientific theories*, 2nd edn. University of Illinois Press, Urbana.

Sutcliffe, W. H. (1953). Further observations on the breeding and migration of the Bermuda spiny lobster, *Panulirus argus*. *Journal of Marine Research* **12**, 173–83.

Sweatman, G. K. (1957). Life history, non-specificity, and revision of the genus *Chorioptes*, a parasitic mite of herbivores. *Canadian Journal of Zoology* **35**, 641–89.

Symons, D. A. (1979). *The evolution of human sexuality.* Oxford University Press, New York.

Szidat, L. (1964). Sobre la evolucion del dimorfismo sexual secondario en isopodos parasitos de la familia Cymothoidae (Crust. Isop.). *Anais Congresso Latino-Americano Zoologia* **2**, (1962), 83–7.

Taberly, G. (1957). Observations sur les spermatophores et leur transfert chez les Oribates (Acariens). *Bulletin de la Société Zoologiques de France* **82**, 139–45.

Tack, P. I. (1941). The life history and ecology of the crayfish *Cambarus immunis* Hagen. *American Midland Naturalist* **25**, 420–46.

Tawfik, M. F. S .and Awadallah, K. T. (1970). The biology of *Pyemotes herfsi* Oudemans and its efficiency in the control of the resting larvae of the pink bollworm, *Pectinophora gossypiella* Sauders, in U.A.R. *Bulletin de la Société Entomologique d'Egypte* **54**, 49–71.

Templeman, W. (1934). Mating in the American lobster. *Contributions to Canadian Biology and Fisheries* **8**, 421–32.

(1936). Further contributions to mating in the American lobster. *Journal of the Biological Board of Canada* **2**, 223–6.

Tessier, G. (1935). Croissance des variants sexuels chez *Maia squinado* L. *Travaux de la Station Biologique de Roscoff* **13**, 93–130.

Thamdrup, H. M. (1935). Beiträge zur Ökologie der Wattenfauna auf experi-

mentaller Grundlage. *Meddelelser fra Kommissionen for Danmarks Fiskeri- og Havundersogelser* serie Fiskeri **10**, No. 2.
Thompson, D. J. and Manning, J. T. (1981). Mate selection by Asellus (Crustacea: Isopoda). *Behaviour* **78**, 178-87.
Thornhill, R. (1976). Reproductive behavior of the lovebug, *Plecia nearctica* (Diptera: Bibionidae). *Annals of the Entomological Society of America* **69**, 843-7.
— (1980). Sexual selection within mating swarms of the lovebug, *Plecia nearctica* (Diptera: Bibionidae). *Animal Behaviour* **28**, 405-12.
Tinbergen, N. (1953). *The herring gull's world*. Collins, London.
— (1959). Comparative studies of the behaviour of gulls: a progress report. *Behaviour* **XV**, 1-70.
Tower, W. L. (1906). An investigation of evolution on chrysomelid beetles of the genus Lepinotarsa. *Carnegie Institution of Washington Publication* No. 48.
Townsend, D. S., Stewart, M. M., Pough, F. H. and Brussard, P. F. (1981). Internal fertilization in an oviparous frog. *Science* **212**, 469-71.
Trilles, J.-P. (1969). Recherches sur les Isopodes *Cymothoidae* des côtes françaises. Aperçu général et comparatif sur la bionomie et la sexualité de ces Crustacés. *Bulletin de la Société Zoologique de France* **54**, 433-45.
Trivers, R. L. (1972). Parental investment and sexual selection. In *Sexual selection and the descent of man* (ed. B. Campbell) pp. 136-79. Heinemann, London.
Türkay, M. (1970). Die Gecarcinidae. *Senckenbergiana Biologica* **51**, 333-54.
— (1973). Die Gecarcinidae Afrikas (Crustacea: Decapoda). *Senckenbergiana Biologica* **54**, 81-103.
— (1974). Die Gecarcinidae Asiens und Ozaniens (Crustacea: Decapoda) *Senckenbergiana Biologica* **55**, 223-59.
Turner, F. B. (1958). Life-history of the western spotted frog in Yellowstone National Park. *Herpetologica* **14**, 96-100.
Unwin, E. E. (1920). Notes upon the reproduction of *Asellus aquaticus*. *Journal of the Linnean Society. Zoology* **34**, 335-42.
Valoušek, B. (1926). Kopulace žabronožky sněžní Chirocephalus grubii Dyb. *Spisy vydávané Přírodovedeckou Fakultou Masarykovy University* No. 75.
Van Beneden, P.-J. (1861). Recherches sur la faune littorale de Belgique. *Mémoires de l'Académie Royale de Belgique* **XXXIII**.
Van Dolah, R. F. (1978). Factors regulating the distribution and population dynamics of the amphipod *Gammarus palustris* in an intertidal salt marsh community. *Ecological Monographs* **48**, 191-217.
Van Gelder, J. J. and Hoedemaekers, H. C. (1971). Sound activity and migration during the breeding period of *Rana temporaria* L., *R. arvalis* Nilsson, *Pelobates fuscus* Laur., and *R. esculenta*. L. *Journal of Animal Ecology* **40**, 559-68.
Van Valen, L. (1973). Are categories in different phyla comparable? *Taxon* **22**, 333-73.
Vandel, A. (1925). Recherches sur la sexualité des Isopodes. Les conditions naturelles de la reproduction chez les Isopodes terrestres. *Bulletin Biologique de la France et de la Belgique* **LIX**, 318-71.
Vernet-Cornubert, G. (1958). Biologie générale de *Pisa tetraodon* (Pennant). *Bulletin de l'Institut Océanographique. Monaco* No. 1113.
— (1958a). Recherches sur la sexualité du crabe *Pachygrapsus marmoratus* (Fabricius). *Archives de Zoologie Expérimentale et Générale* **96**, 101-276.
Verwey, J. (1929). Einiges aus der Biologie von *Talitrus saltator* (Mont.). *X*

Congress International de Zoologie, Budapest, Vol. II, pp. 1156-62.
Veuille, M. (1978). Biologie de la reproduction chez *Jaera* (Isopode Asellote). II.—Evolution des organes reproducteurs femelles. *Cahiers de Biologie Marine* **XIX**, 385-95.
— (1980). Sexual behaviour and evolution of sexual dimorphism in body size in *Jaera* (Isopoda Asellota). *Biological Journal of the Linnean Society* **13**, 89-100.
Vitztum, H. G. (1940-3). Acarina. *H. G. Bronn's Klassen und Ordnungen der Tierreichs* **5**, IV, 5.
Vlasblom, A. G. (1969). A study of Marinogammarus marinus (Leach) in the Oosterschelde. *Netherlands Journal of Sea Research* **4**, 317-38.
Vollrath, F. (1980). Male body size and fitness in the web-building spider *Nephila clavipes*. *Zeitschrift für Tierpsychologie* **53**, 61-78.
Von Bonde, C. (1936). The reproduction, embryology and metamorphosis of the cape crawfish (*Jasus lalandii*) (Milne Edwards) Ortmann. *Investigational Report, Fisheries and Marine Biological Survey Division, Department of Commerce and Industries, South Africa* No. 6.
Von Erlanger, R. (1895). Zur Morphologie und Embryologie eines Tardigraden (*Macrobiotus Macronyx* Duj.). *Biologisches Centralblatt* **15**, 772-77.
Von Kaulbersz, G. J. (1913). Biologische Beobachtungen an Asellus aquaticus. *Zoologischer Jahrbücher. Abteilung für Allgemeine Zoologie* **33**, 287-3.
Von Siebold, C. (1877). Ueber die in München gezüchtet Artemia fertilis aus dem grossen Salzsee von Utah. *Verhandlungen Schweizerischen Naturforschung Gesellschaft* **59**, 267-80.
Wafa, A. K., Zaher, M. A., Soliman, Z. R., and El-Kadi, M. H. (1967). Biology of *Panonychus ulmi* (Koch) in Gaza. *Bulletin de la Société Entomologique d'Egypte* **51**, 131-9.
Walckenaer, C. A. (1837-47). *Histoire Naturelle des Insectes. Aptères*, 4 Vols. and atlas. Vols. III, IV, and atlas with P. Gervais. Libraire Encyclopédique de Roret, Paris.
Wallace, M. M., Pertuit, C. J., and Hvatum, A. R. (1949). Contribution to the biology of the king crab (*Paralithodes camtscatica* Tilesius). *Fishery Leaflet, Fish and Wildlife Service, United States Department of the Interior* No. 340.
Warner, G. F. (1967). The life history of the mangrove tree crab, *Aratus pisoni*. *Journal of Zoology* **153**, 321-35.
Watkin, E. E. (1941). The yearly cycle of the amphipod, *Corophium volutator*. *Journal of Animal Ecology* **10**, 77-93.
Watrous, L. E. and Wheeler, Q. D. (1981). The out-group comparison method of character analysis. *Systematic Zoology* **30**, 1-11.
Watson, J. (1972). Mating behavior in the spider crab, *Chionoecetes opilio*. *Journal of the Fisheries Research Board of Canada* **29**, 447-9.
Watters, F. A. (1912). Size relationships between conjugants and non-conjugants in Blepharisma undulans. *Biological Bulletin* **23**, 195-212.
Weber, E. (1974). Vergleichende Untersuchungen zur Bioakustik von *Discoglossus pictus*, Otth 1837, und *Discoglossus sardus*, Tschudi 1837 (Discoglossidae, Anura). *Zoologischer Jahrbücher. Abteilung für Allgemeine Zoologie und Physiologie der Tiere* **78**, 40-84.
Weismann, A. (1879). Beiträge zur Naturgeschichte des Daphnoiden. VI. Samen und Begattung der Daphnoiden. *Zeitschrift für Wissenschaftliche Zoologie* **33** (1879-80), 55-110.

Wells, K. D. (1977). The social behaviour of anuran amphibians. *Animal Behaviour* 25, 666–93.
— (1978). Courtship and parental behaviour in a Panamanian poison-arrow frog (*Dendrobates auratus*). *Herpetologica* 34, 148–55.
— (1979). Reproductive behavior and male mating success in a neotropical frog, *Bufo typhlonicus*. *Biotropica* 11, 301–7.
— (1980). Social behavior and communication of a dendrobatid frog (*Colostethus trinitatis*). *Herpetologica* 36, 189–99.
— (1980a). Behavioral ecology and social organization of a dendrobatid frog (*Colostethus inguinalis*). *Behavioral Ecology and Sociobiology* 6, 199–209.
Weygoldt, P. and Paulus, H. F. (1979). Untersuchungen zur Morphologie, Taxonomie and Phylogenie der Chelicerata. II. Cladogramme und die Entfaltung der Chelicerata. *Zeitschrift für Zoologische Systematik und Evolutionsforshung* 17, 177–200.
Wharton, G. W. (1941). A typical sand beach animal, the mole crab *Emerita talpoida* (SAY), pp. 199–209 within Pearse, A. S., Human, H. J., and Wharton, G. J. (1941). Ecology of sand beaches at Beaufort, N. C., *Ecological Monographs* 12, 135–90.
White, J. A. (1973). Viable hybrid young from crossmated periodical cicadas. *Ecology* 54, 573–80.
White, M. G. (1970). Aspects of the breeding biology of *Glyptonotus antarcticus* (Eights) (Crustacea, Isopoda) at Signy Island, South Orkney Islands. In *Antarctic ecology*, (ed. M. W. Holdgate) Vol. I, pp. 279–85. Academic Press, London.
Whittington, H. B. and Rolfe, W. D. I. (eds.) (1963). *Phylogeny and evolution of Crustacea*. Museum of Comparative Zoology, Cambridge, Mass.
Wickler, W. and Seibt, U. (1981). Monogamy in Crustacea and man. *Zeitschrift für Tierpsychologie* 57, 215–34.
Wilbur, H., Rubenstein, D. I. and Fairchild, L. (1978). Sexual selection in toads: the roles of female choice and male body size. *Evolution* 32, 264–70.
Wilder, J. (1940). The effects of population density upon growth, reproduction and survival of *Hyalella azteca*. *Physiological Zoology* XIII, 439–61.
Wiley, E. O. (1981). *Phylogenetics*. Wiley, New York.
Wiley, R. H. (1974). Evolution of social organization and life-history patterns among grouse. *Quarterly Review of Biology* 49, 201–27.
Willey, A. (1894). *Amphioxus and the ancestry of vertebrates*. Macmillan, London.
— (1911). *Convergence in evolution*. John Murray, London.
— (1925). Northern Cyclopidae and Canthocamtidae. *Transactions of the Royal Society of Canada*, Section V, Ser. 3, 17.
— (1932). Copepod phenology—observations based on new material from Canada and Bermuda. *Archivio Zoologico Italiano* XVI, 601–17.
Williams, L. W. (1907). The significance of the clasping antennae of harpacticoid copepods. *Science* 25, 225–6.
Williamson, D. I. (1951). On the mating and breeding of some semi-terrestrial amphipods. *Report. Dove Marine Laboratory* (3rd series) 12, 49–62.
Williamson, H. C. (1900). Contribution to the life-history of the edible crab (*Cancer pagurus*, Linn.). *Report of the Fishery Board of Scotland* 18, 77–143.
— (1904). Contributions to the life-histories of the edible crab (*Cancer pagurus*) and of other decapod Crustacea: impregnation: spawning: casting: distribu-

tion: rate of growth. *Report of the Fishery Board of Scotland* **22**, 100–40.
Williamson, M. (1981). *Island populations*. Oxford University Press.
Willoughby, E. J. and Cade, T. J. (1964). Breeding behavior of the American kestrel (sparrow hawk). *Living Bird* **3**, 75–96.
Willoughby, R. R. and Pomerat, C. M. (1932). Homogamy in the toad. *American Naturalist* **LXVI**, 223–34.
Wilson, E. O. (1975). *Sociobiology*. Harvard University Press, Cambridge, Mass.
Winterbottom, J. M. (1929). Studies in sexual phenomena. VI. Communal display in birds. *Proceedings of the Zoological Society of London* 189–95.
Wolf, E. (1905). Vortpflanzungsverhältnisse unserer einheimischen Copepoden. *Zoologischer Jahrbücher. Abteilung für Systematik, Ökologie und geographie der Tiere* **22**, 101–276.
Woodruff, D. S. (1976). Courtship, reproductive rates and mating system in three Austrialian *Pseudophryne* (Amphibia, Anura, Leptodactylidae). *Journal of Herpetology* **10**, 313–18.
Woodward, B. D. (1982). Sexual selection and nonrandom mating patterns in desert anurans (*Bufo woodhousei, Scaphiopus couchi*, and *S. bombifrons*). *Copeia* 351–5.
Wright, A. H. (1914). North American Anura. Life histories of the Anura of Ithaca, New York. *Publication. Carnegie Institute of Washington* No. 197.
Yaldwyn, J. C. (1966). Observations on copulation in the New Zealand grapsid crab *Hemigrapsus crenulatus* (M. Edw.). *Pacific Science* **20**, 384–5.
—— (1968). Notes on the behaviour in captivity of a pair of banded coral shrimps, *Stenopus hispidus* (Olivier). *Australian Zoologist* **17**, 377–89.
Yonge, C. M. (1937). The nature and significance of the membranes surrounding the developing eggs of Homarus vulgaris and other Decapoda. *Proceedings of the Zoological Society of London* **107**, 499–517.
Zaher, M. A. and Soliman, Z. R. (1971). The life-history of the predatory mite, *Cheyletus malaccensis* Oedemans. (*Acarina: Cheyletidae*). *Bulletin de la Société Entomologique d'Egypte* **LV**, 49–53.
—— —— (1971a). Life-history of the predatory mite, *Cheletogenes ornatus* (Canestrini and Fanzago). *Bulletin de la Société Entomologique d'Egypte* **55**, 85–9.
—— —— and El-Safi, G. S. (1974). Biological studies on *Cenopalpus pulcher* (Canestrini and Fanzango) (Acarina: Tenuipalpidae). *Bulletin de la Société Entomologique d'Egypte* **58**, 367–73.
—— Wafa, A. K., and Yousef, A. A. (1969a). Biology of the false spider mite, *Cenopalpus lanceolatisetae* (Attiah) (Acarina: Tenuipalpidae). *Indian Journal of Entomology* **31**, 53–8.
—— —— —— (1969b). Biological studies on *Raoiella indica* Hirst and *Phyllotetranychus aegyptiacus* Sayed infesting date palm trees in U.A.R. (*Acarina-Tenuipalpidae*). *Zeitschrift für Angewandte Entomologie* **63**, 406–11.
Zavattari, E. (1910). Osservazioni etologiche sopra l'Anfipodo tubicolo *Ericthonius brasiliensis* (Dana). *Memoir. R. Com Talass* No. 77.
Zilch, R. (1972). Beiträg zur Verbreitung und Entwicklungsbiologie der Thermosbaenacea. *Internationale Revue der gesamten Hydrobiologie* **57**, 75–107.
Zimmer, C. (1926). Cumacea. *Handbuch der Zoologie* **3**, 1, 651–82.
—— (1941). Cumacea. *H. G. Bronn's Klassen und Ordnungen des Tierreichs* **5**, I, 4.

Index

Taxonomic levels from the family upwards are indexed; lower levels are included only if they are of exceptional interest. Authors are indexed only if their work is discussed beyond a parenthetic citation of authority.

Acari, mites
  classification of 143-5
  moult cycles of 129-31, 144
  precopulas of 145-9
Acaridae, mites 148-9
Acaridei, division of mites 144, 148
adaptation
  hypotheses of 7
  and independent evolution, compared 16, 17-18
  studied by comparison 5-6
  test of 8
Agelenidae, spiders, precopulas of 138
allometry
  anti-adaptationist 5, 30, 32
  Clutton-Brock and Harvey on 13, 15, 17, 32
  of genital size, Pomerat's theory 172, 221
Alpheus, snapping shrimp
  homogamy of 170, 197, 218
  pairs of 82
Ampeliscidae, tube-dwelling amphipods 117
Amphipoda
  classification of 107, 111
  homogamy of 200-2, 217
  precopulas of 52, 110-18, 167
Ampithoidae, tube-dwelling amphipods, precopula of 111
amplexus
  pelvic and pectoral distinguished 150
    adaptive significance 150
  sometime synonym of precopula 52
ancestral characters, adaptations 5, 17-18
Anomura, hermit and sand crabs
  classification of 88-9
  fertilization of 89-90
  homogamy of 198
  precopulas of 90-3
Anostraca, fairy shrimps, precopulas of 70, 127-8
Anura, frogs and toads
  breeding seasons of 61-2, 151-2, 154-9 *passim*
  homogamy of 208-14
  phylogeny of 153-4
  precopulas of 52, 149-59, 167

Aoridae, tube-dwelling amphipods, precopula of 111-12
Arachnida
  classification of 129
  homogamy of 203
  precopulas of 130-49
Araneae, spiders
  moult cycles of 129-31
  precopulas of 132-43
  sexual cannibalism of 132-3
Araneoidea, division of spiders 135
Araneomorpha, division of spiders 134
Aristotle 4
Armadillidae, terrestrial isopods 125
  no precopula 126
Artemia salina, fairy shrimp, precopula of 70
Arthropoda
  homogamy of 186-203
  phylogeny of 37, 64-5
  precopulas of 64-149
Asellidae, isopods
  homogamy of 202
  precopula of 124
Asellus aquaticus
  an examplar 52, 53, 54, 60, 62, 108, 119-20
  homogamy of 202, 217, 220
  precopula of 124
    duration varies with size 175
assortative mating 170n
Atelopodidae, frogs, exceptional precopula? 157, 165
Atypidae, spiders, no precopula 134-5
Aunt Jobisca's theorem, applied? 115, 116, 124

Baker, R. R. (British biologist) (and G. A. Parker) on bird coloration 11-12, 17
barnacles, excluded from precopula study 63, 72
bees and homogamy in *Lythrum* 179
Bibionidae, flies
  no homogamy? 187
    exceptional? 187, 217, 220, 221
  male competition 187

Bibionidae, flies, (cont).
    mating duration 187
biometricians, study homogamy 171-2
birds
    coloration of, comparative trends of 11-12, 40
    homogamy of 214-15
    mate guarding of 63
Birkhead, T. (British ornithologist)
    on *Gammarus* 200, 220
        his theory 201-2
    on magpies 63
Bock, W. (American biologist)
    on adaptation 40
    on comparative method 28-33
Borradaile, L. A. (British carcinologist) his taxonomy 93-4, 95, 106
Bousfield, E. L. (Canadian amphipodologist) his taxonomy 110-11, 113, 117-18
Brachyura, crabs
    chaotic classifications of 93-5
    homogamy of 199
    mechanical constraint on mating? 95-8
    precopulas of 98-105, 167, (summarized) 128, 165-7
Branchiopoda, precopulas of 70-1, 127-8
Branchiura 72
breeding season, anuran, analogous to moult cycle 61-2, 151-3
Brentus, weevil, its mating duration 176, 194
Bristowe, W. S. (British arachnologist) 137-44 *passim*
Brocchi, P. L. A. (French anatomist) 96n
Buffon (*encyclopédiste*) 3-4
Bufo, toad
    an exemplar 60, 150, 152
    homogamy and not 211-14, 217, 219, 221
    seasons and amplexus of 154, 155-7
Bufonidae, toads
    amplectic posture of 150
    and duration 155-7
    homogamy and not 211-14

Caecosphaeorma burgundum, isopod, its long mating interval 123, 168
Cain, A. J. (British biologist)
    ancestral characters and adaptations 5, 18
    criticizes cladism 22-3
    on 'genetic inertia' 17n
Calanoida, copepods
    homogamy of 196
    no precopulas 73-4, 79
Caligoida, parasitic copepods 72

Cambridge, Rev. O. P. (British arachnologist) 135
Cancridae, crabs
    homogamy? 199
    precopula of 98-9
Cantharidae, beetles, homogamous 191
Caprella, an amphipod, precopula? 111
Carcinus maenas, common European crab
    genitalia 96
    homogamous? 199
    precopula? 95, 98-9
cause
    distinguished from coincidence 16, 34-40
    cladistically 37-9
    by experiment 35
    by logic of natural selection 36-7
    other techniques? 39-40
caves, their inhabitants lack precopulas 168-9
Cavolini, Ph. (Neapolitan biologist) on crabs 95
Cerambycidae, beetles, homogamous? 193-4
Cheluridae, wood-boring amphipods, precopulas of 111, 112
Cheyletidae, mites, precopulas of 148
Chomodoris, opisthobranch, homogamous 185
Chorioptes, mange mite 60
    precopula of 149
Chrysomelidae, beetles, homogamous 191-3
cicadas, not homogamous 186
Cichlidae, homogamy of 204-5
Cirolanidae, isopods, their precopulas 123
cladism
    contrasted with method of this book 19
    explained 18-19
    techniques of 21-7, 38-9; *see also* outgroup comparison
Cladocera, no precopulas 71
classification
    of Acari 143-4
    of Amphipoda 110-11
    of Anomura 88-9
    of Anura 153-4
    of Arachnida 128-9
    of Araneae 134
    of Araneomorpha 135
    of Arthropoda 64-5
    of Brachyura 93-5
    of Branchiopoda 70
    of *Bufo* 157
    of Copepoda 72
    of Crustacea 69-70, 128
    of Decapoda 80
    of Flabellifera 121

of Harpacticoida 76
of Isopoda 119
of Malacostraca 79
of Maxillopoda 71-2
of mites 143-4
of Oniscoidea 125
of Peracarida 107
of Reptantia 84-5
of spiders 134
a tradition of comparative biology 3-5
Clubionidae, spiders, precopulas of 139
Clubionoidea, spiders 135
  precopulas of 138-9
Clutton-Brock, T. H. (British biologist) (and P. H. Harvey)
  on allometry 15, 17, 32
  on causes 35
  on phylogenetic inertia 17-18
  on primates 12-15
cockroach 66
Coelopidae, flies, homogamous? 190
cohabitation, sometimes a synonym of precopula 52, 132-3 *passim*
Coleoptera, homogamy of 190-5
Colorado potato beetle, homogamy of 172, 192-3, 221
comparative biology, taxonomic and non-taxonomic contrasted 3-6
comparative method
  analogous to experiment 8
  criticisms of 38-40
  practical procedures 48-51
  trials 8-9, 9-18
Conchostraca, crustaceans, no precopula 71
continuous variables, how studied 11-15, 58
convergence
  evidence of adaptation 5
  minimum estimate of 21, 22-3
Copepoda
  classification of 72
  homogamy of 196
  life cycle of 73
  modified antennae of 73-4
  precopulas of 73-9
Corophium, tube-dwelling amphipod, precopula? 112, 168
Corophoidea, amphipods, precopulas? 112-13
correlation coefficient
  difficulties with hermaphrodites 182, 183, 185
  laboratory estimate 188
  a measure of homogamy 170, 171, 177-9
  spurious 192

Corystidae, crabs, precopula of 98
courtship
  distinguished from precopula 60
  in hermit crabs 91
  in spiders 133
crabs, precopulas of
  anomuran 88-93
  brachyuran 93-107
Crangonycidae, amphipods 116, 118
crayfish
  fertilization of 87
  no precopulas 88, 128, 168
Cribellatae, division of spiders 134, 143
  precopulas in 141-5
Crozier, W. J. (American biologist), on homogamy 172
  and *Gammarus* 199-200
  his observations on molluscs 185
Crustacea
  homogamy of 196-203
  precopulas of 69-128
Cumacea
  classification of 107-8
  precopulas of 118-19
Cyclopidae, copepods, precopula and no precopula 74-6, 79
Cyclops, copepod, precopula and no precopula 74-6, 79
Cymothoidae, parasitic isopods, precopulas? 125-6
Cyprinodontidae, pupfish, not homogamous 207

Darwin, Charles
  on bird coloration 11, 40, 40n
  and convergence 5
  on human heterogamy 170-1
  on sexual selection 47
Decapoda 61
  precopulas of 80-107
Demodicidae, mites, no precopula in? 147-8
Dendrobatidae, frogs, no precopulas in? 155
Dermanyssidae, mites, precopulas of 145
Dictynidae, spiders, precopulas of 141
Dictynoidea, spiders 135
Diogenidae, hermit crabs, precopulas? 90-1
Dioptidae, moths, not homogamous 195
Diptera 67
  homogamy in 187-90
Discoglossidae, frogs
  amplectic posture 150
  no precopula in 154
Drassodes, spider, precopula of 138-9
Drosophila
  in test 217-21 *passim*
  melanogaster homogamous 188

Drosophila, (cont.)
  subobscura not homogamous 189
duration of mating
  in *Alpheus* 82, 197
  in bibionid fly 187
  in *Brentus* 194
  in cantharid beetles 191
  in cerambycid beetles 193
  in cicada 186
  in *Drosophila* 189
  in dungfly 189-90
  in *Leptinotarsa* 192
  long and short distinguished 176-7
  how measured 176
  in molluscs 185
  in moth 195
  in Protozoa 183, 184
  in *Pyrrhocoris* 187
  in *Tetraopes* 194
  varies with size 175-6
  see also precopula
dwarf males
  in copepods 72
  in parasitic isopods 122, 127
  in sand crabs 93
  in species with precopulas 169
  in spiders 136-7
Dysderidae, spiders, lack precopulas 135
Dysderoidea, division of spiders 135

Enock, F. (British arachnologist) 135
Entelegynae, division of spiders 135
Epicaridea, parasitic isopods
  fertilization of 120-1, 127
  pairs of 126-7
epignye, female spider genital organ 130
Eresidae, spiders, no precopula? 141
Eresoidea, division of spiders 135
Erythraeidae, mites, no precopula? 148
Eucarida, crabs
  classification of 79
  homogamy of 196-7
  precopulas of 80-107, 128
Euphausiacea, krill 80, 107, 128
Eylaidae, water mites, precopulas of 148

fairy shrimp, precopula of 70
fecundity, larger in larger females
  in *Alpheus* 197
  in bibionids 187
  not in *Brentus* 194
  in cichlid 103
  in *Drosophila* 188
  in dungfly 189
  in *Eleutherodactylus*? 208
  in *Gammarus* 200

  in geese 214
  not in *Hyla* 209
  in isopods 202-3
  in king crab 198
  in lobster 197
  in molluscs 185
  protozoan analogy 183
  in *Pseudocalanus* 196
  in pupfish 207
  in *Rana* 210
  in salmon 206
  in *Trapezia* 199
  in *Uca* 199
female choice 30, 47-9
  a cause of homogamy? 174, 210-11
fertilization
  of Anura 150
  of Arachnida 130
  of *Artemia* 70
  of *Asellus* 53-4
  of brachyuran crabs 94-8
  of copepods 73
  of crayfish 87
  of *Daphnia* 71
  of *Gammarus* 110
  of hermit crabs 89-90
  of isopods 110-12
  of lobsters 84-6
  of mysids 109
  of natantian shrimps 80-4
  of primitive Peracarida 108
  of tardigrades 68
fiddler crabs, precopulas? 104
fish, homogamy in 204-8
Fisher, R. A. (British geneticist)
  his method of combining significance
    tests 187, 197, 198, 201, 210
    211-212
  his model of female choice 30, 47-9
Flabellifera, a division of isopods, precopulas of 121-4
Formicidae, ants, not homogamous? 195-6
frogs and toads
  homogamy in 208-13
  precopulas of 149-59
  phylogeny of 153-4

Galápagos finch, homogamous 215
Galathea, crab 89, 92
Gammarus
  an exemplar 52, 59, 60, 61
  fertilization of 110, 119
  homogamy of 200
    objections 201-2
  precopula of 115-16, 168
    duration varies with size 175

Gammaridae, amphipods
  homogamy of 200-2
  precopulas of 115-16, 118
Gecarcinidae, crabs
  mechanical constraint on mating? 96
  no precopula 103
geese, homogamy in 214
ghost crab, no precopula 104
Gnaphosidae, spiders, precopula of 138-9
Gnathidiidae, isopods 119, 121
Gnathophyliidae, painted shrimps, pairs of 82
Gonodactylus, mantis shrimp, no precopula 79-80
Gould, S. J. (American palaeontologist)
  on adaptation 7
  on laws 41-2
  on non-adaptive divergence 29
  on phylogenetic inertia 17
Grafen, A. (British biologist)
  his model of precopulas 54-8, 64, 151, 167-8, 176
  statistical advice of 179, 218-19, 222
Grapsidae, crabs, lack precopulas 103-4
Guinot, D. (French carcinologist) her taxonomy 94-5, 105-7

Haplogynae, division of spiders
  lack precopulas 135
  not monophyletic 135
Harpacticoida, copepods
  classification of 76
  precopulas of 76-9
Hartnoll, R. G. (British carcinologist)
  observes crab genitals and mating 96-8, 101-4 *passim*
  his review 50, 166
Harvey, P. H. (British biologist)
  on allometry 15, 17, 32
  association and cause 35
  on comparative method 10, 11, 15, 17
  phylogenetic constraint 17-18
  on primates 12-15
Haustoriidae, interstitial amphipods 116-17
Hemiptera, no homogamy in 189-90
Hennig, W. (German taxonomist) 11, 18n
  jutifies cladism 20, 24
hermit crabs
  classification of 89
  fertilization in 89-90
  homogamy of 198
  precopula? 90-1
Heteropodidae, spiders 138
  precopulas of 139
Heterotremata, division of crabs 94-5
  precopulas of 98-102, 105
Hippidae, sand crabs, precopula? 92-3

Holley, M. (British biologist)
  study of *Artemia* 70
  re-draws copepod 74
Homarus lobster
  fertilization of 85
  homogamy of 197
  precopula of 87-8
homogamy, defined 171
homogamy for size
  biometrical research on 171-2
  caused by mechanical constraint, *see* lock-and-key
  ethology of, *see* Lorenzian evidence
  how measured 177-8; *see also* correlation coefficient
  in purple loosestrife 179
  in protozoans 179-85
  negative evidence
    in some Anura 208-9, 211-13
    in bibionids? 187
    in Cerambycidae 193
    in cicadas 186
    in *Drosophila* 189
    in dungfly 189
    in moth 195
    a null hypothesis 178
    in pupfish 207
    in *Pyrrhocoris* 186
  positive evidence
    in *Alpheus* 170, 197
    in some Anura 210-13
    in cichlids 204-5
    in copepods 196
    in *Drosophila* 188
    in Galápagos finch 214
    Gammarus 200
    in geese 214
    in hermit crab 198
    in isopods 202-3
    in *Leptinotarsa* 192
    in *Limulus* 203
    in lobsters? 197
    in molluscs 185-6
    in salmon 206
  reviewed 171-222
  sexual selection of 173-4
    how tested for 174-8
  spurious evidence of 177
homology, basis of most modern taxonomy 3
Hoplocarida, mantis shrimps, lack precopulas 79-80, 128
Howard, R. (American anurologist)
  observes homogamy 210
  observes precopulas 157-9
humans, proverbially heterogamous 171-2
  not in fact 215-16
Huxley, J. (British biologist) 30

Hyalellidae, amphipods 113
  homogamous 202
  precopulas of 114-15
Hyalidae, amphipods 113
  precopulas of 114-15
Hyas coarctatus, crab, an exception to theory? 101, 105, 166
Hydrachnellae, freshwater mites, precopulas of 148
Hylidae, treefrogs 150
  no homogamy 208-9
  no precopula 155
Hymenoptera
  not homogamous? 195-6
  precopula analogies of 66, 67-8
Hymenosomatidae, crabs, precopulas of 104-5
  exceptional? 105, 166
Hyperiidae, amphipods
  classification of 110-11
  precopula of 113

Idoteidae, isopods, their precopulas 124
independence, statitistical,
  Baker and Parker's method 11-12
  Clutton-Brock and Harvey's 12-15
  necessary in comparative test 9-18
    why 16
    how recognized 18-27
    single taxonomic level methods not independent 15
ingroup comparison 21
  unsound 24-5
insects, precopulas in? 65, 66-8
Iscyroceridae, amphipods, precopulas? 117
isopods
  classification of 107, 119
  fertilization in 120-1
  homogamy in 202-3
  precopulas and their loss 52-4, 121-7, 167, 168-9
Ixodides, ticks 131, 144
  no precopulas 146-9

Jackson, R. R. (arachnologist) on precopulas 51, 131, 134, 140, 141, 142-3
Jaera, isopod
  fertilization of 120-1
  homogamy and its loss 203
  precopulas and their loss 124-5
Janiridae, isopods, precopulas of 124-5
Jasus, rock lobster
  inseminated at moult 86
  internally fertilized 85
  precopula? 87

Jeeves, on human heterogamy 171

king crab
  not homogamous 198, 199
  precopula of 91
  duration varies with size 176
Knowlton, N. (American biologist) 197
  on *Alpheus* 82, 170
Krebs, J. R. (British biologist) (and N. B. Davies) 35
krill 80, 108, 128

Labidognatha, division of spiders 134
laws 8-9
  of homogamy 217-21
  of precopula 164
  a valid concept 40-3
Lepidoptera 66
  not homogamous 195
Leptinotarsa, homogamy in 172, 192-3, 217-21 *passim*
Leptodactylidae, frogs
  no homogamy? 208
  no precopulas? 154-5
Leucosiidae, crabs, lack precopulas 102, 106
Lewontin, R. (American geneticist)
  on adaptation 7
  on non-adaptive divergence 29
  on phylogenetic inertia 17
Ligidae, shore-line isopods, precopulas of 125
Limnoriidae, wood-boring isopods, monogamy in 121, 122
Limulus
  not homogamous 203
  has precopula? 128-9
Linyphiidae, spiders, lack precopulas? 136
lions, mate guarding in 63
Liphistiidae, spiders 134
lobsters
  fertilization of 85
  homogamy of 197
  precopula of 87-8
lock-and-key, a theory of copulation
  causes homogamy? 172, 220-1
  disproved
    in *Asellus* 220
    in *Gammarus* 201, 220
  invoked
    in cerambycids 193
    in crabs 199
    in *Gammarus* 200-1
    in *Leptinotarsa* 193
    in molluscs 185, 186
    in protozoans 183
Locke, J. (philosopher) 4, 5
locust 67

Lorenz, K., 172-3, 204
Lorenzian evidence, of homogamy
   defined 173
   discussed 219
   in falcons 215
   in fish 204
   in lobsters 197
   in shrimps 196-7
Lycosidae, spiders, no precopula in 138
Lycosoidea, spiders 135
   precopulas 138
Lythraceae, flowers, homogamy of 179

Macrura, lobsters and crayfish
   fertilization of 84-7
   precopulas? 87-8
magpie, mate guarding in 63
Majidae, crabs, precopulas of 101-2, 166
Malacostraca
   classification of 70, 79
   precopulas of 79-107, 127-8
male choice
   direct evidence of 194-5, 198, 202, 210
   explains homogamy 183
   indirect evidence of 174
   why rare 48
male competition, overrides sexual selection of homogamy 187, 190, 220
male fitness, increases with size
   in alpheid shrimp 197
   not in ant 196
   in bibionid flies 187
   in brentid beetle 194
   in cerambycid beetle 194
   in cichlid? 205
   in coelopid fly 190
   in *Drosophila*? 188-9
   in dungflies 189
   in some frogs and toads 208-13
   in gammarids 200
   in isopods 202-3
   in king crab 198
   in lobster 198
   in moth 195
   in pupfish 207
   how recognised 175-6
   in salmon 206
   in soldier beetles 191
   in *Trapezia* 199
   see also normalizing selection
mantis shrimp, no precopula in 79-80
Manton, S. (British zoologist)
   arthropod phylogeny 64-5
   on functional criterion of homology 37
mathematician, heroic, needed 178-9
mating, *see* fertilization
   duration of, *see* duration of mating
Maxillopoda, mainly copepods, precopulas of 71-9, 128
mechanical constraint, *see* lock-and-key
Melandryidae, beetles, homogamous? 191
Melitidae, tropical amphipods
   and definition of precopula 59, 166
   precopula of 113
Mesostigmata, mites 144-5
midge 66
midwife toad,
   no precopula 154-8
   pectoral amplectic posture 150
Mimetidae, spiders, no precopula in 135-6
mites
   classification of 143-5
   insemination of 143-4
   moult cycles of 130-1, 144
   precopulas of 145-9
monogamy
   hermit crabs 91
   isopods 121-3, 126-7
   in mudshrimps 88
   a form of precopula 63-4, 167-8
   in natantian shrimps 82-4, 170
   in sand crabs 93
   in trapezid crabs 100
mosquitoes, pupal attendance in 52, 66-7
moth, Californian oak moth, not homogamous 195
moult cycle
   oviposition and mating restricted to moult
      in Arachnida 129-31
      in many Arthropoda 52-169 *passim*
      in *Asellus* 53
      in Brachyura 98
      in copepods 73
      in hermit crabs 98-9
      in insects 67
      in Macrura 84-7
      in natantian shrimps 80-1
      in Peracarida 108
      in tardigrades 68
   oviposition not restricted to moult
      in hermit crabs 90
      in porcellanid crabs 92
      in sand crabs 93
      in some brachyuran crabs 97-8, 102
   relaxed timing of mating, in Peracarida 108, 114-15, 116, 123
mudshrimp, pairs of 88
Mygalomorpha, division of spiders 134
   lack precopulas 134-5
Mysidacea
   classification of 107
   precopula? 109-10, 128, 168

Natantia, prawns and shrimps, pairing of 80-4
Nephila, tropical araneid spider, precopula of 131, 136-7
Nesticidae, spiders, no precopulas 135-6
Niphargidae, amphipods 116, 118
Niphargus orcinus virei, cavernicolous amphipod, no precopula 108, 116, 118, 123, 167, 168-9
non-adaptive convergence 32-3
non-adaptive divergence
    a problem? 28-30, 31-2
    test for 7
normalizing selection, for male size
    in *Leptinotarsa*? 192
    in *Tetraopes* 194
Notostraca, no precopulas 71
numerical taxonomy 20, 24

Ocypodidae, fiddler crabs
    homogamy 199
    lack precopulas 104
Olivella, prosobranch, homogamy in 185
Oliver, J. H. (acarologist) 146, 149
Oniscidae, terrestrial isopods, no precopulas in 126
Oniscoidea, isopods 125
    mating of 119-21
    precopulas and not 125-6
Oonopidae, spiders, no precopula 135
optimal foraging theory, loose analogy with 178
Oribatei, mites 144
Ostracoda, not enough known about 72
outgroup comparison
    applied
        to Amphipoda 117-18
        to Anomura 93
        to Anura 153
        to Araneae 142
        to Arthropoda 64, 65
        to Brachyura 102, 105-7
        to *Bufo* 157
        to Crustacea 127-8
        to the incidence of homogamy 216-21
        to Peracarida 108
        to spiders 142
    and convergence 22-3
    explained 21-6
Oxypodidae, spiders, precopulas? 138

Pachygrapsus crassipes, crab, exceptional? 104, 165
Parker, G. A. (British biologist)
    on bird coloration 11-12, 13-17
    on dungflies 189-90
    on mate guarding 55, 59n, 64-7
Paguridae, hermit crabs
    homogamy in 198
    precopula? 91
Paralithodes, king crab
    no homogamy 198, 199
    precopula of 91
Paramecium
    conjugation of 183-4
    homogamy in 172, 182-4
Parasitengona, mites 144
    precopulas of 148
Parasitidae, mites, precopulas of 145
Parasitiformes, mites 144
    precopulas of 145-6
paternal care, comparative biological exemplar 10, 16-17, 32, 36, 39
Pelobatidae, frogs
    amplectic posture 150
    not homogamous 208
Peracarida crustaceans
    classification of 107
    fertilization of 108-9, 110, 119-21
    precopulas of 108-27, 167
Pholcoidea, spiders 135
Photidae, tube-dwelling amphipods 117
phylogenetic inertia 17-18
phylogeny
    random 33-4
    see also classification
Phytoseiidae, mites, precopulas of 145
Pipidae, toads
    lack precopula 154
    phylogenetic position of 153
Pisauridae, spiders, no precopula? 138
Pliny 3
Podotremata, division of crabs 95
Pomerat, C. M. (American biologist)
    on beetles 190-1
    on *Limulus* 203
    his theory of homogamy 172, 220-1
    on toads 172, 211
Pontoniinae, symbiotic shrimps, pairing of 83-4
Pontoporeia, interstitial amphipod 108
    no precopula? 117, 124
Popper, K. (philosopher) 41-2
Porcellanidae, crabs, pairs and harems of 92
Porcellio scaber, terrestrial isopod
    fertilization of 119, 120
    no precopula 126
Porcellionidae, terrestrial isopods 125
    no precopulas 126
Portunidae, swimming crabs, precopulas of 99, 102
precopulatory mate guarding
    correlated with?

INDEX

cave-dwelling 169
sex ratio
  in mysids 110
  in summary 167-8
  in theory 57
  terrestriality 168
distinction of contact from non-contact 58, 81-4, 86, 114; *see also* tube-dwelling
found in
  amphipods 110-19
  copepods 72-9
  decapods 80-97
  fairy shrimps 70-1
  frogs and toads 154-9
  isopods 119-27
  Limulus 129
  mites 145-9
  mysids? 109-10
  spiders 132-43
  tardigrades 68-9
  Thermosbaenacea? 108
negative evidence of 49
  in general 169
  in mites 147
  in spiders 131-2, 133, 141-3
not courtship 59-60
not found in insects 66-8
part of total mating duration 176
recognition of 58-64
synonyms of 52
theory of 53-8
  confirmed 164
  exceptions destroyed 165-6
primates, trends in societies of 12-15
  consorting in 63
progress, evolutionary, not analysable 7
Prostigmata, mites 144
  precopulas of 147-8
Prussian grenadiers 171
Psoroptidae, mange mites, precopulas of 57, 60, 149
pupfish, not homogamous 207
purple loosestrife, homogamous 179
pyemotid mites
  excluded from precopula study 63
  incestuous matings of 146-7
Pyrrhocoris, hemipteran, not homogamous 186, 217-21 *passim*

Rabaud, E. (French comparative biologist) 8-9
Rana catesbiana, amplexus of 158, 17
Rana esculenta, edible frog, season and amplexus of 158
random phylogenies 33-4

Ranidae, frogs
  amplectic posture 150
  amplexus and seasons of 157-9, 167
  classification of 154
  homogamy of 209-11
Reptantia, lobsters, crabs, etc.
  classification of 80
  precopulas of 84-107
Rhacophoridae, frogs, not much known about 159
riding position
  an amphipodan precopulatory posture 52, 112
  not found if male smaller than female? 116, 117, 169
  not in Tanaidacea 168

salmon, homogamy of 205-6
Salticidae, jumping spiders (formerly Attidae) 130
  classification of 139n
  precopulas of 139-41
Salticoidea, division of spiders 135
sand crabs, precopula? 93
sandhoppers, amphipods, precopulas of 113-14
Sarcoptiformes, mites, precopulas of 148-9
scampi
  fertilization of 85
  lack precopula? 88
Scarabaeidae, beetles, homogamy of 190-1
Scytoidea, spiders, no precopulas 135
season, *see* breeding season
Serolidae, isopods 121
  precopula? 121-2
sexual cannibalism, in spiders 133
sexual selection 11-12
  Fisher's model of 30, 47-9
  theory of homogamy 173-4; *see also* homogamy
Siewing, R. (German zoologist) crustacean phylogeny of 69-70, 71-2, 79, 107
Silberbauer, B. I. (South African zoologist) on fertilization of *Jasus* 85-6, 87
Silentes, crustaceans 85
  fertilization of 85-6
  precopulas 87
Simplson, G. G. (American palaeontologist) 5, 32-3, 41-2
Siphonoecetes, amphipod, mobile harem of 112-13
size, meaning of 178
snapping shrimps
  homogamy of 170, 196
  pairing of 82

Sooglossidae, frogs 163
species, not proper units in comparative tests 10
Sphaeromidae, isopods
　fertilization of 119-20
　homogamy of 202
　precopulas of 123-4
spiders, moult cycles of 130-1
　precopulas of 132-43
Stenopodidae
　homogamous? 196-7
　pairs of 84
Stridentes, crustaceans
　fertilization of 85-6
　lack precopulas 87
Syncarida, crustaceans, mating habits unknown 79
Synesius 4-5

taxonomic artefacts 10, 19
Talitridae, terrestrial amphipods, precopulas lost 113-14, 118
Talitrus, terrestrial amphipod 108
　precopulas lost 113-14, 123, 168-9
Tanaidacea, peracaridan crustaceans
　classification of 107
　precopula? 127
tardigrades
　phylogenetic position of 64-5
　precopulas of 68-9
Tarsenomidae, mites, precopulas of 147
Tarsonemini, mites 144
　precopulas of 146-7
taxonomy, *see* classification
Tenuipalpidae, mites, precopulas of 148
territoriality
　and evolution of paternal care 16-17
　non-adaptively convergent? 32-3
　a form of precopula 68
　in porcellanid crabs 92
test, by comparative method
　analogous to experiment 8
　of homogamy laws 216-21
　of precopula law 164
　single taxonomic level not necessary in 15
　trials must be
　　independent of whether conform 8-9
　　statistically independent 9-18
　units used previously 10-15
Tetragnathidae, spiders, no precopulas 135
Tetranychidae, spider mites, precopulas of 147
Theraphosiidae, spiders, lack precopulas 135
Theridiidae, spiders, precopulas of 136

Thermosbaenacea, crustaceans
　classification of 107
　precopula? 108
Thomisoidea, spider 135
　no precopula 139
Thoraciotremata, a division of crabs 94-5
　precopulas of 103-5
Thornhill, R. (American entomologist) on bibionid flies 187
ticks 131
　no precopulas 146, 149
toads
　homogamy in 172
　precopulas of 149-59
Tower, W. L. (American neo-Lamarckian) on *Leptinotarsa* 191-3
trapezid crabs, symbionts of corals
　homogamous 199
　pairing 101
Trichoniscidae, isopods 125
　evidence inconclusive 126
Trombidiformes, mites
　classification of 144
　precopulas of 147-9
tube-sharing, an amphipodan precopula
　detected 111, 117, 168
　in isopod? 130
　difficult to detect 58, 112
Tylidae, terrestrial isopods, precopulas of 125-6

Uca
　homogamous 199
　lacks precopula 104
Uloboridae, spiders, no precopula? 141
Uropodidae, mites, precopula of 146

Valvifera, isopods, precopulas 124-5
Veuille, M. (French biologist) on *Jaera* 125, 203
Vollrath, F. (international arachnologist) on *Nephila* 131, 137, 137n

Wells, K. D. (American anurologist)
　his observations 151-7 *passim*
　his review 50, 57
Winterbottom, J. (British biologist) on avian mating habits 39

Xanthidae, crabs
　homogamy of 199
　precopulas of 100-1
Xenopus laevis, no precopula 134
Xylotrupes gideon, horned beetle 190

zoological education 50